Emerging Research on Adaptive Plants in Karst Ecosystems

Emerging Research on Adaptive Plants in Karst Ecosystems

Editor

Yanyou Wu

MDPI

Basel • Beijing • Wuhan • Barcelona • Belgrade • Novi Sad • Cluj • Manchester

Editor
Yanyou Wu
State Key Laboratory of
Environmental Geochemistry
Institute of Geochemistry
Guiyang
China

Editorial Office
MDPI
St. Alban-Anlage 66
4052 Basel, Switzerland

This is a reprint of articles from the Special Issue published online in the open access journal *Agronomy* (ISSN 2073-4395) (available at: www.mdpi.com/journal/agronomy/special_issues/adaptable_plants_karst).

For citation purposes, cite each article independently as indicated on the article page online and as indicated below:

Lastname, A.A.; Lastname, B.B. Article Title. *Journal Name* **Year**, *Volume Number*, Page Range.

ISBN 978-3-0365-8663-2 (Hbk)
ISBN 978-3-0365-8662-5 (PDF)
doi.org/10.3390/books978-3-0365-8662-5

Contents

About the Editor

Yanyou Wu

Yanyou Wu is a Professor of Plant Physiology and Ecology at the State Key Laboratory of Environmental Geochemistry, Institute of Geochemistry, Chinese Academy of Sciences, China. He received a BS degree (1986) in Biology from the Department of Biology, Anhui Normal University, China; a master's degree (1989) in Plant Physiology and Biochemistry from the Department of Agronomy, Guizhou University, China; and a doctorate (1994) in Botany from the Department of Biology, Sichuan University, China.

In Dr. Wu's early years, he mainly devoted himself to explaining the fertility of some intergeneric hybrids with "cell fusion - chromosome set segregation" and the extended genetic laws. In recent 25 years, his scholarly activities have been switched to the areas of the adaptive mechanism of karst-adaptable plants, the technology and principle of selection on karst-adaptable plants, the determining the biological and environmental information, the application and determination of electrophysiological information in plants, and biogeochemical action of carbonic anhydrase. He published 15 monographs and more than 250 papers in journals. In addition, he held more than 100 invention patents from the People's Republic of China. His academic contributions appeared in Photosynthesis Research, Frontiers in Plant Science, Geochemistry, Geophysics, Geosystems, Journal of Hazardous Materials, Plant Methods, Chinese Science Bulletin, Scientia Horticulturae, Energy Conversion and Management, BMC Plant Biology, and Physiologiae Plantarum, among others.

agronomy

Editorial

The Diversification of Adaptive Strategies for Karst-Adaptable Plants and the Utilization of Plant Resources in Karst Ecosystems

Yanyou Wu [1,*] and Yansheng Wu [1,2]

1 State Key Laboratory of Environmental Geochemistry, Institute of Geochemistry, Chinese Academy of Sciences, Guiyang 550081, China; wuyansheng@mail.gyig.ac.cn
2 University of Chinese Academy of Sciences, Beijing 100049, China
* Correspondence: wuyanyou@mail.gyig.ac.cn

Citation: Wu, Y.; Wu, Y. The Diversification of Adaptive Strategies for Karst-Adaptable Plants and the Utilization of Plant Resources in Karst Ecosystems. *Agronomy* **2023**, *13*, 2135. https://doi.org/10.3390/agronomy13082135

Received: 2 August 2023
Revised: 11 August 2023
Accepted: 12 August 2023
Published: 15 August 2023

1. Karst Environment

Karst landforms, which account for approximately 15% of the world's total land area, are mainly concentrated in low latitudes, including Southeast Asia, the European Mediterranean, the east coast of North America, the west coast of South America, and the marginal areas of Australia. The concentrated and contiguous karst is mainly distributed in southwest China, northern and southern Europe, and the east of the United States. Strong karstification causes the karst soil environment to be dry, and to have a high content of calcium (magnesium) and bicarbonate, a high pH, a high content of nitrate, and a low content of ammonium. A karst environment with a high spatiotemporal heterogeneity seriously affects the growth and development of plants. Faced with these heterogeneous environments, plants have adopted diversified adaptive strategies. The selected papers published in this Special Issue of *Agronomy* demonstrated some achievements on the physiological and ecological adaption of plants to heterogeneous karst environments, and they also explore how to extend the service period of plant resources in karst regions.

2. Karstification–Photosynthesis Coupling

Karstification–photosynthesis coupling is the most fundamental physiological and ecological mechanism used by plants to adapt to a karst environment, and is the "engine" that drives the biogeochemical cycle in nature. Karst-adaptable plants can "turn waste into treasure", convert bicarbonate formed by carbonate rock dissolution into organic matter, effectively improving the utilization of plant resources, and greatly enhance the biodiversity and carbon sink capacity of the ecosystem [1,2]. The efficient use of bicarbonate by plants is the core mechanism of karst adaptability. Plant utilization of the soluble inorganic carbon from the soil varies with plant species and altitude [3]. The physiological effects of bicarbonate in plants depends on its concentration in the soil, and appropriate levels promote growth and development by enhancing photosynthesis, water, and the nutrient metabolism of *Coix lacryma-jobi* L. [4].

3. The Response of Plant Nitrogen Assimilation

In karst environments, the differential response of plants to inorganic carbon and nitrogen assimilation determines the adaptation to soils with a high content of nitrate and a low content of ammonium. Karst-adaptable plants are more likely to use nitrates, and the positive effects on carbon and nitrogen assimilation are also higher than in the non-karst-adaptable plants [5]. At the same time, karst-adaptable plants can coordinate the distribution of nitrate assimilation between roots and leaves and the balance between internal demand and external supply of nitrogen [6].

4. Diversified Water Metabolisms of Karst Plants

Karst plants respond to water with species specificity. Compared with photosynthetic parameters, plant electrophysiological information can quickly, and in real-time, reflect water metabolism in plant cells [3,7]. In karst environments, the water metabolism patterns of different plant species were clearly different. *Coriaria nepalensis* maintained a high photosynthetic rate, with a high leaf intracellular water-holding capacity and high root water uptake ability. *Broussonetia papyrifera* maintained a stable intracellular water transport and photosynthetic rate by efficiently using the transpired water. *Elaeocarpus decipiens* displayed low transpiration, low photosynthesis, and low instantaneous water-use efficiency [7].

5. Responses of Plants to Karst Soil Environments

Other physical and chemical properties of soils also profoundly affect species diversity and the growth and development of karst plants. Plant diversity and soil seed banks are significantly influenced by soil properties. Orchards have the highest seed density, while primary forests have the highest plant diversity [8]. The Golden *Camellia* species grown on calcareous soils with strong Ca absorption and accumulation displayed a good adaptation to a high content of Ca [9]. The adaptability of plants to the environment changes with plant growth and development. *Zanthoxylum planispinum* 'dintanensis' leaf functional traits change with growth and development, which is also a manifestation of the influence of the soil environment [10]. The synergistic change of the stoichiometric relationship between plants and soil nutrients is also another effective mechanism for plants to adapt to karst environments [11,12].

6. The Adaptability of Karst Plants to the Biological Environment

The interaction between plants and their biological environment is an important aspect of adaptation. *Broussonetia papyrifera* and *Platycladus orientalis* exhibited different adaptation mechanisms to karst environments with different water and nutrient niches and different levels of competition for water and nutrients [13]. Arbuscular mycorrhizal fungi, which are root symbionts, promote nutrient uptake and utilization, improve resource utilization, and promote root morphological development and nutrient utilization in highly heterogeneous soil patches [14]. Soil microorganisms can also regulate litter decomposition and nutrient utilization by acting synergistically with plants [11,12].

7. Extending the Service Period of Plant Resources in Karst Areas

Due to the limited retention and transport of nutrients in karst soil environments, it is necessary to consider extending the service period of plant resources while developing and utilizing karst-adaptable plant resources. The return on investment of an artificial *Zanthoxylum* forest is varies significantly depending on the planting year, and the management of the forest can be used as a reference [10]. Short-term 1-methylcyclopropene (1-MCP) treatment can regulate the metabolism of blueberry fruits and improve the shelf quality of blueberries [15]; and 1-MCP, combined with an SO_2 treatment, can delay postharvest aging of bamboo (*Chimonobambusa quadrangularis*) shoots in karst mountains and extend the shelf life of fresh bamboo shoots [16].

8. Concluding Remarks

In short, the works in this Special Issue can help with understanding the karst-adaptability of plants from multiple perspectives, and they provide a scientific reference for the selection of karst-adaptable plants and the restoration of vegetation in karst areas. Further, they provide theoretical support for organic integration towards the economic, social, and environmental sustainability of karst areas, and the beautiful vision that "green water and green mountains are golden mountains and silver mountains".

Author Contributions: Y.W. (Yanyou Wu) and Y.W. (Yansheng Wu) cooperated to complete this article. All authors have read and agreed to the published version of the manuscript.

Acknowledgments: We gratefully acknowledge all the authors that participated in this Special Issue and the support of the National Natural Science Foundation of China (No. U1612441-2), as well as Plan Projects of Science and Technology of Guizhou Province (No. (2021)YB453).

Conflicts of Interest: The authors declare no conflict of interest.

References

1. Wu, Y. Combined Effect of Bicarbonate and Water in Photosynthetic Oxygen Evolution and Carbon Neutrality. *Acta Geochim.* **2023**, *42*, 77–88. [CrossRef]
2. Wu, Y.; Wu, Y. The Increase in the Karstification–Photosynthesis Coupled Carbon Sink and Its Implication for Carbon Neutrality. *Agronomy* **2022**, *12*, 2147. [CrossRef]
3. Rao, S.; Wu, Y. Species-Specific and Altitude-Induced Variation in Karst Plants' Use of Soil Dissolved Inorganic Carbon. *Agronomy* **2022**, *12*, 2489. [CrossRef]
4. Li, H.; Lv, J.; Su, Y.; Wu, Y. Appropriate Sodium Bicarbonate Concentration Enhances the Intracellular Water Metabolism, Nutrient Transport and Photosynthesis Capacities of *Coix lacryma-jobi* L. *Agronomy* **2023**, *13*, 1790. [CrossRef]
5. Xia, A.; Wu, Y. Differential Responses of Nitrate/Ammonium Use to Bicarbonate Supply in Two *Brassicaceae* Species under Simulated Karst Habitat. *Agronomy* **2022**, *12*, 2080. [CrossRef]
6. Zhang, K.; Zhang, F.; Li, H.; Su, Y.; Wu, Y. Stable Nitrogen Isotopes as an Effective Tool for Estimating the Nitrogen Demand of *Broussonetia papyrifera* (L.) Vent Seedlings under Variable Nitrate Concentrations. *Agronomy* **2023**, *13*, 1663. [CrossRef]
7. Qin, X.; Xing, D.; Wu, Y.; Wang, W.; Li, M.; Solangi, K. Diurnal Variation in Transport and Use of Intracellular Leaf Water and Related Photosynthesis in Three Karst Plants. *Agronomy* **2022**, *12*, 2758. [CrossRef]
8. Guo, Y.; Li, Y.; Li, J.; Li, J.; Wen, S.; Huang, F.; He, W.; Wang, B.; Lu, S.; Li, D.; et al. Comparison of Aboveground Vegetation and Soil Seed Bank Composition among Three Typical Vegetation Types in the Karst Regions of Southwest China. *Agronomy* **2022**, *12*, 1871. [CrossRef]
9. Zhu, X.; Tang, J.; Qin, H.; Bai, K.; Chen, Z.; Zou, R.; Liu, S.; Yang, Q.; Wei, X.; Chai, S. Contrasting Adaptation Mechanisms of Golden *Camellia* Species to Different Soil Habitats Revealed by Nutrient Characteristics. *Agronomy* **2022**, *12*, 1511. [CrossRef]
10. Song, Y.; Yu, Y.; Li, Y. Leaf Functional Traits and Relationships with Soil Properties of *Zanthoxylum planispinum* 'dintanensis' in Plantations of Different Ages. *Agronomy* **2022**, *12*, 1891. [CrossRef]
11. Yu, Y.; Wu, Y.; Song, Y.; Li, Y. Carbon and Nitrogen Stable Isotope Abundance and Soil Stoichiometry of *Zanthoxylum planispinum* var. dintanensis Plantations of Different Ages. *Agronomy* **2022**, *12*, 1248. [CrossRef]
12. Li, Y.; Yu, Y.; Song, Y. Stoichiometry of Soil, Microorganisms, and Extracellular Enzymes of *Zanthoxylum planispinum* var. dintanensis Plantations for Different Allocations. *Agronomy* **2022**, *12*, 1709. [CrossRef]
13. Yao, K.; Wang, Y.; Wu, Y. Competition and Niche Differentiation of Water and Nutrients between *Broussonetia papyrifera* and *Platycladus orientalis* under Prolonged Drought Stress. *Agronomy* **2022**, *12*, 1489. [CrossRef]
14. Li, Q.; Umer, M.; Guo, Y.; Shen, K.; Xia, T.; Xu, X.; Han, X.; Ren, W.; Sun, Y.; Wu, B.; et al. Karst Soil Patch Heterogeneity with Gravels Promotes Plant Root Development and Nutrient Utilization Associated with Arbuscular Mycorrhizal Fungi. *Agronomy* **2022**, *12*, 1063. [CrossRef]
15. Yan, H.; Wang, R.; Ji, N.; Li, J.; Ma, C.; Lei, J.; Ba, L.; Wen, G.; Long, X. Regulation of Cell Wall Degradation and Energy Metabolism for Maintaining Shelf Quality of Blueberry by Short-Term 1-Methylcyclopropene Treatment. *Agronomy* **2023**, *13*, 46. [CrossRef]
16. Xu, J.; Ji, N.; Wang, R.; Ma, C.; Lei, J.; Zhang, N.; Liu, R.; Deng, Y. Study on the Regulation Mechanism of 1-MCP Combined with SO2 Treatment on Postharvest Senescence of Bamboo Shoots (*Chimonobambusa quadrangularis*) in Karst Mountain Area. *Agronomy* **2023**, *13*, 1122. [CrossRef]

agronomy

Review

The Increase in the Karstification–Photosynthesis Coupled Carbon Sink and Its Implication for Carbon Neutrality

Yanyou Wu [1,*] and Yansheng Wu [1,2]

[1] State Key Laboratory of Environmental Geochemistry, Institute of Geochemistry, Chinese Academy of Sciences, Guiyang 550081, China
[2] University of Chinese Academy of Sciences, 100049 Beijing, China
* Correspondence: wuyanyou@mail.gyig.ac.cn

Abstract: Two of the most important CO_2 sequestration processes on Earth are plant photosynthesis and rock chemical dissolution. Photosynthesis is undoubtedly the most important biochemical reaction and carbon sink processes on Earth. Karst geological action does not produce net carbon sinks. Photosynthesis and karstification in nature are coupled. Karstification–photosynthesis coupling can stabilize and increase the capacity of karstic and photosynthetic carbon sinks. Bidirectional isotope tracer culture technology can quantify the utilization of different inorganic carbon sources by plants. Bicarbonate utilization by plants is a driver of karstification–photosynthesis coupling, which depends on plant species and the environment. Carbonic anhydrase, as a pivot of karstification–photosynthesis coupling, can promote inorganic carbon assimilation in plants and the dissolution of carbonate rocks. Karst-adaptable plants can efficiently promote root-derived bicarbonate and atmospheric carbon dioxide use by plants, finally achieving the conjugate promotion of karstic carbon sinks and photosynthetic carbon sinks. Strengthening karstification–photosynthesis coupling and developing karst-adaptable plants will greatly improve the capacity of carbon sinks in karst ecosystems and better serve the "Carbon peak and Carbon neutralization" goals of China.

Keywords: bicarbonate use; carbon sequestration; carbonic anhydrase; karst; karst-adaptable plants; photosynthesis

Citation: Wu, Y.; Wu, Y. The Increase in the Karstification–Photosynthesis Coupled Carbon Sink and Its Implication for Carbon Neutrality. *Agronomy* **2022**, *12*, 2147. https://doi.org/10.3390/agronomy12092147

Academic Editor: Arnd Jürgen Kuhn

Received: 10 August 2022
Accepted: 6 September 2022
Published: 9 September 2022

1. Introduction

Since the Industrial Revolution, atmospheric carbon dioxide (CO_2) concentrations have increased from 280 ppm before the Industrial Revolution to 421 ppm today, an increase of 50% [1]. Since 1850–1900, the global average surface temperature has risen by approximately 1 °C with the increase in atmospheric CO_2 concentrations [2]. Human activities have become one of the driving forces of changes in the Earth system, alongside the sun and the Earth's core, and the resulting global warming phenomenon has become a topic of major public concern. The impact of this phenomenon on the environment in which human beings live has become more direct. To prevent the occurrence of global warming, it is urgent to carry out research into effective and economical atmospheric carbon dioxide capture (carbon sequestration) pathways and take corresponding countermeasures. For this reason, China has made a commitment to reach carbon peak in 2030 and carbon neutrality in 2060, and to fulfill its responsibilities for the realization of global governance and the construction of a community with a shared future for humankind.

Karst is a general term for the geological effects of water on soluble rocks via mainly chemical dissolution, supplemented by mechanical actions such as erosion, latent erosion, and the collapse of flowing water, and the general term for the phenomena produced by these effects [3]. The global karst distribution area is nearly 22 million km^2, accounting for approximately 15% of the planet's land area, and the population living on karst areas is approximately one billion and is mainly concentrated in low latitudes, including Southwest

China, Southeast Asia, Central Asia, the Mediterranean, the east coast of North America, the Caribbean, the west coast of South America, and the south of Australia. Concentrated contiguous karst is mainly distributed in southern Europe, eastern North America and southwestern China. Karst in Southwest China is known for its larger continuous distribution area and for being the complete development type, the beautiful landscape and the fragile ecological environment [4,5]. In the 1,071,400 km^2 geographical range of 451 counties (cities) of Yunnan, Guizhou, Guangxi, Hunan, Guangdong, Sichuan, Chongqing, Hubei, and other provinces (autonomous regions and cities) centered on the Yunnan-Guizhou Plateau, the distribution area of carbonate rock is 450,800 km^2, accounting for 42.08% of the total land area [6].

Two of the most important processes for absorbing carbon dioxide on Earth are the photosynthesis of plants (biological action) and the chemical dissolution of rocks (silicate and carbonate rocks) (geological action). Photosynthesis is undoubtedly the most important biochemical reaction and carbon sink process on Earth. Plants use sunlight energy through photosynthesis to assimilate inorganic carbon and water into organic matter, release oxygen, couple the water and carbon metabolism of plants, and connect the soil–vegetation and vegetation–atmosphere systems. The equation for the dissolution of silicate rocks is as follows: $CO_2 + Ca(Mg)SiO_3 \rightarrow Ca(Mg)CO_3 + SiO_2$, and that for the dissolution of carbonate rocks (karstification) is as follows: $Ca(Mg)CO_3 + H_2O + CO_2 \rightleftarrows Ca(Mg)^{2+} + 2HCO_3^-$. The dissolution of silicate rocks is also undoubtedly a geological carbon sink process, but due to the low dissolution rate of silicate rocks [7], it is believed that silicate rock dissolution is only a carbon sink on a billion-year geological time scale [8], and has little impact on the carbon sink of the short-term time scale that human society is currently concerned about.

The dissolution of carbonate rocks has a profound effect on carbon sinks on short time scales [9–13]. From a short time scale, the development and utilization of fossil fuels released a large amount of carbon dioxide in advance due to human activities, especially the industrial revolution, which breaks the balance between the sequestration of carbon dioxide and the release of carbon dioxide, and conversely affects carbonate rock dissolution. Meanwhile, the lag in plant development and evolution has led to insufficient photosynthetic carbon sequestration. Therefore, the effect of carbonate rocks dissolution on carbon sinks is more profound on a short time scale.

Karstification by itself is a zero-carbon sink geological process, due to the balance of dissolution and the deposition of carbonate rocks on a billion-year-long time scale. Based on the models constructed from the deposition and burial of carbonates and organics, and the continental weathering of silicates, carbonates, and organics, as well as other factors such as volcanoes, diagenesis, and metamorphic degassing, scientists have found that after entering the ocean, HCO_3^- dissolved from carbonate will be transformed into carbonates in seafloor sedimentary rocks while releasing CO_2. From the geological process of carbonate dissolution on land to marine sedimentation, karst geological action does not produce net carbon sinks [14–16].

2. Uncoupled Karstification Limits Photosynthetic Carbon Sequestration

2.1. Karst Drought

Drought is mostly caused by low precipitation, while karst drought are mainly caused by special geological environments, because the average annual rainfall in the karst region of Southwest China is sufficient, up to 1000~1800 mm [17]. Long-term strong karstification has caused the hydrogeological structure in karst areas to be a special spatial structure on the surface and underground. The underground river system in this area is very well developed, and the surface has developed a large number of karst landforms in different landscapes, such as caves, lysophores, lysses, funnels and sinkholes, and skylights. Because most of the rocks are exposed and soil formation is slow and the soil layer is shallow, rainwater quickly leaks into the ground, becoming deeply buried groundwater, forming a pattern of separation between water and soil, and the water covered by the thin soil is quickly evaporated, resulting in soil drought [18]. When torrential rains fall, the underground karst pipes do not provide sufficient drainage, resulting in flooding [19,20]. Karst drought causes the stomata of leaves to close, severely reducing the photosynthesis of plants, and it even causes the death of plants.

2.2. High pH, and High Bicarbonate and Calcium Contents

Most of the bedrock in karst habitats consists of carbonate rocks, and its main chemical components are soluble salts such as $CaCO_3$ and $MgCO_3$. Due to the continuous dissolution of carbonate rocks in these habitats, hydrogen ions were consumed, a large number of bicarbonate ions and calcium and magnesium ions were formed, and the cover soil was modified into an environment with a high pH and high content of calcium and bicarbonate that affected the nutrient uptake by plants [18,21–23].

The average pH of the soil in the karst area of Mulun National Nature Reserve was 6.96, the maximum value was 7.68, and the minimum value reached 5.76 [24]. The pH of the soil developed from limestone was 7.69, and that from dolomite was 7.85, which was significantly higher than that from shale (5.32), sandstone (5.44), and red clay (5.32) [25].

The average exchangeable calcium content in the soil of Puding, Huajiang, Libo, and Luodian in karst areas was 3612.43 $mg \cdot kg^{-1}$, which was seven times higher than the value of 519.33 $mg \cdot kg^{-1}$ found in the non-limestone soils of Fujian, and more than 250 times higher than the average of 14.22 $mg \cdot kg^{-1}$ in the non-limestone soils of hilly and mountainous regions in southern China [26–28].

2.3. Deficiency of Available Nitrogen, Phosphorus, and Nutrient Elements

In karst areas, due to strong dissolution, many effective nutrient elements in the soil are lost. In addition, the high pH value and calcium concentration of the soil further reduce the effective the content of nutrient elements.

Due to the high pH value and high concentration of bicarbonate in soils of karst areas, ammonium nitrogen easily generates ammonia and enters the atmosphere [29]. Therefore, the content of ammonium nitrogen in karst areas is often lower than that in nonkarst areas, while the content of nitrate nitrogen is often higher than that in nonkarst areas. The average content of ammonium nitrogen from 20 soil samples in karst areas was 5.09 $mg \cdot kg^{-1}$, which is less than 70% of that of the Loess Plateau, while the average content of nitrate nitrogen was 13.60 $mg \cdot kg^{-1}$, which is 6.5 times that of the Loess Plateau [30,31]. A low content of ammonium nitrogen and a high content of nitrate nitrogen are typical characteristics of soil in karst areas.

The high pH value and calcium of soil in karst areas make it difficult for phosphorus and other nutrient elements in soil to move, resulting in a lack of available phosphorus and nutrient elements. The average value of available phosphorus in the soil in the limestone area was only 1/3 of that in the sandstone area [31]. The contents of available Zn, Cu, and B in calcareous soil in Guangxi were 1.48, 1.45, and 0.04 mg kg^{-1}, respectively [32], while the contents of available Zn, Cu, and B in the topsoil of paddy soil measured in Yixing city, Jiangsu Province, in 1995 were 3.94, 5.12, and 0.28 mg kg^{-1}, respectively [33]. Clearly, the contents of available Zn, Cu, and B in calcareous soil in Guangxi Province are significantly lower than those in paddy soil in Yixing city, Jiangsu Province.

Therefore, deficiencies in available nitrogen, phosphorus, and other nutrients in karst areas, which limit plant growth and development, are another limiting factor for plant photosynthetic carbon sequestration in karst areas.

3. Karstification–Photosynthesis Coupling Increases Karst Carbon Sinks

Plants (including microalgae) not only directly use carbon dioxide from the atmosphere, but also bicarbonate from soil (water bodies) [34–40], which transform bicarbonates from karstification into photosynthetic carbon sinks [35,41]. Both photosynthetic organisms in the ocean (or other water bodies) and terrestrial plants can "intercept" partial bicarbonate to form a photosynthetic carbon sink. Some bicarbonate dissolved from carbonate becomes sedimentary rock on the seafloor; however, a considerable amount of bicarbonate is involved in photosynthetic assimilation into organic matter. Aquatic photosynthetic organisms and terrestrial plants "intercept" inorganic carbon, which is then replenished with atmospheric CO_2 to capture it. Therefore, the "interception" of bicarbonate by plants stabilizes the apparent karst carbon sink, results in the coupling of karstification and photosynthesis, and yields a net karst carbon sink.

In fact, photosynthesis and karstification in nature are coupled (Figure 1) [42–45]. Karstification–photosynthesis coupling has always constrained atmospheric CO_2 concentrations. In geological history, higher concentrations of atmospheric CO_2 occurred in periods of lagging photosynthesis development, high CO_2 levels in the early Paleozoic era were associated with non-land plant continents, and low CO_2 levels in the Permian-Carboniferous period were associated with the occurrence, development, and flourishing of vascular plants [14,46–48]. Karstification–photosynthesis coupling connects the atmosphere, hydrosphere, and lithosphere through the biosphere, attenuating long-term CO_2 variability and resulting in the atmospheric composition remaining constant from pre-75 Ma to the Holocene (pre-Industrial Revolution) [49–54].

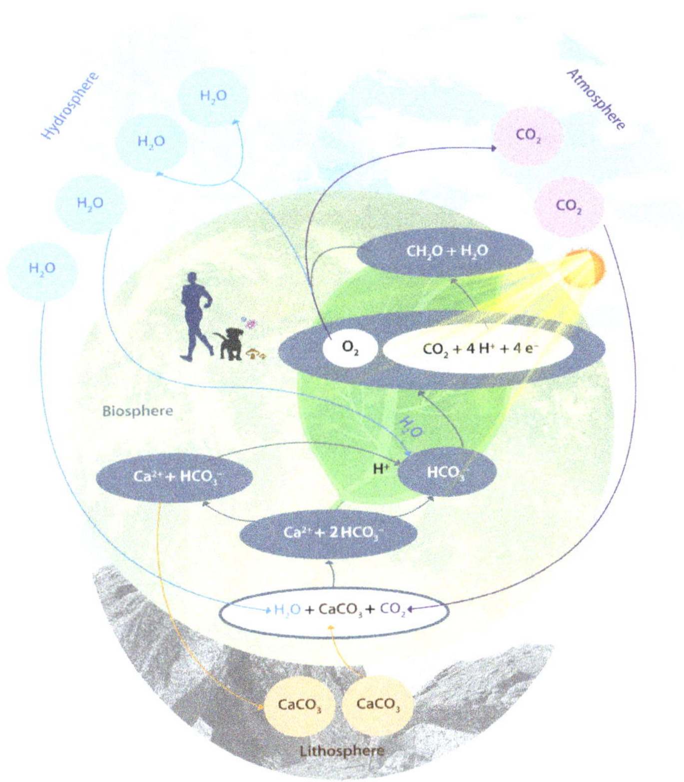

Figure 1. Karstification–photosynthesis coupling processes and their role in the water–carbon balance in nature. Carbonate rocks (lithosphere) are dissolved under the action of water (hydrosphere) and carbon dioxide (atmosphere) to form Ca^{2+} and bicarbonate (biosphere) (carbon sink process). Ca^{2+} combines with bicarbonate to precipitate calcium carbonate ($CaCO_3$) into the lithosphere (carbon source process). Plants split bicarbonate and water to release oxygen and carbon dioxide, then assimilate carbon dioxide to form carbohydrates (CH_2O) (biosphere) (carbon sink process). Finally, organisms utilize oxygen to decompose carbohydrates into carbon dioxide that enters the atmosphere, and water that enters the hydrosphere (carbon source process).

The direct determination of carbonate rock dissolution or changes in bicarbonate seems to reflect the ability of carbonate rock to capture CO_2. However, due to the high variability of carbonate rock dissolution and carbonate rock deposition, as well as the resulting high temporal and spatial heterogeneity of bicarbonate, the application of the carbonate-rock-tablet test, solute load method, and maximum potential dissolution method is limited [11]. Even if the dissolution amount of carbonate can be determined by the carbonate-rock-tablet test, this can only partially reflect the weathering of carbonate (apparent karst carbon sink). In addition, even if changes in the bicarbonate of karstic catchments can be determined via the solute load method or the maximum potential dissolution method, this only partially reflects the instantaneous apparent karst carbon sink capability. The ability of terrestrial and aquatic plants to "intercept" (assimilate) bicarbonates, i.e., net karst carbon sinks, is difficult to obtain using these methods. However, the amount of bicarbonate utilized by plants can represent the net karst carbon sink capacity due to karstification–photosynthesis coupling in nature (Figure 2).

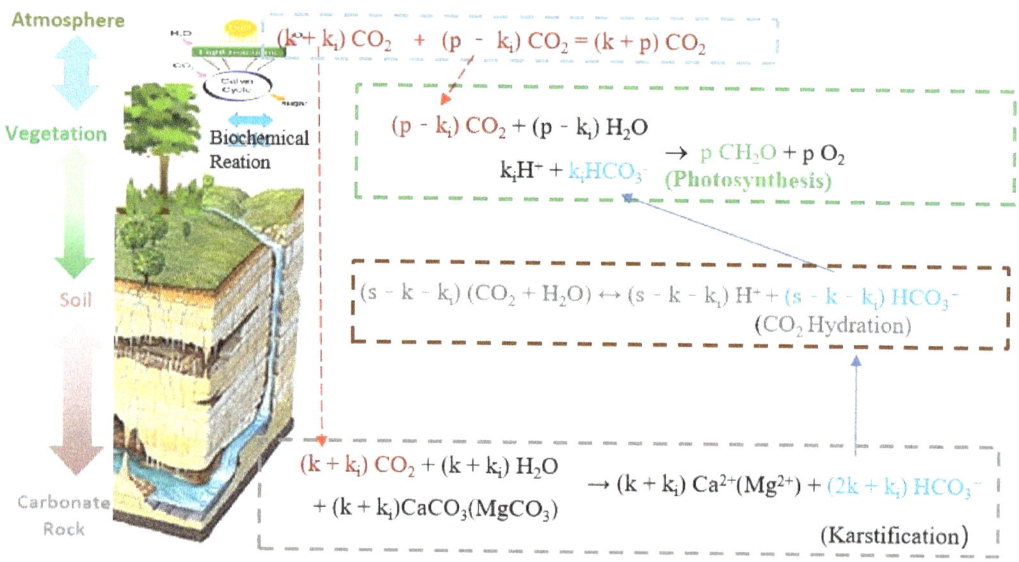

Figure 2. The determination principle of karstification, photosynthesis, and net karst carbon sink under karstification–photosynthesis coupling (example uses are terrestrial plants; aquatic plants replace soil with water bodies), where ki is the amount of root-originated HCO_3^- used by plants, k_i/p is the share of plant utilization of root-originated HCO_3^-, $k_i/(p - k_i)$ is the stoichiometric ratio of root-originated HCO_3^- and atmospheric CO_2 utilized by plants, p is the photosynthesis of plants, and $k + k_i$ is the karstification of carbonate rocks. It should be noted here that from the initial process of carbon sequestration, $p - k_i$ is the photosynthetic carbon sink of plants, $k + k_i$ is the apparent karst carbon sink of carbonate rocks, s is the carbon sink of soil, and $(k + k_i)/(p - k_i)$ is the stoichiometric ratio of karstification and photosynthesis under karstification–photosynthesis coupling. From the substantial process of carbon sequestration, p is the photosynthetic carbon sink of plants, which includes the direct carbon sink $p - k_i$ and the indirect carbon sink k_i, $k + k_i$ is the apparent karst carbon sink of carbonate rock, which includes the apparent karst basic carbon sink k and the intermediate carbon sink k_i from karstification–photosynthesis coupling, and k/p is the stoichiometric ratio of karstification and photosynthesis under karstification–photosynthesis coupling. It can also be seen that the karstification–photosynthesis coupling system not only increases the carbon sink of the system as a whole, but also increases the stability of the carbon sink in karst ecosystems, because the photosynthetic carbon sink is more stable than HCO_3^-.

Stable isotope techniques are widely used in tracing geological and biological processes [55]. Moreover, they can also successfully identify the role of plant photosynthesis, respiration, and water metabolism, and have become an important tool for studying the relationship between water and carbon in ecosystems [56–60]. However, due to the continuous conversion of various inorganic carbons and the continuous exchange of isotopes, conventional stable isotope technology is difficult to trace, and it is more difficult to quantify the utilization information of specific inorganic carbon sources. Therefore, to solve the problem of signal interference during inorganic carbon conversion and isotope exchange, we successfully developed a bidirectional isotope tracer culture technology. Bidirectional isotope tracer culture technology is used to simultaneously label and culture two identical plants with two kinds of bicarbonates at different stable carbon isotope ratios (the difference being more than 10‰), respectively, eliminate signal interference in the process of inorganic carbon transformation and isotope exchange by using parallel isotope difference signals, and finally, analyze and quantify the utilization information of different inorganic carbon sources. Now, bidirectional isotope tracer culture technology has become a promising technique for analyzing and quantifying the utilization of different inorganic carbon sources. Using this technique, combined with metabolic regulation technology (such as applying specific inhibitors), bicarbonate and CO_2 utilization by plants were quantified [34,36–38,40,41]. This technique was even used to successfully quantify the karstification and photosynthesis of microalgae [35,39,61]. Therefore, by comprehensively applying bidirectional isotope tracer culture technology, metabolic regulation technology, and physiological biochemistry, the stoichiometric relationship between karstification and photosynthesis, and HCO_3^- and CO_2 utilization by plants under karstification–photosynthesis coupling can be accurately quantified.

Bicarbonate utilization depends on the plant species and the environment. Isotopic techniques can be used to obtain the share of bicarbonate utilization by plants, and many studies have confirmed that karst-adaptable plants cope with karst adversity by increasing the share of bicarbonate utilization. The effect of sodium bicarbonate (10 mM) on photosynthetic carbon metabolism in karst-adaptable plants (*Broussonetia papyrifera*) using bidirectional isotope tracing culture techniques was studied [34]. The results showed that after 20 days of sodium bicarbonate treatment, the total photosynthetic rate of *Broussonetia papyrifera* was 2.65 $\mu mol(CO_2)\ m^{-2} \cdot s^{-1}$, of which the share of bicarbonate utilization accounted for 30%, while the total photosynthetic rate of mulberry trees (*Morus alba*) was 2.55 $\mu mol(CO_2)\ m^{-2} \cdot s^{-1}$, of which the utilization share of bicarbonate ions accounted for 0. Hang and Wu (2016) quantified the effects of different concentrations of sodium bicarbonate on photosynthetic carbon metabolism in the karst-adaptable plant *Orychophragmus violaceus*, using bidirectional isotope tracing culture techniques. The results showed that when concentrations of bicarbonate were 5, 10, and 15 mM, the share of bicarbonate utilization to the total carbon assimilation in *Orychophragmus violaceus* was 5.4%, 13.5%, and 18.8%, respectively. The bicarbonate utilization of *Brassica juncea* accounts for less than 5% of the total carbon assimilation [36].

When using polyethylene glycol 6000 (PEG 6000) to simulate karst drought to study the inorganic carbon assimilation response of *Orychophragmus violaceus* and *Brassica juncea* to PEG 6000, it was found that when the concentrations of PEG 6000 were 0, 10, 20, and 40 $g \cdot L^{-1}$, the shares of bicarbonate utilization to the total carbon assimilation of *Orychophragmus violaceus* were 6.7%, 13.1%, 17.6%, and 47.7%, respectively, and those of *Brassica juncea* were 2.9%, 7.6%, 7.7%, and 5.9%, respectively [62]. When the concentrations of PEG 6000 were 0, 100, and 200 $g \cdot L^{-1}$, the shares of bicarbonate utilization to the total carbon assimilation of *Camptotheca acuminata* were 10.34%, 20.05%, and 16.60%, respectively [37]. Under a simulated karst environment, after the 180-day growth phase, the shares of bicarbonate utilization to the total carbon assimilation of *Orychophragmus violaceus*, *Brassica juncea*, and *Euphorbia lathyris* were 11.45%, 10.39%, and 9.44%, respectively [38].

The degree of karstification–photosynthesis coupling and the net karst carbon sink capacity of ecosystems varies depending on land use. Table 1 shows the rate of dissolution measured via the carbonate-rock-tablet test method for different land use types. Table 2 shows carbon sink fluxes measured via the solute load method from several different karst catchments.

Table 1. Dissolution of different land use types.

Land Use	Average Dissolution Rate $mg\ cm^{-2}\ y^{-1}$	Karst Carbon Sink Intensity $tCO_2\ km^{-2}\ y^{-1}$	Depth of the Underlying Carbonate-Rock-Tablet cm	Reference
Woodland	1.01	4.69	10	[63]
Dry land	4.25	19.72	60	
Paddy land	0.14	0.65	46	
Shrub	0.61	2.83	50	[64]
Dry land	5.27	24.45	50	
Paddy land	3.05	14.15	50	
Woodland	7.61	35.31	50	[65]
Grassland	8.89	41.25	50	
Vegetable-planted land	6.21	28.81	50	
Tilled land	11.0	51.04	20	[66]
Tilled land	14.9	69.14	50	
Shrub	0.5	2.32	20	
Shrub	2.6	12.06	50	
Woodland	68.7	318.77	20	
Woodland	18.7	86.77	50	
Orchard	87.7	406.93	20	
Orchard	120.1	557.26	50	

Table 2. Carbon sink fluxes (CSFs, $tCO_2\ km^{-2}\ y^{-1}$) from several different karst catchments.

Catchment Name (Location)	CSFs	Reference	Catchment Name (Location)	CSFs	Reference
Banzhai (Libo)	28.84 ± 3.04	[67]	Huangzhou River (Shibing)	36.43	[68]
Huanghou (Libo)	32.81 ± 4.70		Chenqi (Puding)	55.07	[64]
Houzhai (Puding)	39.13 ± 7.56		Houzhai (Puding)	25.70	[69]

Dissolving 1 mol of carbonate rock can consume 1 mol of CO_2 from the atmosphere, contributing to the decline of atmospheric CO_2 concentration. Therefore, the rate of carbonate rock dissolution can reflect the karst carbon sink intensity. As seen from Table 1, the rate of dissolution varies depending on land use, different types of vegetation, or the same vegetation type for different observations. Karst carbon sink intensities range from 0.65 $tCO_2\ km^{-2}\ y^{-1}$ to 557.26 $tCO_2\ km^{-2}\ y^{-1}$, with an average value of 98.60 $tCO_2\ km^{-2}\ y^{-1}$ and a median of 28.81 $tCO_2\ km^{-2}\ y^{-1}$, which is only 2% of the average carbon sink capacity of karst forests (1447.05 $tCO_2\ km^{-2}\ y^{-1}$) [70]. As seen from Table 2, carbon sink fluxes are different at different times and catchments, and even in the same catchments, with an average of 36.33 $tCO_2\ km^{-2}\ y^{-1}$ in the five catchments, which is only 2.5% of the average carbon sink capacity of karst forests [70]. Plants have a much greater ability to "intercept" (assimilate) bicarbonates than the karst carbon sink intensities measured via the carbonate-rock-tablet test method or carbon sink fluxes measured via the solute load method. For example, the average share of bicarbonate utilization by *Camptotheca acuminata* in July was 13.6%, and that in August was 18.8%. The average share of bicarbonate utilization by *Platycarya longipes* in July was 15.3%, and that in August was 14.3% [71]. It can be seen that net karst carbon sinks account for more than 80% of apparent karst carbon sinks.

The positive vegetation succession of the ecosystem represents an increase in karstification–photosynthesis coupling. Table 3 shows the distribution of net primary productivity and carbon sink capacity at various stages of karst vegetation succession in Maolan. Maolan karst vegetation is divided into five successional stages: herbaceous community, shrub–scrub community, shrub community, subclimax community of evergreen–deciduous broadleaved mixed forests, and climax community of evergreen–deciduous broadleaved mixed forests. The strongest karstification–photosynthesis coupling occurs in the climax community of evergreen–deciduous broadleaved mixed forests, and the weakest karstification–photosynthesis coupling occurs in herbaceous communities. The stronger the karstification–photosynthesis coupling between the plant community and carbonate rocks is, the greater the carbon sink in karst ecosystems [72].

Table 3. Distribution of net primary productivity (t hm^{-2} y^{-1}) and carbon sink capacity (tCO_2 km^{-2} y^{-1}) at various stages of karst vegetation succession in Maolan.

Succession Stage	Arbor Layer	Shrub Layer	Herbal Layer	Coarse, Medium Root	Fine Roots	Total	Carbon Sinks [1]
Climax community of evergreen-deciduous broadleaved mixed forests	7.01	0.25	0.04	1.51	4.72	13.58	2240.70
Sub-climax community of evergreen-deciduous broadleaved mixed forests	5.09	0.32	0.09	0.71	2.52	8.73	1440.45
Scrub–shrub community	1.39	0.79	0.06	0.57	3.27	6.08	1003.20
Herb–scrub community		1.03	0.60	0.85	2.10	4.58	755.70
Herb community			1.19	0.12	1.42	2.73	450.45

[1] Note: The conversion factor for converting plant net primary productivity into plant carbon sinks is 165.

4. Carbonic Anhydrase, a Key Pivot of Karstification–Photosynthesis Coupling

Carbonic anhydrase (CA, EC 4.2.1.1) specifically and efficiently catalyzes the reversible conversion of bicarbonate to CO_2 at 10^7 times the rate of the nonenzymatic reaction, being one of the fastest enzymatic reactions. CA is continuously distributed in the rock–soil interface and surface soil–vegetation ecosystems, and is widely present in various organisms in the soil, water, and rock surface layers [73–77]. Different organisms, and even the same organism, have different CA isoenzyme types and activities in different environments; therefore, CA has high heterogeneity [78–81]. Characterized by its continuous distribution, abundance, specificity, and efficient and rapid catalysis of the unique mutual conversion reaction between CO_2 and HCO_3^- in karst ecosystems, carbonic anhydrase is the only key biological enzyme that can closely couple karstification with photosynthesis, that is, inorganic and organic carbon between carbonate rocks–soil–vegetation–atmosphere. On the one hand, CA catalyzes the dissolution of carbonate rocks and the deposition of calcium carbonate to regulate karstification at the rock–soil interface. On the other hand, CA catalyzes the reversible conversion between carbon dioxide and bicarbonate in the soil solution and plants to regulate the photosynthesis and respiration of plants (Figure 3). Carbonic anhydrase is highly spatiotemporally heterogeneous and sensitive to the environment, and is dubbed the karstification "mortise", catalyzing the biogeochemical cycle of water–carbon, and even other elements, regulating the migration and transformation of substances between different interfaces, and maintaining the biodiversity and stability of the system [35,40].

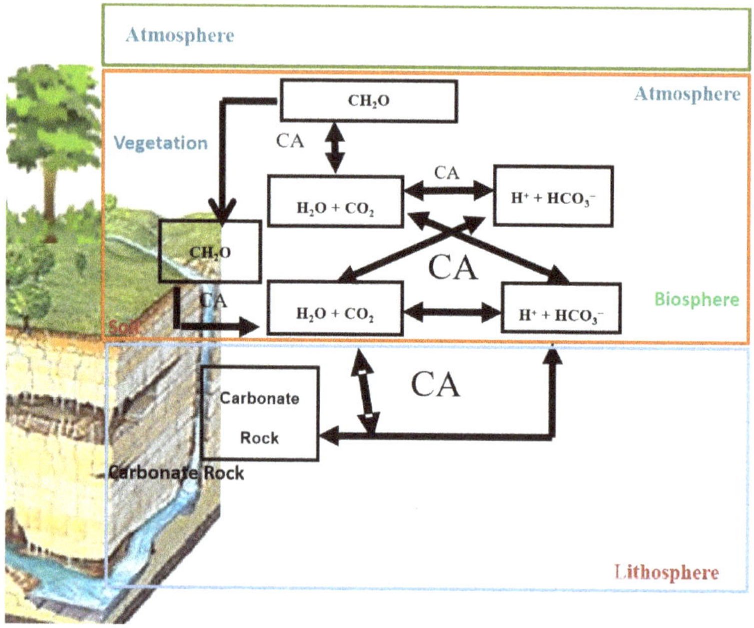

Figure 3. Carbonic anhydrase is pivotal in karstification–photosynthesis coupling. Carbonic anhydrase (CA) catalyzes the dissolution of carbonate rocks at the rock–soil interface, the hydration of CO_2 in the soil solution and soil–atmosphere and soil–vegetation interface, the photosynthesis of plants, etc.

4.1. Carbonic Anhydrase Controls the Transport of Water and Inorganic Carbon at the Carbonate Rocks–Soil–Atmosphere Interface by Catalyzing Carbonate Dissolution and Carbon Dioxide Hydration

The kinetically slow reaction $CO_2 + H_2O \leftrightarrow H^+ + HCO_3^-$ plays an important role in controlling the dissolution and precipitation of carbonate rocks in H_2O-CO_2-$CaCO_3$ systems [82]. Carbonic anhydrase, on the one hand, can quickly catalyze the CO_2 hydration reaction and improve the conversion rate between CO_2 and H^+ and HCO_3^-; on the other hand, the diffusion rate of related ions is affected by catalysis of the use of HCO_3^- and Ca^{2+} by organisms in the soil, thereby catalyzing the dissolution and sedimentation of carbonate rocks [83]. Exogenous CA, algal CA, and extracellular CA of microorganisms can promote the dissolution of carbonate rocks [61,83–85]. The flux of CO_2 in soil is also controlled by CA [86–88].

4.2. Carbonic Anhydrase Promotes Inorganic Carbon Assimilation in Plants

After plants experienced karst adversity, the stomatal conductivity was reduced or the stomata were closed, the atmospheric CO_2 supply was insufficient, and the absorption and transportation of water in roots were blocked. In response to water and CO_2 deficiencies, plants rapidly upregulated the gene expression of specific CA isoenzymes in leaves [40,89,90], increased the activity of corresponding CA isoenzymes [81,91], converted HCO_3^- from soil into water and CO_2, supplemented them to photosynthetic organs, improved the water supply and water efficiency, and increased the utilization of atmospheric CO_2 and HCO_3^- [34,36–38,40].

Carbonic anhydrase catalyzes the dissolution and sedimentation reaction of carbonate rocks at the rock–soil interface, which affects the concentration of HCO_3^- in soil controlled by the CA of plants and soil, and in turn affects the utilization of HCO_3^- by plants. Bicarbonate utilization not only provides plants with metabolic water (intracellular water), increases stomatal opening, and directly increases the assimilation of CO_2, but also increases the photosynthetic area by increasing the adaptability and promoting the growth of plants [40]. In addition, the concentration of bicarbonate in soil can also affect CO_2 fluxes in soil, which are also regulated by CA [77,92].

5. Carbon Sequestration and the Enhancement Effect of Karst-Adaptable Plants during Karstification–Photosynthesis Coupling

After the Industrial Revolution, buried fossil fuels were exploited in large quantities, releasing a large amount of carbon dioxide over a short time, while plant adaptive evolution and development lagged behind, resulting in karstification–photosynthesis decoupling, which eventually led to a global increase in carbon dioxide. After karstification–photosynthesis decoupling, karst-adaptable plants evolve gradually to adapt to karst environments of karst drought, high pH, and high contents of calcium and bicarbonate, and their unique adaptation mechanisms and strategies in turn can accelerate to reach a new degree of karstification–photosynthesis coupling [40,93,94].

Karst-adaptable plants efficiently use bicarbonate, compensatorily absorb nitrate nitrogen, and attain higher photosynthetic carbon sequestration capacity at a lower cost [95], thereby stimulating plant growth, increasing the photosynthetic area, and then promoting root-originated bicarbonate and atmospheric carbon dioxide use by plants, further enhancing photosynthetic carbon sinks. Moreover, karst-adaptable plants reduce the content of bicarbonate in rhizosphere soil during the use of bicarbonate. As a result, the dissolution effect of carbonate rock is further strengthened to supplement bicarbonate into the soil, which further promotes the karst carbon sink capacity, promotes root-originated bicarbonate use by plants, and finally achieves the conjugate promotion of karst carbon sinks and photosynthetic carbon sinks. In fact, karstification that is not coupled with photosynthesis has a limited carbon sink capacity, but if karstification has a high degree of coupling with photosynthesis, its carbon sink capacity is extremely huge. Theoretically, the carbon sink capacity of an 8-year woody plant with karstification–photosynthesis coupling of 10% will be twice that without karstification–photosynthesis coupling. The carbon sink capacity of a 10-year woody plant with a karstification–photosynthesis coupling of 10% is 1.6 times that with a karstification–photosynthesis coupling of 5% ($1.10^{10}/1.05^{10}$); thus, the karst carbon sink capacity is 3.2 times that of the latter.

6. Conclusions and Outlook

Karst geological action by itself is a zero-carbon sink process on a billion-year long time scale. Uncoupled karstification limits photosynthetic carbon sequestration by plants. Karstification–photosynthesis coupling, driven by the bicarbonate utilization of plants and pivoted by carbonic anhydrase, can promote inorganic carbon assimilation in plants and the dissolution of carbonate rocks, thereby stabilizing and increasing the capacity of karst carbon sinks and photosynthetic carbon sinks. Karstification–photosynthesis coupling determines the carbon sink of karst ecosystems. Full use of the adaptation strategies of karst-adaptable plants with high karstification–photosynthesis coupling can maximize carbon sequestration and the enhancement effect of plants in karst areas. Finally, a carbon sequestration system encompassing ecological restoration, rocky desertification control, and the sustainable use of plant resources can be developed.

Author Contributions: Y.W. (Yanyou Wu) and Y.W. (Yansheng Wu) cooperated to complete this article. All authors have read and agreed to the published version of the manuscript.

Funding: This work was supported by the National Natural Science Foundation of China (No. U1612441-2) and Support Plan Projects of Science and Technology of Guizhou Province (No. (2021)YB453).

Institutional Review Board Statement: Not applicable.

Data Availability Statement: All data generated or analyzed during this study are included in this published article.

Acknowledgments: We deeply thank K. Murali, C. Zeng, X. Bai, and Z. Liu for valuable comments. I am grateful to M. Wu for help with Figure 1. Thanks also to the anonymous referees.

Conflicts of Interest: The authors declare no conflict of interest. The funders had no role in the design of the study.

References

1. IPCC. IPCC Sixth Assessment Report. Available online: https://www.ipcc.ch/report/ar6/wg1/, (accessed on 13 July 2022).
2. NOAA. Global Monitoring Laboratory. Available online: https://gml.noaa.gov/ccgg/trends/mlo.html, (accessed on 13 July 2022).
3. Jia, Z.; Li, Z. *Carbonate Sedimentary Facies and Sedimentary Environment*; China University of Geosciences Press: Wuhan, China, 1989; pp. 1–6.
4. Wang, S. The most serious ecological and geological environmental problem in southwest China-Karst rock desertification. *Bull Miner. Pet. Geochem* **2003**, *22*, 120–126.
5. Jiang, Z.; Lian, Y.; Qin, X. Rocky desertification in Southwest China: Impacts, causes, and restoration. *Earth Sci. Rev.* **2014**, *132*, 1–12. [CrossRef]
6. Xiong, P.; Yuan, D.; Xie, S. Progress of research on rocky desertification in South China Karst Mountain. *Carsol Sin* **2010**, *29*, 355–356.
7. Oelkers, E.H.; Declercq, J.; Saldi, G.D.; Gislason, S.R.; Schott, J. Olivine dissolution rates: A critical review. *Chem. Geol.* **2018**, *500*, 1–19. [CrossRef]
8. Gaillardet, J.; Galy, A. Himalaya-Carbon Sink or Source. *Science* **2008**, *320*, 1727–1728. [CrossRef] [PubMed]
9. Yuan, D. Progress in the study on "Karst process and carbon cycle". *Adv. Earth Sci.* **1999**, *14*, 425–432.
10. Larson, C. An unsung carbon sink. *Science* **2011**, *334*, 886–887. [CrossRef]
11. Liu, Z. "Method of maximum potential dissolution" to calculate the intensity of karst process and the relevent carbon sink: With discussions on methods of solute load carbonate-rock-tablet test. *Carsol. Sin.* **2011**, *30*, 379–382.
12. Liu, Z.H. New progress and prospects in the study of rock-weathering-related carbon sinks. *Chin. Sci. Bull.* **2012**, *57*, 95–102. [CrossRef]
13. Chen, C.Y.; Liu, Z.H. The role of biological carbon pump in the carbon sink and water environment improvement in karst surface aquatic ecosystems. *Chin. Sci. Bull.* **2017**, *62*, 3440–3450. [CrossRef]
14. Berner, R.A. Atmospheric carbon dioxide levels over Phanerozoic time. *Science* **1990**, *249*, 1382–1386. [CrossRef] [PubMed]
15. Berner, R.A.; Kothavala, Z. GEOCARB III: A revised model of atmospheric CO_2 over Phanerozoic time. *Am. J. Sci.* **2001**, *301*, 182–204. [CrossRef]
16. Curl, R.L. Carbon shifted but not sequestered. *Science* **2012**, *335*, 655. [CrossRef]
17. Yuan, D. Rock desertification in the subtropical karst of south China. *Z. Geomorph. (Suppl.)* **1997**, *108*, 81–90.
18. Cao, J.; Yuan, D.; Pan, G. Sme soil features in karst ecosystem. *Adv. Earth Sci.* **2003**, *18*, 37–44.
19. Legrand, H.E.; Stringfield, V.T. Karst hydrology:A review. *J. Hydrol.* **1973**, *20*, 97–120. [CrossRef]
20. Qin, X. Characteristics and comprehensive treatment strategy of karst drought in Central Guangxi. *J. Guilin Univ. Technol.* **2005**, *25*, 278–283.
21. Yuan, D. Carbon cycle in Earth system and its effects on environment and resources. *Quatern. Sci.* **2001**, *21*, 223–232.
22. Guo, F.; Jiang, G.; Yuan, D. Major ions in typical subterranean rivers and their anthropogenic impacts in southwest karst areas, China. *Environ. Geol.* **2007**, *53*, 533–541. [CrossRef]
23. Song, T.; Peng, W.; Du, H.; Wang, K.; Zeng, F. Occurrence, spatial-temporal dynamics and regulation strategies of karst rocky desertification in southwest China. *Chin. J. Ecol.* **2014**, *34*, 5328–5341.
24. Liu, L.; Zeng, F.; Song, T.; Peng, W.; Wang, K.; Qin, W.; Tan, W. Spatial heterogeneity of soil nutrients in Karst area's Mulun National Nature Reserve. *Chin. J. Appl. Ecol.* **2010**, *21*, 1667–1673.
25. Dong, L.; He, T.; Liu, Y.; Shu, Y.; Luo, H.; Liu, F. Changes of soil physical- chemical properties derived from different parent materials/rocks in karst mountain. *Chin. J. Soil Sci.* **2008**, *39*, 471–474.
26. Ji, F.; Li, N.; Deng, X. Calcium contents and high calcium adaptation of plants in karst area of China. *Chin. J. Plant Ecol.* **2009**, *33*, 926–935.
27. Xiong, D.; Cai, H.; Zhang, R.; Li, C.; QIan, X. Distribution of soil medium and micro-elements in Fujian tobacco-growing soils. *Chin. J. Eco Agric.* **2007**, *15*, 36–38.
28. Li, Z.; Zou, B.; Cao, Y.; Ren, H.; Liu, J. nutrient characteristics of soils in typical degraded hilly land in Southern China. *Acta. Ecol. Sin.* **2003**, *23*, 1648–1656.
29. Paramasivam, S.; Alva, A.K. Nitrogen recovery from controlled-release fertilizers under intermittent leaching and dry cycles. *Soil Sci.* **1997**, *162*, 447–453. [CrossRef]
30. Wei, X.; Shao, M. The distribution of soil nutrients on sloping land in the gully regiion watersheds of the Loess Plateau. *Acta. Ecol. Sin.* **2007**, *27*, 603–613.
31. Liu, C. *Biogeochemical Processes and Cycling of Nutrients in the Earth's Surface: Cycling of Nutrients in Soil-Plant Systems of Karstic Environments, Southwest China*; Science Press of China: Beijing, China, 2009; pp. 308–353, 401–443.
32. National Soil Census Office of China. *Soil in China*; China Agriculture Press: Beijing, China, 1998; pp. 843–984.

33. Jiang, D.; Cheng, J. Changes of pH and available pool of B, Cu, Zn, Mn in surface rice soils in Yixing, Jiangsu province over the last decade. *J. Nanjing Agric. Univ.* **1997**, *20*, 111–113.
34. Wu, Y.Y.; Xing, D.K. Effect of bicarbonate treatment on photosynthetic assimilation of inorganic carbon in two plant species of *Moraceae*. *Photosynth.* **2012**, *50*, 587–594. [CrossRef]
35. Wu, Y.; Li, H.; Xie, T. *Biogeochemical Action of Microalgal Carbonic Anhydrase*; Science Press of China: Beijing, China, 2015; pp. 75–110.
36. Hang, H.T.; Wu, Y.Y. Quantification of photosynthetic inorganic carbon utilisation via a bidirectional stable carbon isotope tracer. *Acta. Geochim.* **2016**, *35*, 130–137. [CrossRef]
37. Rao, S.; Wu, Y.Y. Root-derived bicarbonate assimilation in response to variable water deficit in *Camptotheca acuminate* seedlings. *Photosyn. Res.* **2017**, *134*, 59–70. [CrossRef] [PubMed]
38. Wang, R.; Wu, Y.; Xing, D.; Hang, H.; Xie, X.; Yang, X.; Zhang, K.; Rao, S. Biomass production of three biofuel energy plants' use of a new carbon resource by carbonic anhydrase in simulated karst soils: Mechanism and capacity. *Energies* **2017**, *10*, 1370. [CrossRef]
39. Xie, T.; Wu, Y. The biokarst system and its carbon sinks in response to pH changes: A simulation experiment with microalgae. *Geochem. Geophys. Geosyst.* **2017**, *18*, 827–843. [CrossRef]
40. Wu, Y.; Xing, D.; Hang, H.; Zhao, K. *Principles and Techniques of Determination on Plants' Adaptation to Karst Environment*; Science Press of China: Beijing, China, 2018; pp. 1–188.
41. Fang, L.; Wu, Y.Y. Bicarbonate uptake experiment show potential karst carbon sinks transformation into carbon sequestration by terrestrial higher plants. *J. Plant Interact.* **2022**, *17*, 419–426. [CrossRef]
42. Wu, Y. Combined effect of bicarbonate and water in photosynthetic oxygen evolution and its implication for carbon neutrality. *Sci. Bull.* **2022**. submitted and revised.
43. He, S.; Yu, S.; Pu, J.; Yuan, Y.; Zhang, C. Dynamics in Riverine Inorganic and Organic Carbon Based on Carbonate Weathering Coupled with Aquatic Photosynthesis in a Karst Catchment, Southwest China. *Water Res.* **2021**, *189*, 116658.
44. Yang, R.; Sun, H.; Chen, B.; Yang, M.; Zeng, Q.; Zeng, C.; Huang, J.; Luo, H.; Lin, D. Temporal variations in riverine hydrochemistry and estimation of the carbon sink produced by coupled carbonate weathering with aquatic photosynthesis on land: An example from the Xijiang River, a large subtropical karst-dominated river in China. *Environ. Sci. Pollut. Res.* **2020**, *27*, 13142–13154. [CrossRef]
45. Cao, J.H.; Wu, X.; Huang, F.; Hu, B.; Groves, C.; Yang, H.; Zhang, C.L. Global significance of the carbon cycle in the karst dynamic system: Evidence from geological and ecological processes. *China Geol.* **2018**, *1*, 17–27. [CrossRef]
46. Berner, R.A. The carbon cycle and carbon dioxide over Phanerozoic time: The role of land plants. *Philos. Trans. R. Soc. B* **1998**, *353*, 75–82. [CrossRef]
47. Moulton, K.L.; Berner, R.A. Quantification of the effect of plants on weathering: Studies in Iceland. *Geology* **1998**, *26*, 895–898. [CrossRef]
48. Knoll, A.H.; Nowak, M.A. The timetable of evolution. *Sci. Adv.* **2017**, *3*, e1603076. [CrossRef] [PubMed]
49. Hart, M.H. The evolution of the atmosphere of the Earth. *Icarus* **1978**, *33*, 23–39. [CrossRef]
50. Des Marais, D.J. When did photosynthesis emerge on Earth? *Science* **2000**, *289*, 1703–1705. [CrossRef]
51. Dismukes, G.C.; Klimov, V.V.; Baranov, S.V.; Kozlov, Y.N.; DasGupta, J.; Tyryshkin, A. The origin of atmospheric oxygen on Earth: The innovation of oxygenic photosynthesis. *Proc. Natl. Acad. Sci. USA* **2001**, *98*, 2170–2175. [CrossRef] [PubMed]
52. Pagani, M.; Caldeira, K.; Berner, R.A.; Beerling, D.J. The role of terrestrial plants in limiting atmospheric CO_2 decline over the past 24 million years. *Nature* **2009**, *460*, 85–88. [CrossRef] [PubMed]
53. Kasting, J.F. Earth's early atmosphere. *Science* **1993**, *259*, 920–926. [CrossRef]
54. Halevy, I.; Bachan, A. The geologic history of seawater pH. *Science* **2017**, *355*, 1069–1071. [CrossRef]
55. Teng, F.Z.; Ma, L. Deciphering isotope signatures of Earth Surface and Critical Zone processes. *Chem. Geol.* **2016**, *445*, 1–3. [CrossRef]
56. Schwender, J.; Goffman, F.; Ohlrogge, J.B.; Shachar-Hill, Y. Rubisco without the Calvin cycle improves the carbon efficiency of developing green seeds. *Nature* **2004**, *432*, 779–782. [CrossRef]
57. Tcherkez, G.; Mahe, A.; Gauthier, P.; Mauve, C.; Gout, E.; Bligny, R.; Cornic, G.; Hodges, M. In folio respiratory fluxomics revealed by ^{13}C isotopic labeling and H/D isotope effects highlight the noncyclic nature of the tricarboxylic acid "Cycle" in illuminated leaves. *Plant Physiol.* **2009**, *151*, 620–630. [CrossRef]
58. Lin, G. Stable isotope ecology: A new branch of ecology resulted from technology advances. *Chin. J. Plant Ecol.* **2010**, *34*, 119–122.
59. Nichols, J.; Booth, R.K.; Jackson, S.T.; Pendall, E.G.; Huang, Y.S. Differential hydrogen isotopic ratios of *Sphagnum* and vascular plant biomarkers in ombrotrophic peatlands as a quantitative proxy for precipitation-evaporation balance. *Geochim. Cosmochim. Acta* **2010**, *74*, 1407–1416. [CrossRef]
60. Broeckx, L.S.; Fichot, R.; Verlinden, M.S.; Ceulemans, R. Seasonal variations in photosynthesis, intrinsic water-use efficiency and stable isotope composition of poplar leaves in a short-rotation plantation. *Tree Physiol.* **2014**, *34*, 701–715. [CrossRef] [PubMed]
61. Xie, T.; Wu, Y. The role of microalgae and their carbonic anhydrase on the biological dissolution of limestone. *Environ. Earth Sci.* **2014**, *71*, 5231–5239. [CrossRef]
62. Hang, H. Expression Encoding Carbonic Anhydrase Gene and Inorganic Carbon Utilization of the Two Plants Species under Karst Environment. Ph.D. Dissertation, The University of Chinese Academy of Sciences, Beijing, China, 2015.

63. Yan, W.; Zeng, C.; Xiao, S.; Lan, J.; Dai, L.; Tai, Z.; He, J.; He, C.; Di, Y. Dissolution rate and karst carbon sink of different land use in typical dolomite watershed with humid subtropical weather. *Earth Environ.* **2021**, *49*, 529–538.
64. Zeng, C.; Zhao, M.; Yang, R.; Liu, Z. Comparison of karst processes-related carbon sink intensity calculated by carbonate rock tablet test and solute load method: A case study in the Chenqi karst spring system. *Hydrogeo Eng. Geol.* **2014**, *41*, 106–111.
65. Lan, J.C.; Fu, W.L.; Peng, J.T.; Zhou, X.P.; Xiao, S.Z.; Yuan, B. Dissolution rate under soil in karst areas and the influencing factors of different land use patterns. *Acta Ecol. Sin.* **2013**, *33*, 3205–3212.
66. Zhang, C. Carbonate rock dissolution rates in different land uses and their carbon sink effect. *Chin. Sci. Bull.* **2011**, *56*, 3759–3765. [CrossRef]
67. Zeng, C.; Liu, Z.; Zhao, M.; Yang, R. Hydrologically-driven variations in the karst-related carbon sink fluxes: Insights from high-resolution monitoring of three karst catchments in Southwest China. *J. Hydrol.* **2016**, *533*, 74–90. [CrossRef]
68. Zeng, C.; He, C.; Xiao, S.; Liu, Z.; Chen, W.; He, J. Karst related inorganic carbon sink flux in a typical humid subtropical dolomite basin. *Earth Sci. Front.* **2022**, *29*, 179–188.
69. Li, S.; Liu, C.; Li, J.; Lang, Y.; Ding, H.; Li, L. Geochemistry of dissolved inorganic carbon and carbonate weathering in a small typical karstic catchment of Southwest China: Isotopic and chemical constraints. *Chem. Geol.* **2010**, *277*, 301–309. [CrossRef]
70. Yu, W.; Dong, D.; Ni, J. Comparison of biomass and net primary productivity of Karst and non-karst forests in mountainous area, Southwestern China. *J. Subtrop. Resour. Environ.* **2010**, *5*, 25–30.
71. Wu, Y.; Rao, S.; Wu, Y.; Fang, L.; Su, Y.; Li, H.; Wang, R.; Wang, S.; Liu, C. *A Method for Obtaining the Share of Bicarbonate Utilization by Plants in Natural Habitats*; ZL2018103132903, Beijing; CN108319820B; State Intellectual Property Office of China: Beijing, China, 2018.
72. Xia, H. Biomass and net primary production in different successional stages of karst vegetation in Maolan, SW China, Guizhou. *For. Sci. Technol.* **2010**, *38*, 1–7.
73. Veitch, F.P.; Blankenship, L.C. Carbonic anhydrase in bacteria. *Nature* **1963**, *197*, 76–77. [CrossRef] [PubMed]
74. Badger, M.R.; Price, G.D. The role of carbonic anhydrase in photosynthesis. *Annu. Rev. Plant Biol.* **1994**, *45*, 369–392. [CrossRef]
75. Smith, K.S.; Ferry, J.G. Prokaryotic carbonic anhydrases. *FEMS Microbiol. Rev.* **2000**, *24*, 335–366. [CrossRef] [PubMed]
76. Gilmour, K.M.; Perry, S.F. Carbonic anhydrase and acid-base regulation in fish. *J. Exp. Biol.* **2009**, *212*, 1647–1661. [CrossRef] [PubMed]
77. Price, G.D.; Long, B.M.; Förster, B. DABs accumulate bicarbonate. *Nat. Microbiol.* **2019**, *4*, 2029–2030. [CrossRef]
78. Ivanov, B.N.; Ignatova, L.K.; Romanova, A.K. Diversity in forms and functions of carbonic anhydrase in terrestrial higher plants. *Russ. J. Plant Physl.* **2007**, *54*, 143–162. [CrossRef]
79. Sun, W.; Wu, Y.; Wen, X.; Xiong, S.; He, H.; Wang, Y.; Lu, G. Different mechanisms of photosynthetic response to drought stress in tomato and violet orychophragmus. *Photosynthetica* **2016**, *54*, 226–233. [CrossRef]
80. DiMario, R.J.; Machingura, M.C.; Waldrop, G.L.; Moroney, J.V. The many types of carbonic anhydrases in photosynthetic organisms. *Plant Sci.* **2018**, *268*, 11–17. [CrossRef]
81. Hang, H.; Wu, Y. Effect of bicarbonate stress on carbonic anhydrase gene expressions from *Orychophragmus violaceus* and *Brassica juncea* seedlings. *Pol. J. Environ. Stud.* **2019**, *28*, 1135–1143. [CrossRef]
82. Liu, Z.; Dreybrod, W. Dissolution kinetics of calcium carbonate minerals in H_2O–CO_2 solutions in turbulent flow: The role of the diffusion boundary layer and the slow reaction $H_2O + CO_2 \rightarrow H^+ + HCO_3^-$. *Geochim. Cosmochim. Acta* **1997**, *61*, 2879–2889. [CrossRef]
83. Li, W.; Yu, L.J.; Wu, Y.; Jia, L.P.; Yuan, D.X. Enhancement of Ca^{2+} release from limestone by microbial extracellular carbonic anhydrase. *Bioresour. Technol.* **2007**, *98*, 950–953. [CrossRef] [PubMed]
84. Liu, Z. Role of carbonic anhydrase as an activator in carbonate rock dissolution and its implication for atmospheric CO_2 sink. *Acta Geol. Sin. (Engl. Ed.)* **2001**, *75*, 275–278.
85. Papamichael, E.M.; Economou, E.D.; Vaimakis, T.C. Dissolution of the carbonate minerals of phosphate ores: Catalysis by carbonic anhydrase II, from bovine erythrocytes, in acid solutions. *J. Colloid Interf. Sci.* **2002**, *251*, 143–150. [CrossRef] [PubMed]
86. Seibt, U.; Wingate, L.; Lloyd, J.; Berry, J.A. Diurnally variable $\Delta^{18}O$ signatures of soil CO_2 fluxes indicate carbonic anhydrase activity in a forest soil. *J. Geophy. Res. Biogeo.* **2006**, *111*, G04005.
87. Wingate, L.; Seibt, U.; Maseyk, K.; Ogee, J.; Almeida, P.; Yakir, D.; Pereira, J.; Mencuccini, M. Evaporation and carbonic anhydrase activity recorded in oxygen isotope signatures of net CO_2 fluxes from a Mediterranean soil. *Glob. Change Biol.* **2008**, *14*, 2178–2193. [CrossRef]
88. Von Sperber, C.; Weiler, M.; Brüggemann, N. The effect of soil moisture, soil particle size, litter layer and carbonic anhydrase on the oxygen isotopic composition of soil-released CO_2. *Eur. J. Soil Sci.* **2015**, *66*, 566–576. [CrossRef]
89. Hu, H.; Boisson-Dernier, A.; Israelsson-Nordström, M.; Böhmer, M.; Xue, S.; Ries, A.; Godoski, J.; Kuhn, J.M.; Schroeder, J.I. Carbonic anhydrases are upstream regulators of CO_2-controlled stomatal movements in guard cells. *Nat. Cell Biol.* **2010**, *12*, 87–93. [CrossRef]
90. Hu, H.; Rappel, W.J.; Occhipinti, R.; Ries, A.; Böhmer, M.; You, L.; Xiao, C.; Engineer, C.B.; Boron, W.F.; Schroeder, J.I. Distinct cellular locations of carbonic anhydrases mediate carbon dioxide control of stomatal movements. *Plant Physiol.* **2015**, *169*, 1168–1178. [CrossRef] [PubMed]

91. Wu, Y.Y.; Liu, C.Q.; Li, P.P.; Wang, J.Z.; Xing, D.; Wang, B.L. Photosynthetic characteristics involved in adaptability to Karst soil and alien invasion of paper mulberry (*Broussonetia papyrifera* (L.) Vent.) in comparison with mulberry (*Morus alba* L.). *Photosynthetica* **2009**, *47*, 155–160. [CrossRef]
92. Nathan, V.K.; Ammini, P. Carbon dioxide sequestering ability of bacterial carbonic anhydrase in a mangrove soil microcosm and its bio-mineralization properties. *Water Air Soil Pollut.* **2019**, *230*, 192. [CrossRef]
93. Wu, Y.; Liu, C.; Wang, S. *Adaptability to Karst of Orychophragmus violaceus*; Guizhou Science and Technology Press: Guiyang, China, 2014; pp. 1–153.
94. Xia, A.; Wu, Y. Joint interactions of carbon and nitrogen metabolism dominated by bicarbonate and nitrogen in *Orychophragmus violaceus* and *Brassica napus* under simulated karst habitats. *BMC Plant Biol.* **2022**, *22*, 264. [CrossRef]
95. Wu, Y. Strategies to increase carbon fixation and sequestration by Karst-adaptable plants. *Carsol. Sin.* **2011**, *30*, 461–465.

agronomy

MDPI

Article

Species-Specific and Altitude-Induced Variation in Karst Plants' Use of Soil Dissolved Inorganic Carbon

Sen Rao [1,2] and Yanyou Wu [2,*]

1 College of Life Sciences and Oceanography, Shenzhen University, Shenzhen 518060, China
2 State Key Laboratory of Environmental Geochemistry, Institute of Geochemistry, Chinese Academy of Sciences, Guiyang 550081, China
* Correspondence: wuyanyou@mail.gyig.ac.cn

Abstract: Root-derived carbon sources supporting photosynthesis have been demonstrated to contribute to plant carbon gain in many laboratory experiments. However, it remains largely unknown whether and to what extent soil dissolved inorganic carbon (DIC) influences leaf photosynthesis in karst habitats characterized by alkaline soils with low water content. We explored this relationship by measuring the concentrations and carbon isotope signals ($\delta^{13}C$) of soil DIC, as well as the $\delta^{13}C$ of water-soluble organic matter (δ_{WSOM}) in leaves of nine woody species across an altitudinal gradient in karst habitats. The δ_{WSOM} varied among species by 7.23‰ and deviated from the $\delta^{13}C$ of photosynthates solely assimilated from atmospheric CO_2 (δ_A) by 0.44–5.26‰, with a mean value of 2.20‰. This systematical discrepancy ($\delta_A - \delta_{WSOM}$) could only be explained by the contribution of soil DIC to leaf total photosynthesis (f_{DIC_soil}). The average values of f_{DIC_soil} considerably varied among the nine species, ranging from 2.48% to 9.99%, and were comparable with or slightly lower than those of previous laboratory experiments. Furthermore, the f_{DIC_soil} of two species significantly increased with altitude, whereas another species exhibited an opposite pattern, suggesting a highly spatial heterogeneity of DIC utilization. The present study improved our understanding of how plants adapt to the alkaline–drought soil conditions of karst habitats and thus acquire additional carbon for growth.

Keywords: soil dissolved inorganic carbon; photosynthesis; water-soluble organic matter; stable isotope; contribution

Citation: Rao, S.; Wu, Y. Species-Specific and Altitude-Induced Variation in Karst Plants' Use of Soil Dissolved Inorganic Carbon. *Agronomy* **2022**, *12*, 2489. https://doi.org/10.3390/agronomy12102489

Academic Editor: Zhengqin Xiong

Received: 28 August 2022
Accepted: 5 October 2022
Published: 12 October 2022

1. Introduction

Karst is a distinctive topography developed in regions with soluble bedrocks, such as limestone, dolomite, and gypsum [1], and covers about 22×10^6 km^2 of the earth's surface [2]. The karst habitats are characterized by thin soil layers, high content of bicarbonate and high pH in soils, and less vegetation cover [3–5]. For plants growing in karst habitats, water demand for maintaining the basic functions is often restricted due to the shallow soil [6]. For example, drought stress can lead to the decline of stomata opening and photosynthesis and downregulated metabolic processes [7]. A prolonged drought will result in hydraulic failure and/or carbon starvation, increasing the risk of mortality [8,9]. On the other hand, the alkaline soil condition in karst habitats is unfavorable for plant uptake of nutrients, such as phosphate, ferrous iron, and zinc [10,11]. As a consequence, the vegetation productivity of the karst ecosystem is usually much less than that in nonkarst regions [12]. Although plants can adopt tolerance and/or avoidance strategies to cope with environmental stress [13,14], it does not fully explain why some species grow well in karst habitats. Hence, other physiological processes of karst plants are suspected to play an important role in regulating their growth under dry and alkaline soil conditions.

Recently, root-derived carbon sources supporting photosynthesis have reemerged as a hotspot in plant ecophysiological studies [15–18]. Research shows that ^{13}C-enriched CO_2

or HCO_3^- labeling in the root zones can be transported upward through the transpiration stream [19] and thus affects the carbon isotope composition ($\delta^{13}C$) of leaf photoassimilates and aboveground tissues [20,21]. Although these results are obtained from the manipulated experiments inside a laboratory, it may be applied to karst plants grown in the field, given the high concentration of bicarbonate contained in karst soils. The additional carbon gain may also compensate for the decrease in the fixation of atmospheric CO_2 induced by drought stress [22,23]. However, to the best of our knowledge, no research has been reported to investigate plants' use of dissolved inorganic carbon (DIC, mainly including dissolved CO_2 and HCO_3^-) in karst habitats. Although many hydroponic experiments were designed to simulate karst soil conditions, for example, bicarbonate stress (usually more than 10 mM) [24,25], osmotic stress (simulating water scarcity) [26], and their interactions [26], their results do not necessarily reflect the utilization of soil DIC by plants due to the differences in the microenvironment of the root zones between soil conditions and hydroponic solutions [27]. Hence, there is a need to fill this knowledge gap between laboratory experiments and field trials.

Appropriate methods and models for quantification are critical to understanding plants' use of soil DIC. Currently, two methods are available to estimate this utilization. Method (i) applies the high abundance ^{13}C labeling and then calculates the ratio of soil DIC fixed in specific tissues or organs to the total carbon gain. The total carbon gain includes soil DIC fixation, apparent photosynthesis measured by commercial infrared gas analyzers, and respiration [20,28]. However, this calculation is not directly linked to photosynthesis. Method (ii) uses a two-source ^{13}C labeling in combination with isotope mixing models to determine the contribution of root-derived DIC to leaf total photosynthesis [22–24,29]. Nevertheless, uncertainties remain associated with the photosynthetic ^{13}C discrimination and post-photosynthesis processes. Additionally, both labeling methods depend on irrigation or hydroponic culture, which will inevitably change the soil conditions, especially in karst environments. The labeling experiment also requires a sealed environment and has a high cost, which does not apply to large-scale field experiments. For these reasons, a new approach needs to be developed to calculate the contribution of soil DIC to leaf total photosynthesis (see Section 2). It has been shown that the natural abundance of ^{13}C has the potential to address a wide range of ecophysiological and biochemical questions [30]. For example, naturally occurring $\delta^{13}C$ signals in plant tissues integrate plant–environment interactions over long periods [31]. In C_3 plants, $\delta^{13}C$ of leaves is mainly controlled by photosynthetic ^{13}C discrimination [32], which can be altered by stomatal control and the activity of the carboxylation enzyme, Rubisco. Furthermore, when other carbon sources are supplying the leaf photosynthesis, for instance, uptake of DIC from the xylem sap of the host by the mistletoe or utilization of root-derived DIC by some plants [26,33], the $\delta^{13}C$ of leaves can also be modified.

In karst habitats, altitude plays an important role in affecting the soil conditions and microclimate [34], which determine the distribution of plant species [35]. Species with different life forms may vary in functional traits, such as root uptake and xylem transport of water and DIC, thus influencing the extent of soil DIC used by plants. In addition, the concentration of soil DIC may change with altitudes, which probably acts on the proportion of soil DIC used by leaf photosynthesis. In this study, the primary objective was to investigate whether and to what extent soil DIC influences the leaf photosynthesis of karst plants at different altitudes. Based on our previous observation [26], we expected to find a large discrepancy between the measured $\delta^{13}C$ of photosynthates and that predicted by a photosynthetic ^{13}C discrimination model, which indicates the involvement of soil DIC in leaf photosynthesis. We hypothesized that the contribution of soil DIC to leaf total photosynthesis was species-specific due to the different physiological responses (e.g., leaf gas exchange) to the environmental factors. Our third hypothesis is the altitude exerts an influence on plants' use of soil DIC, given the differences in soil conditions and microclimate induced by altitude.

2. Materials and Methods

2.1. Study Site

The field experiments were conducted in August 2016 on Mt. Sanmao, Nanming District, Guiyang, southwest China (26°33′57″ N, 106°44′56″ E), with an elevation between 1080 and 1315 m above sea level. In this region, the annual mean precipitation and mean air temperature during 2011 and 2015 were 1129.5 mm and 15.3 °C, respectively. The mean minimum temperature of the coolest month (January) was 4.6 °C, whereas the mean maximum temperature of the warmest month (July) ranged from 25 to 28 °C. Rainfall in August was decreased compared with that in July. The study site was a typical karst landform covered with a secondary forest. Soils under the vegetation examined were alkaline (pH 7.09–7.75), rich in organic matter and bicarbonate, and have low soil water content according to our preliminary survey at the same site [34]. Moreover, these soil properties changed along an altitude gradient. The bedrock was mainly carbonatite, which developed shallow calcareous soil. The thickness of the soil layer varied between 27 and 58 cm. The vegetation community was mixed with evergreen and deciduous broad-leaved forest, with the canopy height in the range of 0.8–7.6 m.

2.2. Experimental Design

Six plots were established along three altitudes: S1 and S2 on the lower altitude, S3 and S4 on the medium altitude, and S5 and S6 on the higher altitude (Figure 1). Each plot had an area of 20 × 20 m^2 and 60–80 m away from the neighboring plot at each altitude. Common species in each plot were recorded in Table 1. Collectively, four tree species (*Ligustrum lucidum* ait., *Broussonetia papyrifera* L., *Platycarya longipes* Wu., and *Zelkova serrata* (Thunb.) Makino) and five shrub species (*Viburnum dilatatum* Thunb., *Ampelopsis delavayana* Planch., *Rosa cymosa* Tratt., *Zanthoxylum armatum* DC., and *Rubus biflorus* Buch.-Ham. ex Smith) were selected for the subsequent measurement and sampling. Species in each plot included four replicates (individuals).

Table 1. Common species in plots of different altitudes. S1–S6 represented six plots established at three altitudes.

Altitudes	Sampling Sites	Elevation (m)	Species	
			Tree	Shrub
Lower	S1	1115	*L. lucidum*, *B. papyrifera*	*V. dilatatum*, *A. delavayana*, *R. cymosa*
	S2	1123	*L. lucidum*, *B. papyrifera*	*A. delavayana*, *R. cymosa*
Medium	S3	1224	*P. longipes*, *Z. serrata*	*V. dilatatum*, *A. delavayana*
	S4	1229	*L. lucidum*, *P. longipes*, *Z. serrata*	*A. delavayana*, *R. cymosa*
Higher	S5	1289	*P. longipes*	*V. dilatatum*, *A. delavayana*, *Z. armatum*, *R. biflorus*
	S6	1292	*L. lucidum*	*A. delavayana*, *Z. armatum*, *R. biflorus*

2.3. Leaf Gas-Exchange Measurements

Leaf gas exchange was measured in the morning (9:00–11:30 a.m.) of three sunny days (17–19 August). The net photosynthetic rate (A), stomatal conductance (g_s), transpiration rate (E), ratio of intercellular to the ambient partial pressure of CO_2 (c_i/c_a), and instantaneous water use efficiency (WUE_i) of the fully expanded leaves were measured using an infrared gas analyzer LI-6400 (Li-Cor, Lincoln, NE, USA), equipped with a clear topped 2 × 3 cm cuvette. The block temperature of the leaf chamber was set to be close to the air temperature. Near-ambient air at the height of 5 m above the ground was supplied to the leaf chamber at a rate of 500 μmol s^{-1} in the sample cell of LI-6400. Leaves were placed inside the cuvette for 2–3 min until steady state, and then the gas exchange parameters were manually logged. In addition, dark respiration of leaves was measured after 30 min of dark adaptation. Three leaves were measured per tree or shrub. At the end of the measurements, leaf samples were immediately collected and stored in a cooler box for further determination.

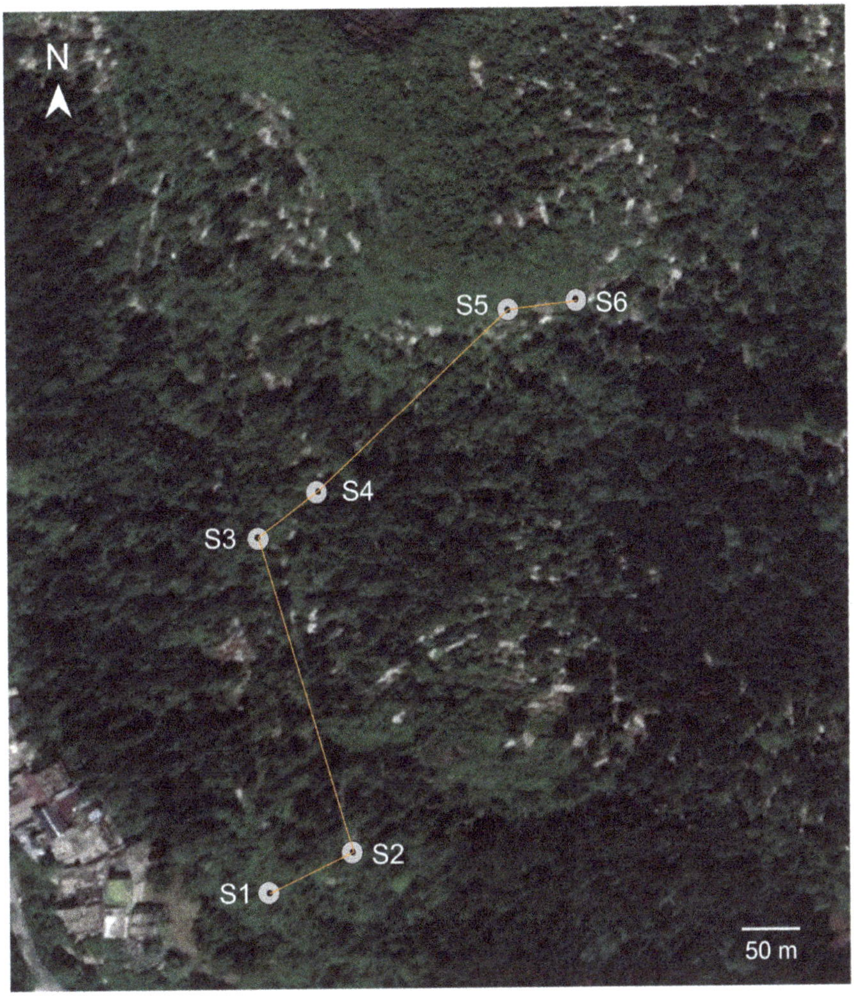

Figure 1. Location of sampling sites (S, circles) near Nanming District, Guiyang, southwest China (adapted from Google Map).

2.4. Air and Soil Sampling

Ambient air at the height of 5 m above the ground was sampled once in each plot during the measurement of leaf gas exchange. The air sample was pumped into gas sampling bags (Dalian Delin Gas Packing Co., Ltd., Dalian, China) for further analysis. Soil air was sampled in the root zone of each plant. Three days prior to our sampling, two holes of 10–15 cm depth with 2 cm diameter were drilled into the soil. Subsequently, two soft gas-piping tubes with filters were inserted into the holes, and then soil air was sucked out. Finally, the end of the tubes was clamped with clips. These processes allowed a long time for soil air to equilibrate between soil porosity and the internal space of tubes. At the day of sampling, the litter layer was removed, and then soil air was pumped into a gas sampling bag with a volume of 30 mL through one of the tubes, waiting for further determination. Another tube was connected to LI-6400, and, thus, the CO_2 concentration of soil air was directly measured when it reached a peak value. Soil samples (approximately

150 g) were collected from the top layer of soil (0 to 20 cm) after removing the litter. They were immediately kept in sealed plastic bags and stored in a cooler box for further analysis.

2.5. Determination of Carbon Isotope Composition

The $\delta^{13}C$ of water-soluble organic matter (WSOM) in leaves instead of bulk leaves was determined to represent the carbon isotope signals of photosynthates. Leaf WSOM has a high turnover rate [36–38] and mainly includes soluble carbohydrates, the hydrolysates of starch, and amino acids [39,40]. The extraction of WSOM referred to the protocols of [41]. Firstly, dried leaf material was ground, and 50 mg sample was put into a 2.5 mL centrifuge tube. After 1 mL of double-deionized water was added, the tube was agitated and then placed in a freezer for 1 h at 4 °C. Secondly, the samples were heated for 10 min at 95 °C, cooled to room temperature (RT), and centrifuged for 10 min at 12,000 g. Thirdly, an aliquot of 25 μL of the supernatant was transferred to a tin capsule and dried at 60 °C for 24 h. The $\delta^{13}C$ of the WSOM was analyzed with an Isotope Ratio Mass Spectrometer (MAT-253, Thermo Fisher Scientific Inc., Waltham, MA, USA). Three leaf samples were measured per plant. The carbon isotopic ratio of the samples was calculated as $\delta^{13}C$ (‰) = $[(R_{sample}/R_{standard}) - 1] \times 1000$, where R_{sample} and $R_{standard}$ are the $^{13}C/^{12}C$ ratio of the sample and Vienna Pee Dee Belemnite (VPDB), respectively. The standard deviation of repeated measurements for MAT253 was 0.1‰ for $\delta^{13}C$.

The $\delta^{13}C$ of atmospheric and soil CO_2 was determined by TG-IsoPrime (GV. Instruments Ltd., Manchester, UK). The method for soil DIC extraction was slightly modified from that of [42]. Briefly, 100 g of soil was placed into a 500 mL glass bottle and then mixed with 200 mL of CO_2-free water. The mixture was homogeneously stirred for 5 min and then incubated at RT for 30 min. Subsequently, the supernatant was centrifuged for 15 min at 3500× g and then filtered through a 0.45 μm water-base membrane. The total alkalinity of the filtrate was detected by an Aquamerck alkalinity kit (MColortest, Merck KGaA, Darmstadt, Germany). At the same time, soil pH was measured using a pH meter (S210-B, Mettler Toledo Group Ltd., Zurich, Switzerland). The remaining filtrate was transferred to a 50 mL bottle and sealed with a rubber plug. The air inside the bottle was drawn by a vacuum pump. Subsequently, the filtrate of DIC was pretreated with an aliquot of 3 mL phosphoric acid to react with DIC and produce CO_2. The generated CO_2 was then processed with a custom-built vacuum extraction system, through which the CO_2 was purified. The $\delta^{13}C$ of the CO_2 generated from soil DIC was then analyzed by MAT-252 (Finnigan, Bremen, Germany). The measurement precision for TG-IsoPrime and MAT252 were 0.1‰ and 0.01‰ for $\delta^{13}C$, respectively.

2.6. Quantification of the Utilization of Soil DIC by Plants

We accounted for two end members, namely atmospheric CO_2 and soil DIC, that might contribute to the constitution of newly formed photosynthates as well as its $\delta^{13}C$ [26]. The $\delta^{13}C$ of photosynthates (δ_A) solely assimilated from atmospheric CO_2 could be predicted with Equation (1) [43]:

$$\delta_A = \frac{\delta_a - \Delta^{13}C_{com}}{1 + \Delta^{13}C_{com}/1000} \quad (1)$$

where δ_a is the carbon isotope composition of atmospheric CO_2, whereas $\Delta^{13}C_{com}$ is the comprehensive discrimination against ^{13}C including the diffusion of CO_2 across the boundary layer, intercellular air space, mesophyllcell, as well as the contributions of respiration and photorespiration. $\Delta^{13}C_{com}$ is calculated using Equation (2) [44]:

$$\Delta^{13}C_{com} = \frac{1}{1-t}\left[\bar{a}\frac{c_a - c_i}{c_a}\right] + \frac{1+t}{1-t}\left[a_m\frac{c_i - c_c}{c_a} + \frac{c_c}{c_a}\left(b_3' - \frac{\alpha_b}{\alpha_e}e\frac{R_d}{V_c} - \frac{\alpha_b}{\alpha_f}f\frac{F}{V_c}\right)\right] \quad (2)$$

where c_a, c_i, and c_c denote the CO_2 partial pressure in the ambient air of the cuvette, intercellular air, and chloroplast, respectively; a_m is the summed $^{12}C/^{13}C$ fractionation during the dissolution of CO_2 and liquid-phase diffusion (a_m = 1.8‰); b_3' is the $^{12}C/^{13}C$

fractionation of Rubisco ($b_3' = 29‰$, [45]); e is the $^{12}C/^{13}C$ fractionation for day respiration ($e = 0$, [46]); and f is the $^{12}C/^{13}C$ fractionation during photorespiration ($f = 11‰$, [47]); R_d, V_c, and F are nonphotorespiratory CO_2 released in the dark, Rubisco carboxylation rate for the fixed atmospheric CO_2, and photorespiratory rate, respectively; $\alpha_b = 1 + b_3'$, $\alpha_e = 1 + e$, and $\alpha_f = 1 + f$; t and $\bar{a}$ are the ternary correction factor and the weighed fractionation for CO_2 diffusion across the boundary layer and stomata, respectively, which are calculated as [32]

$$t = \frac{\alpha_{ac} E}{2 g_{ac}} \tag{3}$$

$$\bar{a} = \frac{a_b(c_a - c_s) + a_s(c_s - c_i)}{c_a - c_i} \tag{4}$$

where $\alpha_{ac} = 1 + \bar{a}$, E is the transpiration rate; g_{ac} is the combined boundary layer and stomatal conductance to CO_2; a_b and a_s are the $^{12}C/^{13}C$ fractionation for CO_2 diffusion across the boundary layer and through the stomata ($a_b = 2.9‰$, $a_s = 4.4‰$), respectively; and c_s is the CO_2 partial pressure at the leaf surface. c_c was not directly measured by LI-6400, but it was calculated with Equation (5) [37]:

$$A = g_i(c_i - c_c) \tag{5}$$

where g_i is an internal CO_2 conductance. In this study, we assumed g_i of 0.5 mol m^{-2} s^{-1} for woody species according to [48]. V_c and F are calculated as [44]

$$V_c = A + R_d + F \tag{6}$$

where

$$F = \frac{\Gamma^*(A + R_d)}{c_c - \Gamma^*} \tag{7}$$

where Γ^* is the CO_2 compensation point in the absence of R_d, which was fitted using Equation (8) [49]:

$$\Gamma^* = e^{\left(c - \frac{\Delta H * 1000}{(R * (273 + T_{leaf}))}\right)} \tag{8}$$

where c, ΔH, and R are the scaling constant (13.49), energy of activation (24.46 kJ K^{-1} mol^{-1}), and molar gas constant (0.008314 kJ J^{-1} mol^{-1}), respectively.

When accounting for the involvement of soil DIC in leaf photosynthesis, the carbon isotopic composition of photosynthate (δ_A') originating from the assimilation of soil DIC is given by

$$\delta_A' = \frac{\delta_{DIC} - \Delta^{13}C_{DIC}}{1 + \Delta^{13}C_{DIC}/1000} \tag{9}$$

where δ_{DIC} is the carbon isotope composition of soil DIC collected from root zones of each plant; $\Delta^{13}C_{DIC}$ is the photosynthetic discrimination against ^{13}C combining the carboxylation of DIC and the effects of respiration and photorespiration, which is expressed as

$$\Delta^{13}C_{DIC} = b_3' - \frac{\alpha_b}{\alpha_e} e \frac{R_d}{V_c'} - \frac{\alpha_b}{\alpha_f} f \frac{F}{V_c'} \tag{10}$$

where V_c' is the rate of Rubisco carboxylation for DIC and given by

$$V_c' = A_{DIC\_soil} + R_d + F \tag{11}$$

where A_{DIC_soil} is the net photosynthetic rate for soil DIC. Empirically, A_{DIC} is less than 10% of A in many species. Here, we assumed an initial value for A_{DIC_soil} (A_{DIC_soil0}) to be $A/10$. Although this assumption introduced some errors, it had little effect on the calculation of $\Delta^{13}C_{DIC}$ and the contribution of DIC to leaf total photosynthesis. Considering bicarbonate was the dominant species in soil DIC under high pH (7.09–7.75), δ_{DIC} in Equation (9) is derived from the $\delta^{13}C$ of bicarbonate minus the discrimination related to the conversion between bicarbonate and CO_2 (e_b', 9‰ at 25 °C).

However, studies showed that root-respired CO_2 ($C_{\_R}$) can also dissolve in the xylem sap [50]. $C_{\_R}$ is primarily reported to range from 0.01 to 10.98 mM and varies with species, height, season, etc. [51,52]. Therefore, we proposed a ratio of the concentration of soil CO_2 or DIC ($C_{\_DIC}$) to $C_{\_R}$ in the xylem, $f_{s/x} = C_{\_DIC}/(C_{\_DIC} + C_{\_R})$. In the present study, we directly measured $C_{\_DIC}$ and adopted the mean value of 5 mM for $C_{\_R}$. Thus, the $\delta^{13}C$ of the mixed DIC in the xylem sap (δ_{DIC_xylem}) is calculated as

$$\delta_{DIC\_xylem} = f_{s/x} \cdot \delta_{DIC} + (1 - f_{s/x})\, \delta_R \quad (12)$$

where δ_R is the carbon isotope composition of root-respired CO_2 dissolved in the xylem sap. We assumed that the respiration of living tissues came from the recently fixed C, such as WSOM, which would carry the enriched isotopic signal to roots due to post-photosynthetic fractionation (−2‰, [53]) and then result in −1−−3‰ of respiratory fractionation in roots [54–56]. Here, we used the intermediate value −2‰ for respiratory fractionation; thus, δ_R is calculated as $\delta_R = \delta_{WSOM} + 4‰ - a_m$.

Considering that both atmospheric CO_2 and soil DIC could be used for leaf photosynthesis, the mixture of photosynthates from two substrates determined the value of δ_{WSOM}, which is expressed as a two-end-member mixing model:

$$\delta'_A \cdot f_{DIC\_xylem} + \delta_A \left(1 - f_{DIC\_xylem}\right) = \delta_{WSOM} \quad (13)$$

where f_{DIC_xylem} is the proportion of xylem DIC contributing to δ_{WSOM}; δ_A' is recalculated with Equation (9) but in which δ_{DIC} is replaced by δ_{DIC_xylem}. Finally, the contribution of soil DIC to the total leaf photosynthesis (f_{DIC_soil}) is calculated as

$$f_{DIC\_soil} = f_{s/x} \cdot f_{DIC\_xylem} \quad (14)$$

2.7. Statistical Analysis

The comparison of mean values between species and altitudes was determined by one-way ANOVA with a *t*-test. Linear regression was used to investigate the relationship between f_{DIC_soil} and δ_A-δ_{WSOM}. Pearson's correlation was conducted to evaluate the relationships between physiological or environmental factors and f_{DIC_soil}. All statistical analyses were performed using SPSS statistical software for Windows 21 (SPSS Inc., Chicago, IL, USA). The tests were considered significant at the $p < 0.05$ level. All data were expressed as the mean ± 1SE.

3. Results

3.1. Leaf Gas Exchange

All gas-exchange parameters exhibited significant differences within nine species (Table 2). However, no clear trend was observed between two life forms (tree and shrub). The maximum and minimum values of A were 18.24 ± 0.92 μmol m^{-2} s^{-1} in *B. papyrifera* and 9.81 ± 0.92 μmol m^{-2} s^{-1} in *R. cymosa*, respectively. The g_s of nine species ranged from 0.14 to 0.28 mol m^{-2} s^{-1}, suggesting mild-to-moderate drought stress. In contrast, there were small variations of E, c_i/c_a, and WUE_i among nine species, with mean values of 3.91 mol H_2O m^{-2} s^{-1}, 0.67, and 3.27 μmol H_2O mol^{-1}, respectively.

Table 2. Leaf gas exchange of nine species across six plots. *A*, net photosynthetic rate; g_s, stomatal conductance; *E*, transpiration rate; c_i/c_a, ratio of intercellular to ambient partial pressure of CO_2; WUE_i, instantaneous water use efficiency. Values represent mean ± SE. N = 4 for each species in the specific plot. Capital letters indicate significant differences ($p < 0.05$) among species.

Species	*A* (μmol m^{-2} s^{-1})	g_s (mol m^{-2} s^{-1})	*E* (mol H_2O m^{-2} s^{-1})	c_i/c_a	WUE_i (μmol H_2O mol^{-1})
L. lucidum	11.07 (0.72) C	0.14 (0.01) D	3.38 (0.20) CD	0.60 (0.01) C	3.30 (0.14) BCD
B. papyrifera	18.24 (0.92) A	0.28 (0.01) A	3.67 (0.19) BCD	0.67 (0.01) AB	4.98 (0.13) A
P. longipes	9.99 (0.59) C	0.19 (0.02) BCD	3.81 (0.16) ABCD	0.72 (0.02) A	2.71 (0.23) D
Z. serrata	13.91 (0.27) B	0.21 (0.01) BC	3.95 (0.25) ABCD	0.67 (0.01) AB	3.61 (0.22) B
V. dilatatum	10.58 (0.47) C	0.18 (0.01) CD	3.59 (0.11) CD	0.70 (0.01) AB	2.96 (0.13) CD
A. delavayana	13.39 (0.47) B	0.24 (0.01) AB	4.58 (0.26) A	0.70 (0.01) AB	3.11 (0.18) BCD
R. cymosa	9.81 (0.50) C	0.16 (0.02) CD	3.23 (0.22) D	0.65 (0.03) BC	3.10 (0.11) BCD
Z. armatum	14.07 (1.03) B	0.21 (0.02) BC	4.18 (0.27) ABC	0.66 (0.02) AB	3.42 (0.24) BC
R. biflorus	13.54 (0.41) B	0.21 (0.01) BC	4.46 (0.13) AB	0.68 (0.01) AB	3.04 (0.10) BCD

3.2. Characteristics of Soil DIC and CO_2

The average concentrations of DIC (or $C_{\_DIC}$) in the rhizosphere of four tree species were similar (around 9.26 mM), whereas in five shrub species, they exhibited large variations, ranging from 8.10 mM in *R. cymosa* to 10.50 mM in *Z. armatum* (Figure 2A). In contrast, the mean concentrations of soil CO_2 varied from 3919.91 ppm in *P. longipes* to 5279.83 ppm in *B. papyrifera*, higher than the mean value of 3454.42 ppm in all shrub species. According to Henry's law, the partial pressure of CO_2 over a solution was proportional to the concentration of the CO_2 in the solution. The calculated quantity of CO_2 dissolved in soil water ([CO_2^*]) was in the range of 0.55 to 0.92 mM at pH 7 and 25 °C across nine species, much lower than the mean value of [CO_2^*] in xylem sap (or $C_{\_R}$) investigated in many species. Thus, it was unlikely that soil CO_2 could be fixed by plant roots due to a CO_2 gradient from roots to the soil. In addition, the interspecies differences in the $\delta^{13}C$ of soil DIC and CO_2 were both less than 2‰ (Figure 2B). The mean values of soil DIC and CO_2 among nine species were −10.19‰ and −20.01%, respectively.

3.3. $\delta^{13}C$ of Photosynthates

The $\delta^{13}C$ of the newly formed photosynthates in leaves could be determined by measuring the isotope signals of WSOM, or predicted with Eqn. 1 that only accounted for the assimilation of atmospheric CO_2 (Figure 3A). δ_{WSOM} varied all among species by about 3–4‰ and was generally lower than δ_A in all species. The discrepancy between δ_A and δ_{WSOM} (δ_A-δ_{WSOM}) was species-specific. Figure 3B displayed the distribution of δ_A-δ_{WSOM} in leaves of nine species under varying altitudes. The highest frequency (approximate 30%) of δ_A-δ_{WSOM} was located in the range of 1 to 2‰. Furthermore, half of the δ_A-δ_{WSOM} was distributed between 2‰ and 6‰. This phenomenon clearly illustrated the fact that there were other carbon sources, for example, soil DIC, which participated in leaf photosynthesis. The photosynthates originating from the mixture of soil DIC and xylem [CO_2^*] carried more ^{13}C-depleted signal (δ_A', see Supplementary Data), thus resulting in δ_{WSOM} systematically lower than δ_A.

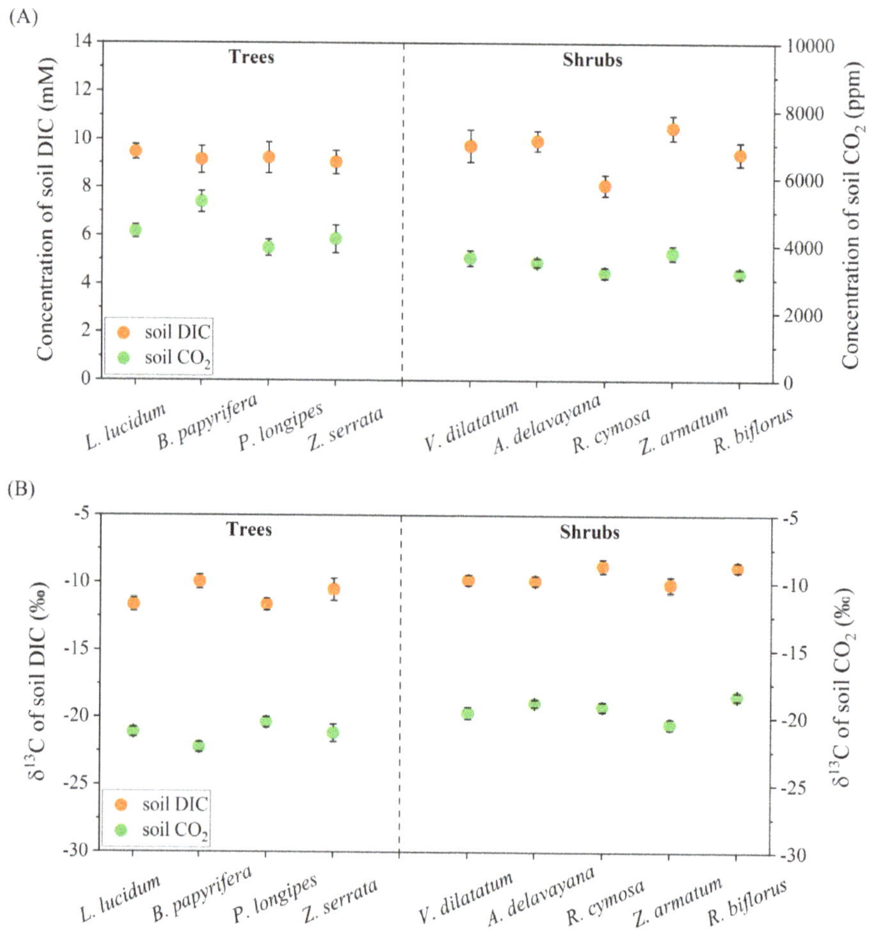

Figure 2. Concentrations (**A**) and $\delta^{13}C$ (**B**) of soil DIC and CO_2 in the root zones of nine species across three altitudes. N = 4 for each species in the specific plot.

3.4. Interspecies Difference in f_{DIC_soil}

The potential contribution of soil DIC to leaf total photosynthesis (f_{DIC_soil}) considerably varied with species across the six plots (Figure 4). In tree species, the average values of f_{DIC_soil} in *L. lucidum* and *B. papyrifera* were 8.93% and 9.54%, respectively, significantly higher than those in *P. longipes* and *Z. serrata* (both less than 2.2%). By contrast, the mean values of f_{DIC_soi} within the shrub species varied from 2.48% to 9.99%, with the highest value in *Z. armatum* and the lowest value in *R. biflorus*. Furthermore, f_{DIC_soil} strongly correlated with δ_A-δ_{WSOM} ($p < 0.001$), with the coefficient of determination (R^2) as 0.99. For the nine species, a higher value of δ_A-δ_{WSOM} generally indicated a higher value of f_{DIC_soil} (Figure 5). This tight relation further demonstrated the plants' use of soil DIC in the karst habitats. Additionally, f_{DIC_soi} significantly correlated with *A*, c_i/c_a, and WUE_i (Table 3).

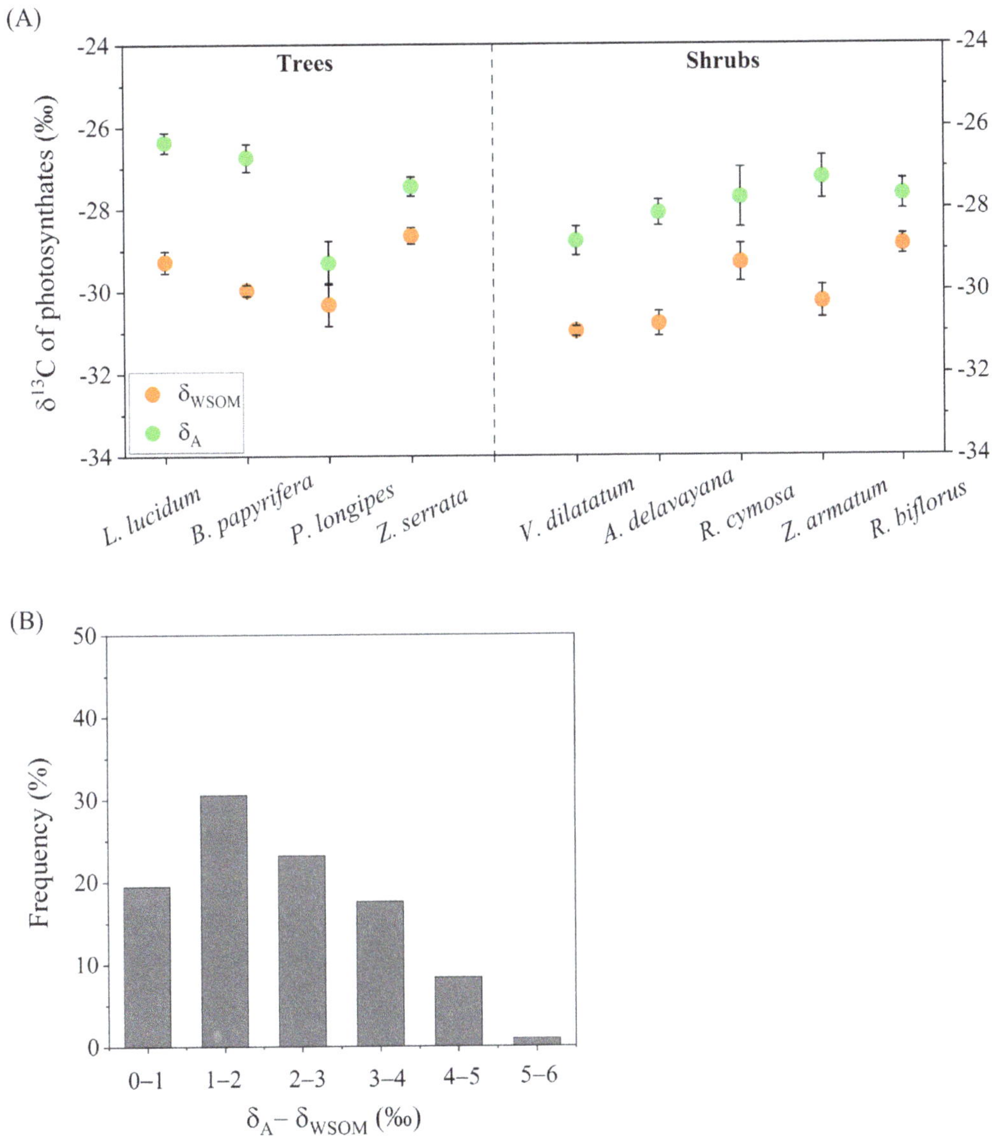

Figure 3. $\delta^{13}C$ of photosynthates determined by measuring isotope signals of leaf WSOM (orange closed circle) or predicted with Equation (1) that only accounted for assimilation of atmospheric CO_2 (green closed circle) of nine species across three altitudes (**A**); frequency distribution of the discrepancy between δ_A and δ_{WSOM} (**B**).

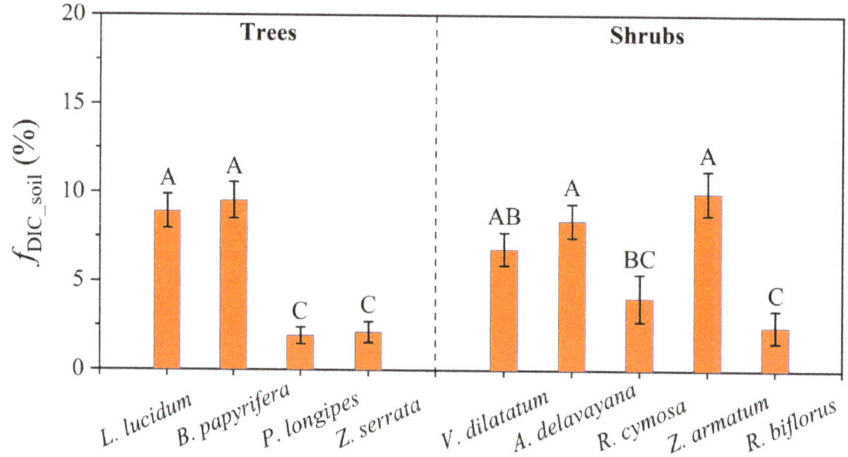

Figure 4. Contribution of soil DIC to leaf total photosynthesis (f_{DIC_soil}) in nine species across three altitudes. N = 4 for each species in the specific plot. Capital letters on the error bar indicate significant differences ($p < 0.05$) among species.

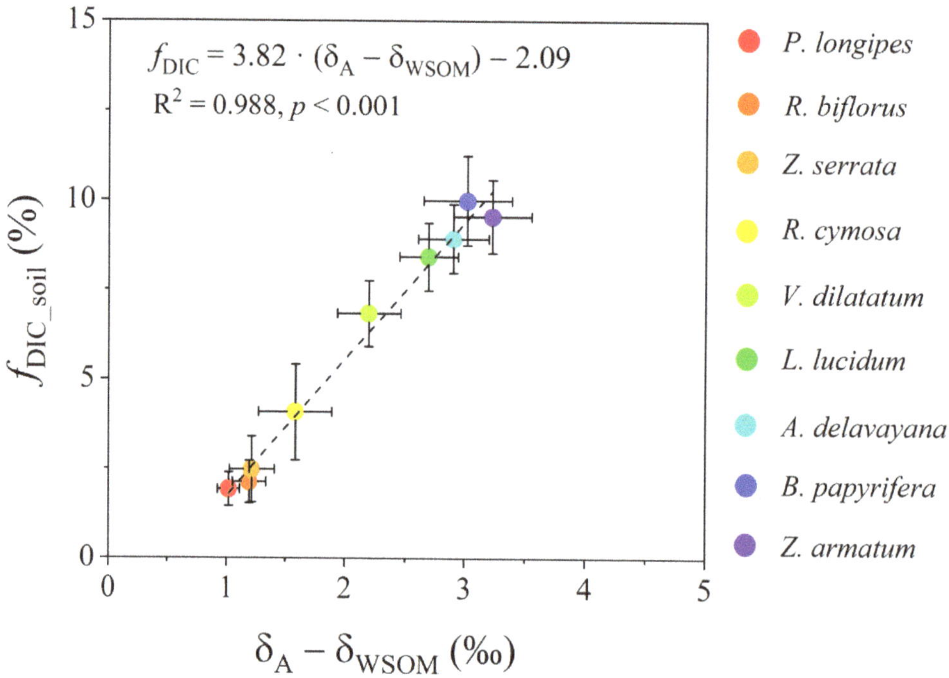

Figure 5. Relationship between (δ_A-δ_{WSOM}) and f_{DIC_soil} among nine species across three altitudes. N = 4 for each species in the specific plot.

Table 3. Relationships between physiological or environmental factors and f_{DIC_soil} analyzed by Pearson's correlation. N = 108.

Factors	f_{DIC_soil}	
	r	*p*
A	0.287	0.003
g_s	−0.122	0.210
E	−0.096	0.325
c_i/c_a	−0.486	<0.001
WUE_i	0.350	<0.001
$C\__{DIC}$	0.053	0.586
δ_{DIC}	−0.100	0.306
δ_{WSOM}	−0.155	0.105

3.5. Variations in f_{DIC_soil} at Different Altitudes

The impact of altitude on f_{DIC_soil} was investigated in one tree (*L. lucidum*) and two shrub species (*V. dilatatum* and *A. delavayana*), which all appeared at three altitudes (Table 1; Figure 6). In *L. lucidum* and *A. delavayana*, f_{DIC_soil} tended to increase from lower to higher altitude ($p < 0.05$), whereas in *V. dilatatum*, the pattern was opposite.

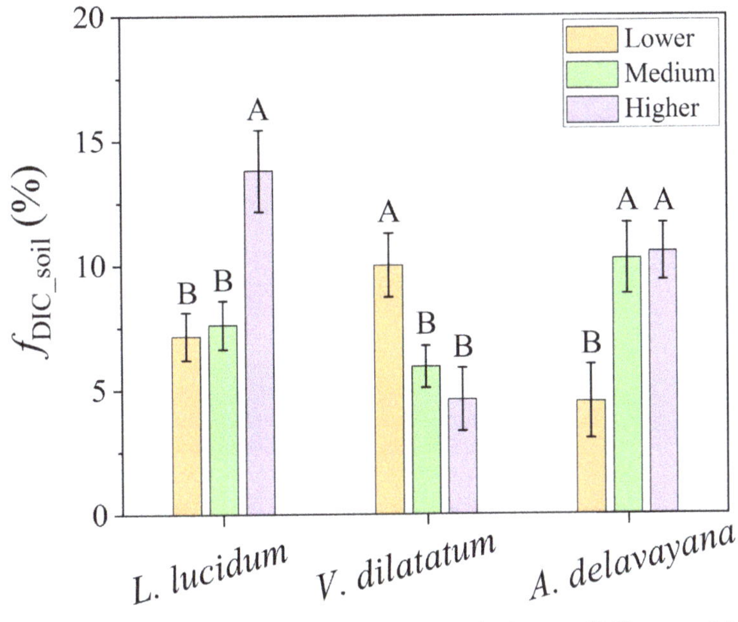

Figure 6. f_{DIC_soil} of one tree species (*L. lucidum*) and two shrub species (*V. dilatatum* and *A. delavayana*) at different altitudes. N = 4 for each species in the specific plot. Capital letters on the error bar indicate significant differences ($p < 0.05$) among altitudes.

4. Discussion

4.1. Isotope Evidence for the Utilization of Soil DIC by Karst Plants

Two issues needed to be addressed in field trials: (1) whether the natural abundance of ^{13}C signals of soil DIC could be detected in leaves, and (2) what the major carbon source supplying the plant roots was. A few studies showed that even if the plants were irrigated with $^{13}CO_2$ or $H^{13}CO_3^-$ solution, the C isotope signals were too weak to be detected [20,57]. This is probably due to a low concentration of the labeled species dissolved in the solution (usually less than 1 mM), resulting in less $^{13}CO_2$ or $H^{13}CO_3^-$

taken up by roots as the diffusion against the concentration gradient from roots to the soil was minimal [15]. However, in the karst habitats, the high concentration of soil DIC made it possible to be fixed by plants. In addition, labeling experiments usually have a single type of carbon source [16,17,20], whereas, in field conditions, the major form of carbon source depends on soil pH, humidity, temperature, root respiration, microbial activity, etc. [58,59]. For this reason, we measured both the concentrations of soil DIC and CO_2 (Figure 2). The result suggested that soil DIC rather than soil CO_2 was the main source supplying the roots. However, direct evidence is needed to support this idea, especially in field conditions. Then we examined whether soil DIC could affect the δ_{WSOM}. The experimental data confirmed our speculation that δ_{WSOM} largely deviated from the predicted $\delta^{13}C$ of photosynthates (δ_A) solely assimilated from atmospheric CO_2 across nine species (Figure 3), consistent with our previous finding [26]. As the values of δ_A-δ_{WSOM} were all positive and half of them were larger than 2‰ (Figures 3B and 5), the isotope measurement error (less than 0.1‰) and diurnal variations in δ_{WSOM} (less than 1‰) [60] are not sufficient to explain the discrepancy. When combined with the very negative ^{13}C signal of soil DIC (δ_A', see Supplementary Data), this large and positive deviation of δ_A from δ_{WSOM} suggests the participation of soil DIC in leaf photosynthesis. Therefore, both atmospheric CO_2 and soil DIC contributed to the newly assimilated carbon (i.e., leaf WSOM). The contribution of soil DIC (f_{DIC_soil}) to leaf total photosynthesis could be quantified with a two-end-member mixing model (Equation (13)). We further showed that δ_A-δ_{WSOM} was closely linked to f_{DIC_soil} among all species (Figure 5). That is, the more the soil DIC was involved in photosynthesis, the more δ_A deviated from δ_{WSOM}, and the larger the proportion of soil DIC contributed to leaf WSOM.

However, uncertainties remain concerning the estimation of f_{DIC_soil}. For example, we considered the tissues-respired CO_2 dissolved in the xylem sap in the calculation of f_{DIC_soil}. Although we assigned a mean value of 5 mM for the $C_{\_R}$, species might differ in $C_{\_R}$ due to their different physiological statues and anatomical characteristics [15,52,61]. We also assumed that the source of respired CO_2 came from the newly assimilated carbon and used the value of δ_{WSOM} to infer the $\delta^{13}C$ of CO_2 respired by living tissues (δ_R). Nevertheless, recent studies [62–65] showed that vegetation pools respired carbon of a wide range of ages in many species. For instance, the respired carbon in stems and roots was, on average, older than 1 year [65,66]. When plants were exposed to stress, respiration almost completely relied on the old carbon [64]. Therefore, in this study, there was a high probability that the investigated species used reserved carbon for respiration under the condition of frequent water limitation in karst habitats. To the best of our knowledge, the $\delta^{13}C$ of reserved carbon has a close relationship with that of respired CO_2 [56]. Our unpublished data showed that the maximum range of $\delta^{13}C$ in the storage carbon of some karst-adaptable species was 4.13‰ during the growing season, which implied a similar or less variation in the range of $\delta^{13}C$ for respired CO_2 in stems and roots. Some studies also showed that seasonal or annual variability of $\delta^{13}C$ in respired CO_2 in stems and roots ranged within 2.0–4.5‰ [67–69]. In this study, taking *L. lucidum* for example, a change of 4.5‰ in $\delta^{13}C$ of leaf WSOM only led to 0.5% of the change in f_{DIC_soil}, indicating a limited influence of the source of carbon used for respiration on the quantification of plants' utilization of soil DIC.

4.2. Interspecies Difference in Plants' Use of Soil DIC

An assessment of f_{DIC_soil} between two plant life forms, namely trees and shrubs, showed no clear trend (Figure 4), indicating that the variation of f_{DIC_soil} was mainly resulted from the interspecies differences. In most species, the values of f_{DIC_soil} were comparable with those reported in the laboratory experiments [24–26], suggesting that soil DIC was easily accessible to plants in the karst habitats. In this study, nine species were all native, most of which were deciduous trees or shrubs. Although species differed in height, biomass, ages, etc., our measurements and sampling were all conducted with newly expanded leaves and within the root zones of the same soil layer. As leaves and roots control the uptake of atmospheric CO_2 and soil DIC, respectively, the species-specific variation in f_{DIC_soil} might be related to leaf gas-exchange traits (Table 2) and soil conditions (Figure 2).

f_{DIC_soil} positively correlates with A ($p = 0.003$) and WUE_i ($p < 0.001$) but negatively with c_i/c_a ($p < 0.001$). At first sight, it was surprising that f_{DIC_soil} increased with an increase in A, which was inconsistent with the results reported by our previous study [26]. For instance, when plants were confronted with moderate or severe water limitation, it would drastically reduce A and g_s [7,29] and thus increased the proportion of root or soil-derived DIC to support photosynthesis [26,29]. The fact was that the change of A was not proportional to that of A_{DIC_soil} or E [26], thus arithmetically promoting the value of f_{DIC_soil}. In this study, the level of g_s indicated mild drought stress in many species (Table 2). Among these species, higher A corresponded to higher g_s and E, which benefited the long-distance transport of DIC in the xylem sap [70,71]. Similarly, higher WUE_i also implied a higher contribution of soil DIC to leaf photosynthesis, although the correlation between E and f_{DIC_soil} was not significant (Table 3). The negative correlation between f_{DIC_soil} and c_i/c_a corresponded to the reverse change of f_{DIC_soil} and the contribution of atmospheric CO_2 to leaf total photosynthesis (1-f_{DIC_soil}) because higher c_i/c_a usually suggested higher $\Delta^{13}C_{com}$ (Equation (2)) and lower A and E [32]. In addition, there was no significant correlation between f_{DIC_soil}, C_DIC, and δ_{DIC}. This could be explained by less variation of these indicators in comparison with that of leaf gas-exchange parameters among species (Figure 2; Table 2).

4.3. Effect of Altitude on Plants' Use of DIC

As the altitude increases, the number of tree species reduces and that of shrub species increases (Table 1), suggesting a change in the plant community. In this study, only three species presented in all plots and thus were available for evaluating the effect of altitude on f_{DIC_soil} (Figure 6). Among these species, f_{DIC_soil} of *L. lucidum* and *A. delavayana* increased from the lower to higher altitude, whereas in *V. dilatatum*, f_{DIC_soil} decreased along the altitudes. The impact of altitude on f_{DIC_soil} was a little bit confusing. However, altitude did not directly affect f_{DIC_soil}, but through imposing influences on site-characteristic related microclimate and soil conditions [35,72]. For example, the altitude could constrain daily mean temperature and vapor pressure deficit [72], and produce considerable variations in soil moisture and nutrient availability [35]. However, in the present study, the air temperature and relative humidity recorded by LI-6400 did not linearly change with altitude (see Supplementary Data) but exhibited a high degree of spatial heterogeneity. The reasons might be that the altitude gradient was not large enough, and some occasional factors, e.g., wind, shading, and vegetation coverage, could redistribute these resources, e.g., light and vapor [72,73]. Our previous study reported decreasing patterns for soil water content, organic matter, and some nutrients with the altitudes in the same area and same season of 2015 [34]. However, the increasing trend of f_{DIC_soil} along the altitude gradient in *L. lucidum* and *A. delavayana* was opposite to that in *V. dilatatum* (Figure 6), implying that the differences in soil conditions could not solely explain the variation of f_{DIC_soil}. Therefore, we speculated that all these variabilities were combined to influence the leaf gas-exchange and subsequent estimation of f_{DIC_soil}. Future studies may choose a large range of environmental gradients and exclude the influences of some occasional factors.

5. Conclusions

In the present study, we aimed to investigate whether soil DIC could be assimilated through leaf photosynthesis in the karst habitats, and whether this process, if taken place, varied between species and altitudes. Firstly, we showed that all plots in the study sites had a high content of soil DIC in the root zones of nine species. This implied a high possibility that soil DIC could be fixed by karst plants through photosynthesis. Secondly, we observed that there were large discrepancies between the measured and predicted $\delta^{13}C$ of newly formed photosynthates (δ_A-δ_{WSOM}), and this systematic difference could not be explained by measurement errors or diurnal variations in δ_{WSOM} alone. Therefore, we accounted for the involvement of soil DIC in photosynthesis and thus deviating δ_{WSOM} from δ_A. Thirdly, we applied a two-end-member mixing model to estimate the contribution of soil DIC to leaf total photosynthesis (f_{DIC_soil}). The values of f_{DIC_soil} largely differed among species and some of which were influenced by altitudes. The present study improved our understanding of how plants adapted to alkaline soil conditions of karst habitats and acquired additional carbon for growth.

Supplementary Materials: The following supporting information can be downloaded at: https://www.mdpi.com/article/10.3390/agronomy12102489/s1.

Author Contributions: Conceptualization, S.R. and Y.W.; experimentation, S.R.; writing, S.R.; data analysis, S.R. and Y.W.; review and editing, Y.W. All authors have read and agreed to the published version of the manuscript.

Funding: This study was supported by Support Plan Projects of Science and Technology of Guizhou Province (No. (2021)YB453) and the National Key Research and Development Program of China (No. 2021YFD1100300).

Institutional Review Board Statement: Not applicable.

Informed Consent Statement: Not applicable.

Data Availability Statement: All data related to this article can be found in the online version.

Acknowledgments: We deeply thank W. Lin, two anonymous reviewers for their valuable comments and suggestions that greatly improved the manuscript.

Conflicts of Interest: The authors declare no conflict of interest.

References

1. Bonacci, O.; Pipan, T.; Culver, D.C. A framework for karst ecohydrology. *Environ. Geol.* **2009**, *56*, 891–900. [CrossRef]
2. Goldscheider, N.; Chen, Z.; Auler, A.S.; Bakalowicz, M.; Broda, S.; Drew, D.; Hartmann, J.; Jiang, G.; Moosdorf, N.; Stevanovic, Z.; et al. Global distribution of carbonate rocks and karst water resources. *Hydrogeol. J.* **2020**, *28*, 1661–1677. [CrossRef]
3. Yuan, D.X. On the karst ecosystem. *Acta Geol. Sin.-Engl. Ed.* **2001**, *75*, 336–338.
4. Wang, S.; Liu, Q.; Zhang, D. Karst rocky desertification in southwestern China: Geomorphology, landuse, impact and rehabilitation. *Land Degrad. Dev.* **2004**, *15*, 115–121. [CrossRef]
5. Cao, J.; Yuan, D.; Tong, L.; Azim, M.; Yang, H.; Huang, F. An overview of karst ecosystem in southwest china: Current state and future management. *J. Resour. Ecol.* **2015**, *6*, 247–256.
6. Yu, F.; Huang, X.; Liang, Q.; Yao, P.; Li, X.; Liao, Z.; Duan, C.; Zhang, A.; Shao, H. Ecological water demand of regional vegetation: The example of the 2010 severe drought in Southwest China. *Plant Biosyst.* **2015**, *149*, 100–110. [CrossRef]
7. Liu, C.; Liu, Y.; Guo, K.; Li, G.; Zheng, Y.; Yu, L.; Yang, R. Comparative ecophysiological responses to drought of two shrub and four tree species from karst habitats of southwestern China. *Trees* **2011**, *25*, 537–549. [CrossRef]
8. Sevanto, S.; McDowell, N.G.; Dickman, L.T.; Pangle, R.; Pockman, W.T. How do trees die? A test of the hydraulic failure and carbon starvation hypotheses. *Plant Cell Environ.* **2014**, *37*, 153–161. [CrossRef]
9. Rowland, L.; da Costa, A.C.; Galbraith, D.R.; Oliveira, R.S.; Binks, O.J.; Oliveira, A.A.; Pullen, A.M.; Doughty, C.E.; Metcalfe, D.B.; Vasconcelos, S.S.; et al. Death from drought in tropical forests is triggered by hydraulics not carbon starvation. *Nature* **2015**, *528*, 119–122. [CrossRef]
10. Msilini, N.; Attia, H.; Bouraoui, N.; M'rah, S.; Ksouri, R.; Lachaâl, M.; Ouerghi, Z. Responses of Arabidopsis thaliana to bicarbonate-induced iron defciency. *Acta Physiol. Plant.* **2009**, *31*, 849–853. [CrossRef]
11. Du, Y.; Pan, G.; Li, L.; Hu, Z.; Wang, X. Leaf N/P ratio and nutrient reuse between dominant species and stands: Predicting phosphorus deficiencies in karst ecosystems, southwestern China. *Environ. Earth Sci.* **2011**, *64*, 299–309. [CrossRef]
12. Jiang, Z.; Liu, H.; Wang, H.; Peng, J.; Meersmans, J.; Green, S.; Quine, T.A.; Wu, X.; Song, Z. Bedrock geochemistry influences vegetation growth by regulating the regolith water holding capacity. *Nat. Commun.* **2020**, *11*, 2392. [CrossRef]
13. Wang, R.; Wu, Y.; Xing, D.; Hang, H.; Xie, X.; Yang, X.; Zhang, K.; Rao, S. Biomass production of three biofuel energy plants' use of a new carbon resource by carbonic anhydrase in simulated karst soils: Mechanism and capacity. *Energies* **2017**, *10*, 1370. [CrossRef]
14. Wu, Y.; Hong, W.; Chen, Y. Leaf physiological and anatomical characteristics of two indicator species in the limestone region of southern China under drought stress. *Pak. J. Bot.* **2018**, *50*, 1335–1342.
15. Bloemen, J.; Teskey, R.O.; McGuire, M.A.; Aubrey, D.P.; Steppe, K. Root xylem CO_2 flux: An important but unaccounted-for component of root respiration. *Trees* **2016**, *30*, 343–352. [CrossRef]

16. Shimono, H.; Kondo, M.; Evans, J.R. Internal transport of CO_2 from the root-zone to plant shoot is pH dependent. *Physiol. Plant.* **2019**, *165*, 451–463. [CrossRef]
17. Tarvainen, L.; Wallin, G.; Linder, S.; Nasholm, T.; Oren, R.; Löfvenius, M.O.; Rantfors, M.; Tor-Ngern, P.; Marshall, J.D. Limited vertical CO_2 transport in stems of mature boreal *Pinus sylvestris* trees. *Tree Physiol.* **2020**, *41*, 63–75. [CrossRef] [PubMed]
18. Salomon, R.L.; De Roo, L.; Bodé, S.; Boeckx, P.; Steppe, K. Efflux and assimilation of xylem-transported CO_2 in stems and leaves of tree species with different wood anatomy. *Plant Cell Environ.* **2021**, *44*, 3494–3508. [CrossRef] [PubMed]
19. Aubrey, D.P.; Teskey, R.O. Root-derived CO_2 efflux via xylem stream rivals soil CO_2 efflux. *New Phytol.* **2009**, *184*, 35–40. [CrossRef] [PubMed]
20. Ford, C.R.; Wurzburger, N.; Hendrick, R.L.; Teskey, R.O. Soil DIC uptake and fixation in *Pinus taeda* seedlings and its C contribution to plant tissues and ectomycorrhizal fungi. *Tree Physiol.* **2007**, *27*, 375–383. [CrossRef] [PubMed]
21. Rao, S.; Wu, Y.; Wang, R. Bicarbonate stimulates non-structural carbohydrate pools of *Camptotheca acuminata*. *Physiol. Plant.* **2019**, *165*, 780–789. [CrossRef]
22. Liang, F.; Yang, W.; Xu, L.; Ji, L.; He, Q.; Wu, L.; Ran, Y.; Yan, S. Closing extra CO_2 into plants for simultaneous CO_2 fixation, drought stress alleviation and nutrient absorption enhancement. *J. CO2 Util.* **2020**, *42*, 101319. [CrossRef]
23. Simkin, A.J.; Faralli, M.; Ramamoorthy, S.; Lawson, T. Photosynthesis in non-foliar tissues: Implications for yield. *Plant J.* **2020**, *101*, 1001–1015. [CrossRef]
24. Wu, Y.; Xing, D. Effect of bicarbonate treatment on photosynthetic assimilation of inorganic carbon in two plant species of Moraceae. *Photosynthetica* **2012**, *50*, 587–594. [CrossRef]
25. Hang, H.; Wu, Y. Quantification of photosynthetic inorganic carbon utilisation via a bidirectional stable carbon isotope tracer. *Acta Geochim.* **2016**, *35*, 130–137. [CrossRef]
26. Rao, S.; Wu, Y. Root-derived bicarbonate assimilation in response to variable water deficit in *Camptotheca acuminata* seedlings. *Photosynth. Res.* **2017**, *134*, 59–70. [CrossRef]
27. Valipour, M.; Khoshgoftarmanesh, A.H.; Baninasab, B. Physiological responses of hawthorn (*Crataegus persica* Pojark.) and quince (*Cydonia oblonga* Mill.) rootstocks to bicarbonate-induced iron deficiency in nutrient solution. *J. Plant Nutr. Soil Sci.* **2018**, *181*, 905–913. [CrossRef]
28. Han, X.; Jing, Y.; Xu, C.; Gao, L.; Li, M.; Liu, Y.; Qi, H. Root-zone CO_2 concentration affects partitioning and assimilation of carbon in oriental melon seedlings. *Int. J. Mol. Sci.* **2022**, *23*, 10694. [CrossRef]
29. Wu, Y.; Wu, Y. The increase in the karstification–photosynthesis coupled carbon sink and its implication for carbon neutrality. *Agronomy* **2022**, *12*, 2147. [CrossRef]
30. Göttlicher, S.; Knohl, A.; Wanek, W.; Buchmann, N.; Richter, A. Short-term changes in carbon isotope composition of soluble carbohydrates and starch: From canopy leaves to the root system. *Rapid Commun. Mass Spectrom.* **2006**, *20*, 653–660. [CrossRef]
31. Yang, Y.; Siegwolf, R.T.; Körner, C. Species specific and environment induced variation of $\delta^{13}C$ and $\delta^{15}N$ in alpine plants. *Front. Plant Sci.* **2015**, *6*, 423. [CrossRef]
32. Cernusak, L.A.; Ubierna, N.; Winter, K.; Holtum, J.A.; Marshall, J.D.; Farquhar, G.D. Environmental and physiological determinants of carbon isotope discrimination in terrestrial plants. *New Phytol.* **2013**, *200*, 950–965. [CrossRef]
33. Marshall, J.D.; Dawson, T.E.; Ehleringer, J.R. Integrated nitrogen, carbon, and water relations of a xylem-tapping mistletoe following nitrogen fertilization of the host. *Oecologia* **1994**, *100*, 430–438. [CrossRef]
34. Rao, S.; Wu, Y. Comparison of physiological and morphological traits of *Platycarya longipes* in different slope locations in karst area. *Earth Environ.* **2017**, *45*, 10–17.
35. Zhang, Z.H.; Hu, G.; Ni, J. Effects of topographical and edaphic factors on the distribution of plant communities in two subtropical karst forests, southwestern China. *J. Mt. Sci.* **2013**, *10*, 95–104. [CrossRef]
36. Brandes, E.; Kodama, N.; Whittaker, K.; Weston, C.; Rennenberg, H.; Keitel, C.; Adams, M.A.; Gessler, A. Short-term variation in the isotopic composition of organic matter allocated from the leaves to the stem of *Pinus sylvestris*: Effects of photosynthetic and postphotosynthetic carbon isotope fractionation. *Glob. Chang. Biol.* **2006**, *12*, 1922–1939. [CrossRef]
37. Gessler, A.; Tcherkez, G.; Peuke, A.D.; Ghashghaie, J.; Farquhar, G.D. Experimental evidence for diel variations of the carbon isotope composition in leaf, stem and phloem sap organic matter in Ricinus communis. *Plant Cell Environ.* **2008**, *31*, 941–953. [CrossRef]
38. Ghashghaie, J.; Badeck, F.W.; Girardin, C.; Sketriené, D.; Lamothe-Sibold, M.; Werner, R.A. Changes in $\delta^{13}C$ of dark respired CO_2 and organic matter of different organs during early ontogeny in peanut plants. *Isot. Environ. Health Stud.* **2015**, *51*, 93–108. [CrossRef]
39. Yousfi, S.; Serret, M.D.; Araus, J.L. Comparative response of $\delta^{13}C$, $\delta^{18}O$ and $\delta^{15}N$ in durum wheat exposed to salinity at the vegetative and reproductive stages. *Plant Cell Environ.* **2013**, *36*, 1214–1227. [CrossRef]
40. Studer, M.S.; Siegwolf, R.T.; Leuenberger, M.; Abiven, S. Multi-isotope labelling of organic matter by diffusion of $^{2}H/^{18}OH_2O$ vapour and ^{13}C-CO_2 into the leaves and its distribution within the plant. *Biogeosciences* **2015**, *12*, 1865–1879. [CrossRef]
41. Ruehr, N.K.; Offermann, C.A.; Gessler, A.; Winkler, J.B.; Ferrio, J.P.; Buchmann, N.; Barnard, R.L. Drought effects on allocation of recent carbon: From beech leaves to soil CO_2 efflux. *New Phytol.* **2009**, *184*, 950–961. [CrossRef]
42. Zuo, Y.; Ren, L.; Zhang, F.; Jiang, R. Bicarbonate concentration as affected by soil water content controls iron nutrition of peanut plants in a calcareous soil. *Plant Physiol. Biochem.* **2007**, *45*, 357–364. [CrossRef] [PubMed]

43. Farquhar, G.D.; O'Leary, M.H.; Berry, J.A. On the relationship between carbon isotope discrimination and the intercellular carbon-dioxide concentration in leaves. *Funct. Plant Biol.* **1982**, *9*, 121–137. [CrossRef]
44. Ubierna, N.; Holloway-Phillips, M.M.; Farquhar, G.D. Using stable carbon isotopes to study C_3 and C_4 photosynthesis: Models and calculations. In *Photosynthesis*; Humana Press: New York, NY, USA, 2018; pp. 155–196.
45. Evans, J.R.; von Caemmerer, S. Temperature response of carbon isotope discrimination and mesophyll conductance in tobacco. *Plant Cell Environ.* **2013**, *36*, 745–756. [CrossRef]
46. Ubierna, N.; Cernusak, L.A.; Holloway-Phillips, M.M.; Busch, F.A.; Cousins, A.B.; Farquhar, G.D. Critical review: Incorporating the arrangement of mitochondria and chloroplasts into models of photosynthesis and carbon isotope discrimination. *Photosynth. Res.* **2019**, *141*, 5–31. [CrossRef] [PubMed]
47. Tcherkez, G. How large is the carbon isotope fractionation of the photorespiratory enzyme glycine decarboxylase? *Funct. Plant Biol.* **2006**, *33*, 911–920. [CrossRef]
48. Gillon, J.; Yakir, D. Influence of carbonic anhydrase activity in terrestrial vegetation on the ^{18}O content of atmospheric CO_2. *Science* **2001**, *291*, 2584–2587. [CrossRef]
49. Bernacchi, C.J.; Portis, A.R.; Nakano, H.; von Caemmerer, S.; Long, S.P. Temperature response of mesophyll conductance. Implications for the determination of Rubisco enzyme kinetics and for limitations to photosynthesis in vivo. *Plant Physiol.* **2002**, *130*, 1992–1998. [CrossRef]
50. Teskey, R.O.; McGuire, M.A. Measurement of stem respiration of sycamore (*Platanus occidentalis* L.) trees involves internal and external fluxes of CO_2 and possible transport of CO_2 from roots. *Plant Cell Environ.* **2007**, *30*, 570–579. [CrossRef]
51. Levy, P.E.; Meir, P.; Allen, S.J.; Jarvis, P.G. The effect of aqueous transport of CO_2 in xylem sap on gas exchange in woody plants. *Tree Physiol.* **1999**, *19*, 53–58. [CrossRef] [PubMed]
52. Teskey, R.O.; Saveyn, A.; Steppe, K.; McGuire, M.A. Origin, fate and significance of CO_2 in tree stems. *New Phytol.* **2008**, *177*, 17–32. [CrossRef]
53. Bowling, D.R.; Pataki, D.E.; Randerson, J.T. Carbon isotopes in terrestrial ecosystem pools and CO_2 fluxes. *New Phytol.* **2008**, *178*, 24–40. [CrossRef] [PubMed]
54. Gessler, A.; Keitel, C.; Kodama, N.; Weston, C.; Winters, A.J.; Keith, H.; Grice, K.; Leuning, R.; Farquhar, G.D. $\delta^{13}C$ of organic matter transported from the leaves to the roots in Eucalyptus delegatensis: Short-term variations and relation to respired CO_2. *Funct. Plant Biol.* **2007**, *34*, 692–706. [CrossRef]
55. Kodama, N.; Barnard, R.L.; Salmon, Y.; Weston, C.; Ferrio, J.P.; Holst, J.; Werner, R.A.; Saurer, M.; Rennenberg, H.; Buchmann, N.; et al. Temporal dynamics of the carbon isotope composition in a Pinus sylvestris stand: From newly assimilated organic carbon to respired carbon dioxide. *Oecologia* **2008**, *156*, 737–750. [CrossRef]
56. Bathellier, C.; Badeck, F.W.; Ghashghaie, J. Carbon isotope fractionation in plant respiration. In *Plant Respiration: Metabolic Fluxes and Carbon Balance*; Springer: Cham, Switzerland, 2017; pp. 43–68.
57. Ubierna, N.; Kumar, A.S.; Cernusak, L.A.; Pangle, R.E.; Gag, P.J.; Marshall, J.D. Storage and transpiration have negligible effects on $\delta^{13}C$ of stem CO_2 efflux in large conifer trees. *Tree Physiol.* **2009**, *29*, 1563–1574. [CrossRef] [PubMed]
58. Shahabi, A.; Malakouti, M.J.; Fallahi, E. Effects of bicarbonate content of irrigation water on nutritional disorders of some apple varieties. *J. Plant Nutr.* **2005**, *28*, 1663–1678. [CrossRef]
59. Poschenrieder, C.; Fernández, J.A.; Rubio, L.; Pérez, L.; Terés, J.; Barceló, J. Transport and use of bicarbonate in plants: Current knowledge and challenges ahead. *Int. J. Mol. Sci.* **2018**, *19*, 1352. [CrossRef] [PubMed]
60. Dubbert, M.; Rascher, K.G.; Werner, C. Species-specific differences in temporal and spatial variation in $\delta^{13}C$ of plant carbon pools and dark-respired CO_2 under changing environmental conditions. *Photosynth. Res.* **2012**, *113*, 297–309. [CrossRef]
61. Werner, C.; Gessler, A. Diel variations in the carbon isotope composition of respired CO_2 and associated carbon sources: A review of dynamics and mechanisms. *Biogeosciences* **2011**, *8*, 2437–2459. [CrossRef]
62. Sierra, C.A.; Ceballos-Núñez, V.; Hartmann, H.; Herrera-Ramírez, D.; Metzler, H. Ideas and perspectives: Allocation of carbon from net primary production in models is inconsistent with observations of the age of respired carbon. *Biogeosciences* **2022**, *19*, 3727–3738. [CrossRef]
63. Gao, D.; Joseph, J.; Werner, R.A.; Brunner, I.; Zürcher, A.; Hug, C.; Wang, A.; Zhao, C.; Bai, E.; Meusburger, K.; et al. Drought alters the carbon footprint of trees in soils—Tracking the spatio-temporal fate of ^{13}C-labelled assimilates in the soil of an old-growth pine forest. *Glob. Chang. Biol.* **2021**, *27*, 2491–2506. [CrossRef]
64. Huang, J.; Hammerbacher, A.; Gershenzon, J.; van Dam, N.M.; Sala, A.; McDowell, N.G.; Chowdhury, S.; Gleixner, G.; Trumbore, S.; Hartmann, H. Storage of carbon reserves in spruce trees is prioritized over growth in the face of carbon limitation. *Proc. Natl. Acad. Sci. USA* **2021**, *118*, e2023297118. [CrossRef]
65. Hilman, B.; Muhr, J.; Helm, J.; Kuhlmann, I.; Schulze, E.D.; Trumbore, S. The size and the age of the metabolically active carbon in tree roots. *Plant Cell Environ.* **2021**, *44*, 2522–2535. [CrossRef]
66. Muhr, J.; Trumbore, S.; Higuchi, N.; Kunert, N. Living on borrowed time–Amazonian trees use decade-old storage carbon to survive for months after complete stem girdling. *New Phytol.* **2018**, *220*, 111–120. [CrossRef]
67. Brændholt, A.; Ibrom, A.; Ambus, P.; Larsen, K.S.; Pilegaard, K. Combining a quantum cascade laser spectrometer with an automated closed-chamber system for $\delta^{13}C$ measurements of forest soil, tree stem and tree root CO_2 fluxes. *Forests* **2019**, *10*, 432. [CrossRef]

68. Kuptz, D.; Fleischmann, F.; Matyssek, R.; Grams, T.E. Seasonal patterns of carbon allocation to respiratory pools in 60-yr-old deciduous (*Fagus sylvatica*) and evergreen (*Picea abies*) trees assessed via whole-tree stable carbon isotope labeling. *New Phytol.* **2011**, *191*, 160–172. [CrossRef] [PubMed]
69. Diao, H.; Wang, A.; Yuan, F.; Guan, D.; Dai, G.; Wu, J. Environmental effects on carbon isotope discrimination from assimilation to respiration in a coniferous and broad-leaved mixed forest of Northeast China. *Forests* **2020**, *11*, 1156. [CrossRef]
70. Bloemen, J.; McGuire, M.A.; Aubrey, D.P.; Teskey, R.O.; Steppe, K. Assimilation of xylem-transported CO_2 is dependent on transpiration rate but is small relative to atmospheric fixation. *J. Exp. Bot.* **2013**, *64*, 2129–2138. [CrossRef] [PubMed]
71. Stutz, S.S.; Hanson, D.T. Contribution and consequences of xylem-transported CO_2 assimilation for C_3 plants. *New Phytol.* **2019**, *223*, 1230–1240. [CrossRef]
72. Jucker, T.; Hardwick, S.R.; Both, S.; Elias, D.M.; Ewers, R.M.; Milodowski, D.T.; Swinfield, T.; Coomes, D.A. Canopy structure and topography jointly constrain the microclimate of human-modified tropical landscapes. *Glob. Chang. Biol.* **2018**, *24*, 5243–5258. [CrossRef]
73. Burgess, A.J.; Retkute, R.; Preston, S.P.; Jensen, O.E.; Pound, M.P.; Pridmore, T.P.; Murchie, E.H. The 4-dimensional plant: Effects of wind-induced canopy movement on light fluctuations and photosynthesis. *Front. Plant Sci.* **2016**, *7*, 1392. [CrossRef]

agronomy

MDPI

Article

Appropriate Sodium Bicarbonate Concentration Enhances the Intracellular Water Metabolism, Nutrient Transport and Photosynthesis Capacities of *Coix lacryma-jobi* L.

Haitao Li [1], Jiamei Lv [1], Yue Su [1,*] and Yanyou Wu [2,*]

1 Engineering Technology Research Center for Protection and Detection of Germplasm Resources of Karst-Adaptable Crops, Guizhou Vocational College of Agriculture, Qingzhen 551400, China; lisea02@126.com (H.L.); jiameilv123@126.com (J.L.)

2 State Key Laboratory of Environmental Geochemistry, Institute of Geochemistry, Chinese Academy of Sciences, Guiyang 550081, China

* Correspondence: suyue09136@163.com (Y.S.); wuyanyou@mail.gyig.ac.cn (Y.W.); Tel.: +86-1808-5105-941 (Y.S.); +86-0851-8439-1746 (Y.W.)

Abstract: Karst ecological stresses are harmful to plant growth, especially high bicarbonate concentrations, drought, high pH, etc. In this study, the effects of 0, 2.0, 7.0 and 12.0 mmol L^{-1} sodium bicarbonate concentrations on the biomass, electrophysiological properties, intracellular water metabolism, nutrient transport, photosynthesis and chlorophyll fluorescence of *Coix lacryma-jobi* L. were investigated. The results show that 2.0 mmol L^{-1} sodium bicarbonate effectively improved the biomass formation of *Coix lacryma-jobi* L., notably increased its intrinsic capacitance (IC) and decreased its intrinsic resistance (IR), intrinsic impedance (IZ), intrinsic capacitive reactance (IXc) and intrinsic inductive reactance (IXL) as well as reliably enhanced its intracellular water metabolism, nutrient transport and photosynthetic capacities. However, 7.0 and 12.0 mmol L^{-1} sodium bicarbonate concentrations exhibited marked inhibitory effects on the plants' photosynthetic rate, stomatal conductance, transpiration rate and dry weight, whereas they did not significantly change the intracellular water metabolism or the nutrient transport capacity of *Coix lacryma-jobi* L. This study highlights that appropriate bicarbonate levels could enhance the intracellular water metabolism, nutrient transport, photosynthesis and growth of *Coix lacryma-jobi* L., which can be rapidly monitored by the plant's electrophysiological properties. Importantly, plant electrophysiological measurement is significantly superior to photosynthesis measurement. In the future, plant electrophysiological measurement can be used as a means to quickly and effectively evaluate the physiological response of plants to the external environment.

Keywords: dissolved inorganic carbon; karst; *Coix* seed rice; electrical signal; substance transport

Citation: Li, H.; Lv, J.; Su, Y.; Wu, Y. Appropriate Sodium Bicarbonate Concentration Enhances the Intracellular Water Metabolism, Nutrient Transport and Photosynthesis Capacities of *Coix lacryma-jobi* L.. *Agronomy* **2023**, *13*, 1790. https://doi.org/10.3390/agronomy13071790

Academic Editor: Guanfu Fu

Received: 5 June 2023
Revised: 27 June 2023
Accepted: 30 June 2023
Published: 3 July 2023

1. Introduction

Strong karstification leads to high bicarbonate concentrations, drought, high pH, high calcium and magnesium levels and low nutrition in karst soil environments [1]. In karst ecological systems, stable bicarbonate (HCO_3^-) reservoirs in soil and water can be formed by the dissolution of limestone and dolomite by water, as their HCO_3^- concentrations are notably higher than those of non-karst areas [2]. For instance, in the present literature, it was reported that the HCO_3^- concentration in calcareous soils is usually 1~5 mmol L^{-1} [3,4]; Zhang et al., showed that the HCO_3^- concentration in karst rivers and lakes is usually about 4.5 mmol L^{-1} [5], and Hussner et al., reported that the HCO_3^- concentration was usually several times higher than that of carbon dioxide if the ambient pH value was above 7 [6]. Plants grown in karst areas often suffer from various degrees of karst ecological stresses, such as high bicarbonate levels, drought, high pH, etc. [7]. Consequently, plants' adaptive mechanisms to karst environments is a permanent research hot spot.

Previous studies have reported that karstification has many negative effects on plant growth and metabolism. It affects the growth and photosynthesis of plants by inducing HCO_3^- stress in *Broussonetia papyrifera* L. [8] and tomatoes [9] as well as through decreasing the activity of key enzymes in wheat under drought conditions [10]. Alkali stress affects the growth and physiological characteristics of melon seedlings [11]. High concentrations of HCO_3^- can reduce the uptake of Fe in plants by limiting the expression of Fe acquisition genes, leading to Fe deficiency [12,13]. Fe deficiency strongly limits the biosynthesis of chlorophyll [14], which will result in a significant decrease in photosynthesis. In fact, plants are very tenacious, and they have developed a variety of unique adaptive mechanisms to adversity [15]. Some reports show that HCO_3^- not only provides short-term carbon and water sources for plants under drought stress, but also promotes stomatal opening and restores photosynthesis [16–20]. Meanwhile, HCO_3^- can promote carbon and nitrogen metabolism by regulating the activities of key enzymes involved in carbon and nitrogen metabolism and by participating in the regulation of complex physiological processes, including carbon assimilation and nitrogen reduction in plants [15,21]. Zhou et al., reported that 3~4 mmol L^{-1} sodium bicarbonate concentration added to cultures was conducive to growth and the improvement of carbonic anhydrase activity in macroalgae [22]. Additionally, HCO_3^- can also change the activity of glycolysis and pentose phosphate pathways via the distribution of their substrates to enhance the stress resistance of plants and obtain survival opportunities [8]. Wu et al., systematically summarized the karst-adaptive mechanisms of plants in terms of morphology, ecology, photosynthetic capacity, utilization of inorganic nutrients, carbonic anhydrase activity, biodiversity, calcium regulation root organic acid exudation, etc. [1].

Coix lacryma-jobi L., a typical karst and oldest grass crop, is rich in protein, oil and medicinal components, such as fatty acids, amino acids, triterpenes, vitamins and various minerals [23,24]. Due to its high nutritional and medical value, it is widely used in improving immunity, inhibiting tumor angiogenesis, invigorating stomach and spleen tissue and regulating blood sugar; in addition, it possesses anti-tumor, anti-convulsant and apocenosis effects [25,26]. As a promising economic crop and natural medicine, *Coix lacryma-jobi* L. is widely planted in Southwestern China, especially in Guizhou Province. Currently, the planting area and yield of *Coix lacryma-jobi* L. in China is estimated to be around 73,000 ha and 0.22 million tons, respectively [24,27]. Recently, Guizhou's *Coix lacryma-jobi* L. industry has developed rapidly, becoming the largest producer and distribution center for *Coix lacryma-jobi* L. in Southeast Asia [24]. Moreover, it has made great contributions in alleviating poverty, revitalizing rural areas and controlling rocky desertification, and its planting area reached over 48,600 ha by 2021 [27,28]. In Guizhou Province, however, *Coix lacryma-jobi* L. is mostly planted in karst soil environments, and its growth and development are influenced by many adversities such as high bicarbonate content, drought and high pH levels. Therefore, it is worth exploring whether high levels of bicarbonate can promote the growth of *Coix lacryma-jobi* L. Moreover, to date, there are no data available about the adaptive mechanisms of *Coix lacryma-jobi* L. to high bicarbonate levels.

The electrophysiological activities of plants, including charge separation, electron movement, proton and dielectric transport, etc., control almost all of their biological processes [29–33]. Alterations in plants' electrical parameters are considered to be their fastest response to environmental stresses [34–37]. Numerous findings have shown that the passive electric properties of plants, such as their capacitance (C), impedance (Z), resistance (R), capacitive reactance (Xc) and inductive reactance (XL), can be used to evaluate their physiological status [33–45]. For instance, Zhang et al. found that the theoretical relationships among the clamping force (F) and leaf Z, R, Xc or XL were revealed to be Z, R, Xc or XL $= y + ke^{-bF}$ based on bioenergetics. Then, the intrinsic R (IR), intrinsic Z (IZ), intrinsic Xc (IXc) and intrinsic XL (IXL) of plant leaves were successfully described as IR, IZ, IXc or IXL $= y + k$ when the clamping force is 0 (F = 0) [44,45]. Meanwhile, they also developed intracellular water metabolism and nutrient transport parameters based on the IR, IZ, IXc, IXL and intrinsic capacitance (IC) of plants, which accurately revealed the intracellular water metabolism and nutrient transport strategies of plants and allowed them to be evaluated [33,43–45]. Thus, whether plants' electrophysiological properties can be faster and more effective to determine the appropriate bicarbonate level for *Coix lacryma-jobi* L. growth compared to common photosynthesis indicators is worth further attention.

In this study, the effects of sodium bicarbonate on the growth of *Coix lacryma-jobi* L. were evaluated. Moreover, the effects of sodium bicarbonate on the electrophysiological properties, intracellular water metabolism, nutrient transport, photosynthesis and chlorophyll fluorescence of *Coix lacryma-jobi* L. were also investigated. This work provides a scientific basis for rapidly screening the appropriate bicarbonate level for *Coix lacryma-jobi* L. grown in karst environments.

2. Materials and Methods

2.1. Materials

Sodium bicarbonate (AR, ≥99.8%) was provided by Rhawn Reagent Chemical Technology Co., Ltd. (Shanghai, China). All reagents were analytical grade. Three-leaf seedlings of *Coix lacryma-jobi* L. 'Yizhu 1' with similar growth were used as experimental materials. The seedlings were grown in hydroponic conditions for 14 days. The weights of the aboveground parts were as follows: 0.38 ± 0.01 (fresh); 0.04 ± 0.01 (dry). The weights of the root were as follows: 0.17 ± 0.03 (fresh); 0.02 ± 0.01 (dry). Hoagland nutrient solution contained the following (in mmol L^{-1}): 3.0 mmol KNO_3, 0.125 mmol $NH_4H_2PO_4$, 0.1875 mmol $(NH_4)_2SO_4$, 1.0 mmol $MgSO_4$, 2.0 mmol $Ca(NO_3)_2$, 2.0 μmol KCl, 25 μmol H_3BO_3, 2 μmol $MnSO_4$, 2 μmol $ZnSO_4$, 0.1 μmol $CuSO_4$, 0.1 μmol Na_2MoO_4, 0.043 μmol $CoCl_2$ and 50 μmol Fe-EDTA.

2.2. Sodium Bicarbonate Hydroponic Experiment Using Coix lacryma-jobi L.

The hydroponic method was used for studying the growth effects of sodium bicarbonate on *Coix lacryma-jobi* L. The culture solution was 1/2 Hogland nutrient solution containing different sodium bicarbonate concentrations. Based on the preliminary experimental results and the sodium bicarbonate concentration (1.2~3.3 mmol L^{-1}) in field planting soils of *Coix lacryma-jobi* L., four treatments were designed with 0 (CK), 2.0, 7.0 and 12.0 mmol L^{-1} sodium bicarbonate. Each treatment had three replicates, and each replicate contained nine plants as biological replicates. Three plants were randomly selected for each replicate. The original culture solution was replaced with fresh culture solution with a pH of 8.00 ± 0.10 each day, and the whole treatment cycle was 14 days. The culture temperature, illumination time, illumination intensity and relative humidity of *Coix lacryma-jobi* plants were 25.0 ± 2.0/19.0 ± 2.0 °C (light/dark), 12 h/12 h (light/dark), 500.0 ± 25.0 μmol m^{-2} s^{-1} and 55.4~60.6%, respectively.

2.3. Determination of Biomass Parameters, Electrophysiological and Photosynthetic Characteristics and Chlorophyll Fluorescence

The height and weight of *Coix lacryma-jobi* L. plants were measured. The plant height of *Coix lacryma-jobi* L. was measured using a ruler. The fresh weights of the aboveground parts and root were measured by weighing them, and their dry weights were determined after drying them to a constant weight. The fresh weight of a whole *Coix lacryma-jobi* L. plant was the sum of the fresh weights of its aboveground parts and roots, and its whole dry weight was the sum of the dry weights of the aboveground parts and roots.

The intracellular water metabolism and nutrient transport parameters of *Coix lacryma-jobi* L. were determined as described by Zhang et al. [33,44,45]. The fresh second fully expanded leaves of *Coix lacryma-jobi* L. were sampled after measuring their photosynthetic parameters and chlorophyll fluorescence. Leaf samples were immediately soaked in distilled water for 30 min, and then the water on the leaf surface was removed. The electrophysiological parameters of the plant leaves were determined using a LCR-6300 tester (Gwinstek, Taiwan, China) in parallel connection mode with a tested voltage and frequency of 1.5 V and 3 kHz, respectively. The leaf was first put between the two electrodes of a self-made parallel-plate capacitor, and then the leaf's passive electric properties, including C, R and Z under different clamping forces (1.139, 2.149, 3.178, 4.212 and 5.245 N), were continuously recorded, yielding 11~13 data points for each clamping force. Different clamping forces were obtained by adding iron blocks. Subsequently, leaf Xc and XL were calculated based on Equations (1) and (2), respectively.

$$\mathrm{Xc} = \frac{1}{2\pi f C} \tag{1}$$

$$\frac{1}{-\mathrm{XL}} = \frac{1}{\mathrm{Z}} - \frac{1}{\mathrm{R}} - \frac{1}{\mathrm{Xc}} \tag{2}$$

where π is 3.1416 and f is frequency.

The fitting equations for the clamping force and Z, R, Xc, XL and C of the leaves were constructed as Equations (3)–(7):

$$\mathrm{Z} = \mathrm{y_1} + \mathrm{k_1 e^{-b_1 F}} \tag{3}$$

$$\mathrm{R} = \mathrm{y_2} + \mathrm{k_2 e^{-b_2 F}} \tag{4}$$

$$\mathrm{Xc} = \mathrm{y_3} + \mathrm{k_3 e^{-b_3 F}} \tag{5}$$

$$\mathrm{XL} = \mathrm{y_4} + \mathrm{k_4 e^{-b_4 F}} \tag{6}$$

$$\mathrm{C} = \mathrm{y_0} + \mathrm{hF} \tag{7}$$

Furthermore, the real-time intrinsic Z, R, Xc, Xl, C and specific effective thickness (d) of the leaves (F = 0 N) were calculated using the corresponding parameters from Equations (3)–(7):

$$\mathrm{IZ} = \mathrm{y_1} + \mathrm{k_1} \tag{8}$$

$$\mathrm{IR} = \mathrm{y_2} + \mathrm{k_2} \tag{9}$$

$$\mathrm{IX_C} = \mathrm{y_3} + \mathrm{k_3} \tag{10}$$

$$\mathrm{IXL} = \mathrm{y_4} + \mathrm{k_4} \tag{11}$$

$$\mathrm{IC} = \frac{1}{2\pi \mathrm{fIXC}} \tag{12}$$

$$\mathrm{d} = \frac{\mathrm{U^2 h}}{2} \tag{13}$$

where π is 3.1416, f is frequency and U is the tested voltage.

Subsequently, the intracellular water metabolism parameters of *Coix lacryma-jobi* L. including intracellular water-holding capacity (IWHC), intracellular water use efficiency (IWUE), intracellular water-holding time (IWHT) and water transfer rate (WTR) were calculated according to Equations (14)–(17):

$$\mathrm{IWHC} = \sqrt{(\mathrm{IC})^3} \tag{14}$$

$$\mathrm{IWUE} = \frac{\mathrm{d}}{\mathrm{IWHC}} \tag{15}$$

$$\mathrm{IWHT} = \mathrm{IC} \times \mathrm{IZ} \tag{16}$$

$$\mathrm{WTR} = \frac{\mathrm{IWHC}}{\mathrm{IWHT}} \tag{17}$$

Additionally, the intracellular water metabolism parameters of *Coix lacryma-jobi* L., such as nutrient flux per unit area (UNF), active transport flow of nutrient (NAF), nutrient transfer rate (NTR), nutrient transport capacity (NTC) and nutrient active transport capacity (NAC), were calculated according to Equations (18)–(22):

$$\text{UNF} = \frac{\text{IR}}{\text{IXc}} + \frac{\text{IR}}{\text{IXL}} \quad (18)$$

$$\text{NAF} = \frac{\text{IXc}}{\text{IXL}} \quad (19)$$

$$\text{NTR} = \text{WTR} \quad (20)$$

$$\text{NTC} = \text{UNF} \times \text{NTR} \quad (21)$$

$$\text{NAC} = \text{UAF} \times \text{NTR} \quad (22)$$

The chlorophyll content of the second fully expanded leaf of each *Coix lacryma-jobi* L. specimen was measured using a SPAD-502Plus chlorophyll meter (Konica Minolta Inc., Tokyo, Japan). The net photosynthetic rate (Pn), stomatal conductance (Gs) and transpiration rate (Tr) of the second fully expanded leaf of each *Coix lacryma-jobi* L. specimen was determined using a portable Li-6800 photosynthesis measurement system (LI-COR Inc., Lincoln, NE, USA) between 8:00 and 10:00 a.m. The measurement conditions were a flow rate of gas of 500 mmol s^{-1}, photosynthetic active radiation of 500 μmol m^{-2} s^{-1}, leaf temperature of 27 °C and CO_2 concentration of 400 μmol mol^{-1}. The chlorophyll fluorescence parameters of *Coix lacryma-jobi* L. leaves were also measured using a portable Li-6800 photosynthesis measurement system after dark adaptation for 1 h. First, the initial fluorescence (Fo) and maximum fluorescence (Fm) of *Coix lacryma-jobi* L. leaves were determined under dark adaptation conditions, and the maximum photochemical efficiency (Fv/Fm) of PSII was calculated according to Equation (23). Moreover, the maximum fluorescence (Fm′) and stable fluorescence (Fs) of *Coix lacryma-jobi* L. leaves were determined under light adaptation conditions, and the actual photosynthetic efficiency (Φp) of PSII was calculated according to Equation (24).

$$\frac{F_v}{F_m} = \frac{F_m - F_0}{F_m} \quad (23)$$

$$\Phi p = \frac{F_{m'} - F_s}{F_{m'}} \quad (24)$$

2.4. Statistical Analyses

Data are displayed as means ± standard deviation (SD) of three replicates, and all analyses were carried out using SPSS 18.0 (SPSS Inc., Chicago, IL, USA). The data were subjected to analysis of variance (ANOVA) tests. The means of the different groups were compared via Tukey's test ($p < 0.05$). Small letters indicate significant differences between treatments ($n = 3$; $p < 0.05$).

3. Results

3.1. Responses of the Growth in Coix lacryma-jobi L. to Sodium Bicarbonate

The effects of sodium bicarbonate on the growth of *Coix lacryma-jobi* L. are shown in Figure 1. A 2.0 mmol L^{-1} concentration of sodium bicarbonate significantly ($p < 0.05$) promoted plant growth, shoot fresh weight and plant fresh weight compared to CK. A 7.0 mmol L^{-1} concentration of sodium bicarbonate decreased the plant height, aboveground part fresh weight and root fresh weight of *Coix lacryma-jobi* L., whereas the difference between 7.0 mmol L^{-1} sodium bicarbonate and CK did not reach a significant level. A 12.0 mmol L^{-1} concentration of sodium bicarbonate further reduced *Coix lacryma-jobi* L. growth compared with 7.0 mmol L^{-1} sodium bicarbonate.

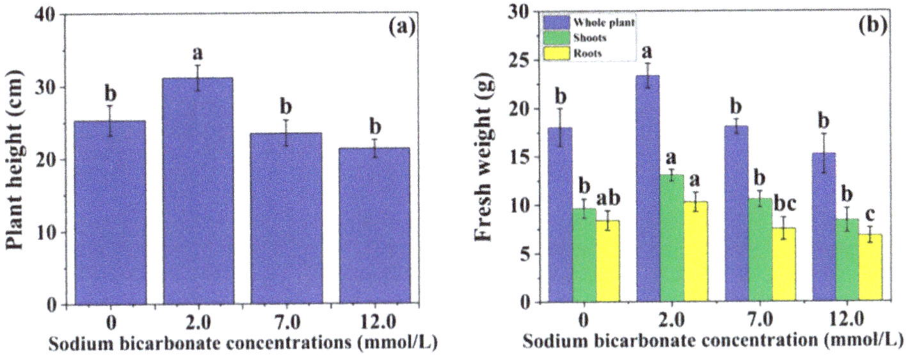

Figure 1. Effects of sodium bicarbonate on the growth of *Coix lacryma-jobi* L. Small letters indicate significant differences between treatments ($n = 3$; $p < 0.05$ according to Tukey's HSD). (**a**) plant height; (**b**) fresh weight.

3.2. Responses of Electrophysiological, Intracellular Water Metabolism and Nutrient Transport Parameters of Coix lacryma-jobi L. to Sodium Bicarbonate

The effects of sodium bicarbonate on the electrophysiological parameters of *Coix lacryma-jobi* L. are shown in Table 1. The plant treated with 2.0 mmol L^{-1} sodium bicarbonate became vigorous, its IC value was higher and its IR, IZ, IXc and IXL values were lower. Sodium bicarbonate effectively increased the IC of *Coix lacryma-jobi* L. and decreased its IR, IZ, IXc and IXL. A 2.0 mmol L^{-1} concentration of sodium bicarbonate significantly ($p < 0.05$) enhanced the IC of *Coix lacryma-jobi* L. and decreased its IR, IZ, IXc and IXL compared to CK. Although 7.0 and 12.0 mmol L^{-1} sodium bicarbonate also decreased the IC of *Coix lacryma-jobi* L. and increased its IR, IZ, IXc and IXL, these electrophysiological parameters were not significantly ($p < 0.05$) different compared to the 0 mmol L^{-1} sodium bicarbonate treatments. These findings emphasize that 2.0 mmol L^{-1} sodium bicarbonate can effectively enhance the IC of *Coix lacryma-jobi* L., decrease its IR, IZ, IXc and IXL and promote its healthy growth.

Table 1. Effects of sodium bicarbonate on the electrophysiological parameters of *Coix lacryma-jobi* L.

$[HCO_3^-]$ (mmol L^{-1})	IC (pF)	IR (MΩ)	IZ (MΩ)	IXc (MΩ)	IXL (MΩ)
0	80.35 ± 22.18 ^{b}	0.79 ± 0.39 ^{a}	0.52 ± 0.20 ^{a}	0.70 ± 0.20 ^{a}	1.26 ± 0.46 ^{a}
2.0	359.70 ± 44.70 ^{a}	0.07 ± 0.02 ^{b}	0.06 ± 0.01 ^{b}	0.15 ± 0.02 ^{b}	0.19 ± 0.03 ^{b}
7.0	108.20 ± 13.75 ^{b}	0.41 ± 0.04 ^{ab}	0.31 ± 0.02 ^{ab}	0.50 ± 0.06 ^{ab}	0.77 ± 0.04 ^{ab}
12.0	97.49 ± 36.47 ^{b}	0.37 ± 0.22 ^{ab}	0.32 ± 0.17 ^{ab}	0.61 ± 0.27 ^{a}	0.83 ± 0.41 ^{ab}

$[HCO_3^-]$: $NaHCO_3$ concentration added to the culture medium. The data are shown as the mean ± SD. The data were subjected to analysis of variance (ANOVA) tests. The means of the different groups were compared using Tukey's test. Small letters indicate significant differences between treatments ($n = 3$; $p < 0.05$).

The effects of sodium bicarbonate on the intracellular water metabolism characteristics of *Coix lacryma-jobi* L. are shown in Table 2. A 2.0 mmol L^{-1} concentration of sodium bicarbonate significantly ($p < 0.05$) increased the IWHC and WTR of *Coix lacryma-jobi* L. compared to the 0, 7.0 and 12.0 mmol L^{-1} sodium bicarbonate treatments. Moreover, 2.0 mmol L^{-1} sodium bicarbonate significantly ($p < 0.05$) decreased the IWHT of *Coix lacryma-jobi* L. compared to CK. At the same time, the d, IWUE, IWHT and WTR of *Coix lacryma-jobi* L. did not show significant ($p < 0.05$) differences between the 0, 7.0 and 12.0 mmol L^{-1} sodium bicarbonate treatments. This indicates that there was little difference in the intracellular water-use efficiency of leaves in the 0, 7.0 and 12.0 mmol L^{-1} sodium bicarbonate treatments. The present findings show that 2.0 mmol L^{-1} sodium bicarbonate could effectively improve the IWHC and WTR of *Coix lacryma-jobi* L. and promote its intracellular water metabolism.

Table 2. Effects of sodium bicarbonate on the intracellular water metabolism characteristics of *Coix lacryma-jobi* L.

$[HCO_3^-]$ (mmol L^{-1})	d	IWHC	IWUE	IWHT	WTR
0	153.57 ± 99.08 [a]	734.05 ± 298.17 [b]	0.27 ± 0.28 [a]	38.52 ± 7.86 [a]	20.51 ± 12.48 [b]
2.0	407.40 ± 63.75 [a]	6848.49 ± 1263.66 [a]	0.06 ± 0.01 [a]	21.87 ± 1.96 [b]	316.87 ± 80.77 [a]
7.0	238.29 ± 176.15 [a]	1130.00 ± 217.07 [b]	0.20 ± 0.11 [a]	33.86 ± 3.24 [ab]	33.22 ± 3.83 [b]
12.0	181.56 ± 40.55 [a]	997.09 ± 524.67 [b]	0.22 ± 0.11 [a]	27.80 ± 4.85 [ab]	36.93 ± 19.39 [b]

$[HCO_3^-]$: $NaHCO_3$ concentration added to the culture medium. The data are shown as the mean ± SD. The data were subjected to analysis of variance (ANOVA) tests. The means of the different groups were compared using Tukey's test. Small letters indicate significant differences between treatments (n = 3; $p < 0.05$).

The effects of sodium bicarbonate on the nutrient transport characteristics of *Coix lacryma-jobi* L. are shown in Table 3. Compared to CK, 2.0 mmol L^{-1} sodium bicarbonate significantly ($p < 0.05$) increased the NAC of *Coix lacryma-jobi* L., significantly ($p < 0.05$) enhanced its NAF and NTC and significantly ($p < 0.05$) decreased its UNF. Compared to the 7.0 and 12.0 mmol L^{-1} sodium bicarbonate treatments, 2.0 mmol L^{-1} sodium bicarbonate also significantly ($p < 0.05$) increased the NAC of *Coix lacryma-jobi* L. Moreover, the UNF, NAF, NAC and NTC of *Coix lacryma-jobi* L. did not show significant ($p < 0.05$) differences between the 0, 7.0 and 12.0 mmol L^{-1} sodium bicarbonate treatments. These findings here emphasize that 2.0 mmol L^{-1} sodium bicarbonate could effectively improve the NAF, NAC and NTC of *Coix lacryma-jobi* L. and promote its nutrient transport processes.

Table 3. Effects of sodium bicarbonate on the nutrient transport characteristics of *Coix lacryma-jobi* L.

$[HCO_3^-]$ (mmol L^{-1})	UNF	NAF	NAC	NTC
0	1.70 ± 0.55 [a]	0.58 ± 0.12 [b]	12.67 ± 10.15 [b]	36.39 ± 4.65 [b]
2.0	0.84 ± 0.08 [b]	0.77 ± 0.02 [a]	245.58 ± 70.30 [a]	43.53 ± 0.73 [a]
7.0	1.36 ± 0.20 [ab]	0.64 ± 0.06 [ab]	21.30 ± 1.70 [b]	39.13 ± 2.03 [ab]
12.0	1.03 ± 0.19 [ab]	0.75 ± 0.05 [ab]	28.07 ± 15.51 [b]	42.71 ± 1.64 [ab]

$[HCO_3^-]$: $NaHCO_3$ concentration added to the culture medium. The data are shown as the mean ± SD. The data were subjected to analysis of variance (ANOVA) tests. The means of the different groups were compared using Tukey's test. Small letters indicate significant differences between treatments (n = 3; $p < 0.05$).

3.3. *Responses of the Photosynthetic Characteristics of* Coix lacryma-jobi *L. to Sodium Bicarbonate*

The effects of sodium bicarbonate on the photosynthetic characteristics of *Coix lacryma-jobi* L. are shown in Figure 2. A 2.0 mmol L^{-1} concentration of sodium bicarbonate effectively increased the chlorophyll content of *Coix lacryma-jobi* L. compared to CK. Plants treated with 2.0 mmol L^{-1} sodium bicarbonate possessed a superior photosynthetic rate of leaves of 20.86 $\mu mol \cdot CO_2 \cdot m^{-2} \cdot s^{-1}$, which was significantly ($p < 0.05$) higher than that of the plants treated with 0, 7.0 and 12.0 mmol L^{-1} sodium bicarbonate (1.15-, 1.19- and 1.8-fold higher, respectively). Compared to CK, 7.0 mmol L^{-1} sodium bicarbonate significantly ($p < 0.05$) increased the chlorophyll content of *Coix lacryma-jobi* L. while significantly ($p < 0.05$) decreasing its stomatal conductance and transpiration rate; 12.0 mmol L^{-1} sodium bicarbonate significantly ($p < 0.05$) decreased its photosynthetic rate, stomatal conductance and transpiration rate. These results indicate that 2.0 mmol L^{-1} sodium bicarbonate can enhance the photosynthesis of *Coix lacryma-jobi* L. plants and promote their growth, and that 7.0~12.0 mmol L^{-1} sodium bicarbonate inhibits the growth of *Coix lacryma-job* L.

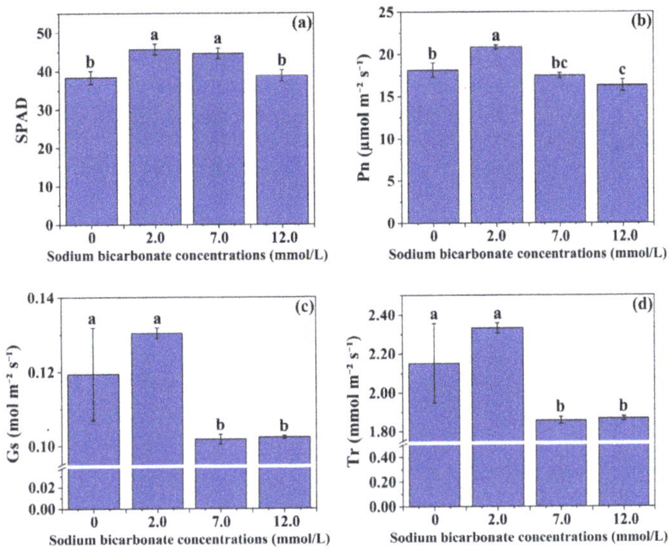

Figure 2. Effects of sodium bicarbonate on the chlorophyll content (SPAD), net photosynthetic rate (Pn), stomatal conductance (Gs) and transpiration rate (Tr) of *Coix lacryma-jobi* L. Small letters indicate significant differences between treatments ($n = 3$; $p < 0.05$ according to Tukey's HSD). (**a**) the chlorophyll content (SPAD); (**b**) net photosynthetic rate (Pn); (**c**) stomatal conductance (Sc); (**d**) transpiration rate (Tr).

3.4. Responses of the Chlorophyll Fluorescence of *Coix lacryma-jobi* L. to Sodium Bicarbonate

The effects of sodium bicarbonate on the chlorophyll fluorescence characteristics of *Coix lacryma-jobi* L. are shown in Figure 3. There were no significant ($p < 0.05$) differences in the Fv/Fm of *Coix lacryma-jobi* L. between the 0, 2.0, 7.0 and 12.0 mmol L^{-1} sodium bicarbonate treatments. A 2.0 mmol L^{-1} concentration of sodium bicarbonate effectively increased the Φp of *Coix lacryma-jobi* L. compared to CK, although the difference was not significant. Concentrations of 7.0 and 12.0 mmol L^{-1} decreased the Φp of *Coix lacryma-jobi* L. compared with 0 mmol L^{-1} sodium bicarbonate, and there was a significant ($p < 0.05$) difference between the 0 and 12.0 mmol L^{-1} sodium bicarbonate treatments. The present findings show that 2.0 mmol L^{-1} sodium bicarbonate can increase the Φp of *Coix lacryma-jobi* L., thereby enhancing its photosynthetic capacity.

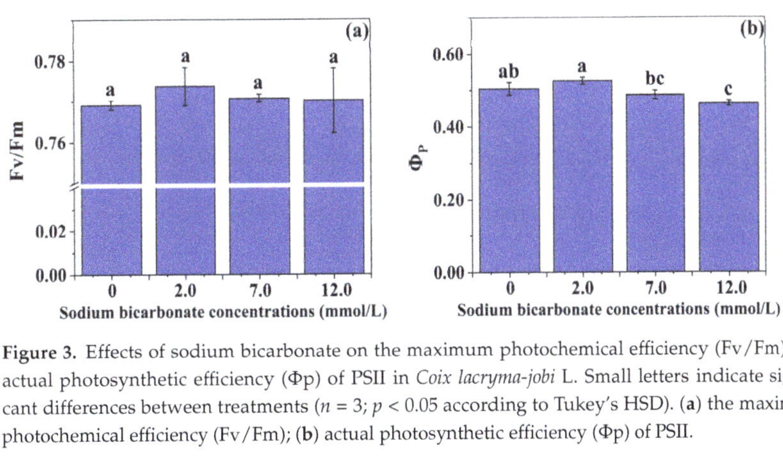

Figure 3. Effects of sodium bicarbonate on the maximum photochemical efficiency (Fv/Fm) and actual photosynthetic efficiency (Φp) of PSII in *Coix lacryma-jobi* L. Small letters indicate significant differences between treatments ($n = 3$; $p < 0.05$ according to Tukey's HSD). (**a**) the maximum photochemical efficiency (Fv/Fm); (**b**) actual photosynthetic efficiency (Φp) of PSII.

4. Discussion

Sodium bicarbonate can not only increase the pH of the environment, but it also affects plants' absorption of water and mineral elements [46]. At the appropriate concentrations of sodium bicarbonate, the positive effect is greater than the negative effect, such as being conducive to carbon assimilation, stimulating stomatal opening and improving drought resistance, etc. In contrast, at excessively high levels of sodium bicarbonate, it can bring anionic toxicity, causing effects such as physiological drought, stomata closure, reducing metabolite synthesis and transport rates, etc. It has been reported that high concentrations of HCO_3^- can reduce the uptake of Fe in plants by limiting the expression of Fe acquisition genes, leading to Fe deficiency [12]. Generally, salt stress occurs when the concentration of sodium is in the range of 25–200 mM [47,48]. Hence, the *Coix lacryma-jobi* L. seedlings did not suffer from salt stress because the concentration of sodium bicarbonate was far below 25 mM in this study. The results here demonstrate that 2.0 mmol L^{-1} sodium bicarbonate can effectively enhance the growth of *Coix lacryma-jobi* L., which is probably derived from the long-term domestication of *Coix lacryma-jobi* L. with the appropriate sodium bicarbonate concentration. With increasing HCO_3^- concentrations, the pH of the culture solution increased gradually, which led to a reduction of the absorption of water and inorganic salts by plant roots, thereby affecting the plants' normal growth and development [12,49]. Rao and Wu found similar results in other plant species (*Camptotheca acuminate* seedlings) [15]. The water potential of plant leaves decreased gradually with the increase in sodium bicarbonate concentration, and the transpiration rate of plants under bicarbonate stress decreased significantly. These effects could explain how the 12.0 mmol L^{-1} sodium bicarbonate treatment on *Coix lacryma-job* L. affected its growth. We also found that the sodium bicarbonate concentration in the soils of Guizhou's *Coix lacryma-jobi* L. planting areas was about 1.2~3.3 mmol L^{-1}.

Plants' physiological activities are vigorous in the healthy state, and its cells store more charge (ions, ion groups and electric dipoles), which can be considered as a generalized charging phenomenon [33]. Thus, the stronger the growth of plants, the higher its charge and C value and the lower its R, Z, Xc and XL values. The findings here emphasize that 2.0 mmol L^{-1} sodium bicarbonate can effectively enhance the IC of *Coix lacryma-jobi* L., decrease its IR, IZ, IXc and IXL and promote its healthy growth, whereas these electrophysiological information parameters were not significantly ($p < 0.05$) different between the 0, 7.0 and 12.0 mmol L^{-1} sodium bicarbonate treatments. These results demonstrate that the electrical parameters of *Coix lacryma-jobi* L. rapidly responded to bicarbonate stress, and the appropriate bicarbonate concentration can enhance the growth of *Coix lacryma-jobi* L.

The intracellular water metabolism and nutrient transport statuses of plants strongly reflects their growth and development. Zhang et al. reported that *Broussonetia papyrifera* plants grown in agricultural soils had higher IC, d, IWHC, WTR (or NTR), NTC and water content values and lower IZ and IXc values than those grown in moderately rocky, desertified soils [44,45]. The present findings show that 2.0 mmol L^{-1} sodium bicarbonate can effectively improve the IWHC, WTR (or NTR), NAF, NAC and NTC values of *Coix lacryma-jobi* L. and promote its intracellular water metabolism and nutrient transport. Although the d, IWUE, IWHT, WTR (or NTR), IWUE UNF, NAF, NAC and NTC values of *Coix lacryma-jobi* L. were not significantly ($p < 0.05$) different between the 0, 7.0 and 12.0 mmol L^{-1} sodium bicarbonate treatments, the 7.0 and 12.0 mmol L^{-1} sodium bicarbonate treatments can improve the IWHC, WTR (or NTR), NAF, NAC and NTC of *Coix lacryma-jobi* L. Zhang et al. also reported four nutrient transport strategies in the tested plants: (1) low UNF, high NTR and high NTC (e.g., *Broussonetia papyrifera* grown in agricultural soils, *Ipomoea batatas* L., *Solanum tuberosum* L.); (2) high UNF, low NTR and low NTC (e.g., *Broussonetia papyrifera* grown in moderately rocky desertified soils, *Senecio scandens* Buch.-Ham. ex D., *Capsicum annuum* L.); (3) high UNF, low NTR and high NTC (e.g., *Rhus chinensis* Mill.); (4) low UNF, high NTR and low NTC (e.g., *Toona sinensis*) [45]. In this study, three nutrient transport strategies were found: (1) high UNF, low NTR and low NTC (0 mmol L^{-1}); (2) low UNF, high NTR and high NTC (2.0 mmol L^{-1}); (3) high UNF, low NTR and high

NTC (7.0 and 12.0 mmol L^{-1}). From this study, we can conclude that plant leaves with high intracellular water-holding capacity (IWHC) have a faster water transfer rate (WTR). These results also show that the appropriate bicarbonate concentration can improve the intracellular water metabolism and nutrient transport of *Coix lacryma-jobi* L., and that *Coix lacryma-jobi* L. can adapt to bicarbonate environments by improving its intracellular water metabolism and nutrient transport capacities. In addition, the inorganic carbon metabolism of *Coix lacryma-jobi* L., which is dependent on both atmospheric and soil inorganic carbon, can be captured by measuring the plant's electrophysiological parameters.

Shahsavandia et al., found that high HCO_3^- concentrations can reduce the chloroplast content of plants by affecting their reduced iron ion amounts [49]. Li et al., reported that high HCO_3^- concentration is a type of abiotic stress, which could affect the water use of plants and thereby inhibit their photosynthetic capacity [50]. Rao and Wu found that the water potential of plant leaves decreased gradually with the increase in sodium bicarbonate concentration, and the transpiration rate of plants under sodium bicarbonate stress decreased significantly [15]. In this work, 7.0 and 12.0 mmol L^{-1} sodium bicarbonate had inhibitory effects on the chlorophyll content, photosynthetic rate, stomatal conductance and transpiration rate of *Coix lacryma-job* L. The intensification of osmotic stress and the decrease in transpiration rate caused by high HCO_3^- concentrations might be the main reasons for the decrease in photosynthetic capacity of *Coix lacryma-job* L. Nevertheless, plants treated with 2.0 mmol L^{-1} sodium bicarbonate possessed a superior photosynthetic rate of 20.86 $\mu mol \cdot CO_2 \cdot m^{-2} \cdot s^{-1}$, which was significantly ($p < 0.05$) higher than those of plants that received the 0, 7.0 and 12.0 mmol L^{-1} sodium bicarbonate treatments. These positive effects were probably related to the utilization of HCO_3^- from the soils by *Coix lacryma-jobi* L.; this HCO_3^- is kept in the reaction center of PSII, and it plays an important role in the transfer of electrons in photosynthesis [1,8,15,51–53]. Our previous work has shown that in karst environments, plants can alternatively use bicarbonate from soil and atmospheric carbon dioxide, and when plants encounter some adversity, their stomatal conductance decreases, carbonic anhydrase activity increases, and their use of bicarbonate also increases, thereby improving their intercellular water status, which in turn increases the use of atmospheric carbon dioxide by these plants [1,15,16,20]. Bicarbonates can stimulate plants to release oxygen, which is called the bicarbonate effect [54]. Recently, Wu proposed that sodium bicarbonate may directly participate in photosynthetic O_2 evolution, resulting in the chemical equilibria, $HCO_3^- + H^+ \rightarrow 1/2O_2 + 2e^- + 2H^+ + CO_2$, which provides electrons and concentrating CO_2 in the Calvin cycle, which is involved in photosynthetic carbon assimilation [51,53].

There are many studies on the impact of plants' photosynthetic characteristics [55–57]. Giovanna et al. reported that added HCO_3^- is beneficial for plant growth and improving photosynthetic efficiency [58]. The present findings show that there was no significant ($p < 0.05$) difference in the Fv/Fm of *Coix lacryma-jobi* L. in the 0, 2.0, 7.0 and 12.0 mmol L^{-1} sodium bicarbonate treatments, which indicates that sodium bicarbonate treatment does not damage the reaction center of PSII. Moreover, 2.0 mmol L^{-1} sodium bicarbonate increased the Φp of PSII in *Coix lacryma-jobi* L. and thereby enhanced photosynthesis, which further demonstrates that the appropriate bicarbonate concentration can enhance the photosynthesis and growth of *Coix lacryma-jobi* L.

Plants grown in karst areas often suffer from various degrees of karst ecological stresses including high bicarbonate concentrations, drought and high pH levels [1,7]. In this work, the results show that 2.0 mmol L^{-1} sodium bicarbonate effectively enhanced the photosynthetic rate and Φp of *Coix lacryma-jobi* L., notably increased its IWHC, WTR (or NTR), NAF, NAC and NTC and improved its intracellular water metabolism and nutrient transport, thereby markedly enhancing its growth and biomass formation. Meanwhile, 12.0 mmol L^{-1} sodium bicarbonate exhibited inhibitory effects on the chlorophyll content, photosynthetic rate, stomatal conductance, transpiration rate and biomass of *Coix lacryma-job* L. Generally, the utilization of bicarbonate from soil by plants is difficult to measure. The results indicate that the inorganic carbon used by *Coix lacryma-jobi* L., which comes from both atmosphere and soil, can be studied using measurements of plants' electrophysiological parameters. This work highlights that the appropriate bicarbonate concentration can improve the intracellular water metabolism, nutrient transport, photosynthetic capacity and growth of *Coix lacryma-jobi* L., and plants' electrophysiological parameters can be used for determining the appropriate bicarbonate level.

5. Conclusions

In conclusion, the present study indicates that 2.0 mmol L^{-1} sodium bicarbonate effectively improves the biomass formation of *Coix lacryma-jobi* L. and notably increases its IC and decrease its IR, IZ, IXc and IXL as well as reliably enhances its intracellular water metabolism, nutrient transport and photosynthesis capacity. Moreover, 7.0 and 12.0 mmol L^{-1} sodium bicarbonate exhibited inhibitory effects on the chlorophyll content, photosynthetic rate, stomatal conductance, transpiration rate and biomass of *Coix lacryma-job* L. This study highlights that the appropriate level of bicarbonate can enhance the intracellular water metabolism, nutrient transport, photosynthetic capacity and growth of *Coix lacryma-jobi* L., and its appropriate bicarbonate level can be rapidly obtained using plants' electrophysiological parameters.

Author Contributions: Y.S. constructed the project; Y.W. and H.L. designed the experiments; J.L. and H.L. performed the experiments; H.L. and J.L. analyzed the data; H.L., Y.W. and Y.S. wrote the paper. All authors have read and agreed to the published version of the manuscript.

Funding: This work was jointly supported by the Science and technology innovation talent project of Guizhou Province [(2016)5672], the Support Plan Projects of Science and Technology of Guizhou Province [(2021)YB453], and the Key project of Guizhou Vocational College of Agriculture [YJKT2020-9].

Institutional Review Board Statement: Not applicable.

Data Availability Statement: Not applicable.

Conflicts of Interest: We declare that we do not have any commercial or associative interest that represent conflicts of interest in connection with the work submitted.

Abbreviations

HCO_3^-	sodium bicarbonate
C	capacitance
Z	impedance
R	resistance
F	force
d	specific effective thickness
Xc	capacitive reactance
XL	inductive reactance
IC	intrinsic capacitance
IR	intrinsic resistance
IZ	intrinsic impedance
IXc	intrinsic capacitive reactance
IXL	intrinsic inductive reactance
IWHC	intracellular water-holding capacity
IWUE	intracellular water use efficiency
IWHT	intracellular water-holding time
WTR	water transfer rate
UNF	nutrient flux per unit area
NAF	active transport flow of nutrient
NTR	nutrient transfer rate
NTC	nutrient transport capacity
NAC	nutrient active transport capacity

References

1. Wu, Y.; Xing, D.; Hang, H.; Zhao, K. *Principles and Technology of Determination on Plant' Adaptation to Karst Environment*; Science Press: Beijing, China, 2019; pp. 1–88.
2. Stokes, T.R.; Griffiths, P.A. An overview of the Karst Areas in British Columbia, Canada. *Geosci. Can.* **2019**, *46*, 49–66. [CrossRef]
3. Luo, L.; Wu, Y.; Li, H.; Xing, D.; Zhou, Y.; Xia, A. Drought Induced Dynamic Traits of Soil Water and Inorganic Carbon in Different Karst Habitats. *Water* **2022**, *14*, 3837. [CrossRef]
4. McCray, J.M.; Matocha, J.E. Effects of soil water levels on solution bicarbonate, chlorosis and growth of sorghum. *J. Plant Nutr.* **1992**, *15*, 1877–1890. [CrossRef]
5. Zhang, C.; Wang, J.; Pu, J.; Yan, J. Bicarbonate daily variations in a karst river: The carbon sink effect of subaquatic vegetation photosynthesis. *Acta Geol. Sin.-Engl.* **2012**, *86*, 973–979. [CrossRef]
6. Hussner, A.; Mettler-Altmann, T.; Weber, A.; Sand-Jense, K. Acclimation of photosynthesis to supersaturated CO_2 in aquatic plant bicarbonate users. *Freshw. Biol.* **2016**, *61*, 1720–1732. [CrossRef]
7. Xing, D.; Chen, X.; Wu, Y.; Xu, X.; Chen, Q.; Li, L.; Zhang, C. Rapid prediction of the re-watering time point of *Orychophragmus violaceus* L. based on the online monitoring of electrophysiological indexes. *Sci. Hortic.* **2019**, *256*, 108642. [CrossRef]
8. Yao, K.; Wu, Y. Rhizospheric bicarbonate improves glucose metabolism and stress tolerance of *Broussonetia papyrifera* L. seedlings under simulated drought stress. *Russ. J. Plant Physl.* **2021**, *68*, 126–135. [CrossRef]
9. Gong, B.; Wen, D.; Vandenlangenberg, K.; Wei, M.; Yang, F.; Shi, Q.; Wang, X. Comparative effects of NaCl and $NaHCO_3$ stress on photosynthetic parameters, nutrient metabolism, and the antioxidant system in tomato leaves. *Sci. Hortic.* **2013**, *157*, 1–12. [CrossRef]
10. Zhang, X.; Pu, P.; Tang, Y.; Zhang, L.; Lv, J. C_4 photosynthetic enzymes play a key role in wheat spike bracts primary carbon metabolism response under water deficit. *Plant Physiol. Biochem.* **2019**, *142*, 163–172. [CrossRef]
11. Sun, X.; Dong, X.; Li, X.; Chen, L.; Liu, J. Response of growth and physiological characteristics of melon seedlings under alkali stress. *North Hortic.* **2020**, *9*, 51–59. [CrossRef]
12. García, M.J.; García-Mateo, M.J.; Lucena, C.; Romera, F.J.; Rojas, C.L.; Alcántara, E.; Pérez-Vicente, R. Hypoxia and bicarbonate could limit the expression of iron acquisition genes in Strategy I plants by affecting ethylene synthesis and signaling in different ways. *Physiol. Plant.* **2014**, *150*, 95–106. [CrossRef]
13. Ding, W.; Clode, P.L.; Lambers, H. Effects of pH and bicarbonate on the nutrient status and growth of three Lupinus species. *Plant Soil.* **2020**, *447*, 9–28. [CrossRef]
14. Kallala, N.; M'sehli, W.; Jelali, K.; Kais, Z.; Mhadhbi, H. Inoculation with efficient nitrogen fixing and indoleacetic acid producing bacterial microsymbiont enhance tolerance of the model legume *Medicago truncatula* to iron deficiency. *Biomed Res. Int.* **2018**, *2018*, 9134716. [CrossRef] [PubMed]
15. Rao, S.; Wu, Y. Root derived bicarbonate assimilation in response to variable water deficit in *Camptotheca acuminata* seedlings. *Photosynth. Res.* **2017**, *134*, 59–70. [CrossRef] [PubMed]
16. Wu, Y. Strategies to increase carbon fixation and sequestration by karst-adaptable plants. *Carsol. Sin.* **2011**, *30*, 461–465. [CrossRef]
17. Tang, Y.; Lian, B. Diversity of endolithic fungal communities in dolomite and limestone rocks from Nanjiang canyon in Guizhou karst area, China. *Can. J. Microbiol.* **2012**, *58*, 685–693. [CrossRef] [PubMed]
18. Wang, R.; Wu, Y.; Xing, D.; Hang, H.; Liu, Y.; Zhang, K.; Yan, K. Physiological characteristics and inorganic carbon usage capacity of three biomass plants under simulative karst adversity (bicarbonate stress). *Earth Environ.* **2015**, *43*, 21–30. [CrossRef]
19. Hang, H.; Wu, Y. Quantification of photosynthetic inorganic carbon utilisation via a bidirectional stable carbon isotope tracer. *Acta Geochim.* **2016**, *35*, 130–137. [CrossRef]
20. Wang, R.; Wu, Y.; Xing, D. Biomass production of three biofuel energy plants'use of a new carbon resource by carbonic anhydrase in simulated karst soils: Mechanism and capacity. *Energies* **2017**, *10*, 1370. [CrossRef]
21. Banerjee, G.; Ghosh, I.; Kim, C.J.; Debus, R.J.; Brudvig, G.W. Bicarbonate rescues damaged proton-transfer pathway in photosystem II. *BBA-Bioenerg.* **2019**, *1860*, 611–617. [CrossRef]
22. Zhou, W.; Sui, Z.; Wang, J.; Hu, Y.; Kang, K.H.; Hong, H.R.; Niaz, Z.; Wei, H.; Du, Q.; Peng, C.; et al. Effects of sodium bicarbonate concentration on growth, photosynthesis, and carbonic anhydrase activity of macroalgae *Gracilariopsis lemaneiformis, Gracilaria vermiculophylla,* and *Gracilaria chouae* (Gracilariales, Rhodophyta). *Photosynth. Res.* **2016**, *128*, 259–270. [CrossRef]
23. Fu, Y.; Yang, C.; Meng, Q.; Liu, F.; Shen, G.; Zhou, M.; Ao, M. Genetic diversity and structure of *Coix lacryma-jobi* L. from its world secondary diversity center, southwest China. *Int. J. Genom.* **2019**, *2019*, 9815697. [CrossRef]
24. Miao, G.; Qin, Y.; Guo, J.; Zhang, Q.; Bao, Y. Transcriptome characterization and expression profile of *Coix lacryma-jobi* L. in response to drought. *PLoS ONE* **2021**, *16*, e0256875. [CrossRef] [PubMed]
25. Lee, Y.J.; Park, H.J.; Cho, S.J.; Park, S.A.; Kim, M.L. The effect of yulmu (*Coix lacryma-jobi* L.)-sunsik on obesity and hyperlipidemia in mice fed high-fat diet. *J. Korean Soc. Food Sci. Nutr.* **2021**, *50*, 664–671. [CrossRef]
26. Yu, Q.; Ye, G.; Lei, Z.; Yang, R.; Chen, R.; He, T.; Huang, S. An isolated compound from stems and leaves of *Coix lacryma-jobi* L. and its anticancer effect. *Food Biosci.* **2021**, *42*, 101047. [CrossRef]
27. Diao, X. Production and genetic improvement of minor cereals in China. *Crop. J.* **2017**, *5*, 103–114. [CrossRef]

28. Li, X.; Lu, X.; Pan, H.; Wei, X.; Shi, M.; Su, Y. Research status of germplasm resource and genetic breeding in *Coix lacryma-jobi* L. *Guizhou Agri. Sci.* **2022**, *50*, 8–15. [CrossRef]
29. Fromm, J.; Lautner, S. Electrical signals and their physiological significance in plants. *Plant Cell Environ.* **2007**, *30*, 249–257. [CrossRef]
30. Volkov, A.G. *Plant Electrophysiology*; Theory and Methods; Springer: Berlin/Heidelberg, Germany, 2006.
31. Szechyńska-Hebda, M.; Lewandowska, M.; Karpiński, S. Electrical signaling, photosynthesis and systemic acquired acclimation. *Front. Physiol.* **2017**, *8*, 684. [CrossRef] [PubMed]
32. Sukhov, V. Electrical signals as mechanism of photosynthesis regulation in plants. *Photosynth. Res.* **2016**, *130*, 373–387. [CrossRef]
33. Zhang, C.; Su, Y.; Wu, Y.; Li, H.; Zhou, Y.; Xing, D. Comparison on the nutrient plunder capacity of *Orychophragmus violaceus* and *Brassica napus* L. Based on electrophysiological properties. *Horticulturae* **2021**, *7*, 206. [CrossRef]
34. Sukhov, V.; Gaspirovich, V.; Mysyagin, S.; Vodeneev, V. High-temperature tolerance of photosynthesis can be linked to local electrical responses in leaves of Pea. *Front. Physiol.* **2017**, *8*, 763. [CrossRef] [PubMed]
35. Sukhov, V.; Surova, L.; Sherstneva, O.; Bushueva, A.; Vodeneev, V. Variation potential induces decreased PSI damage and increased PSII damage under high external temperatures in pea. *Funct. Plant Biol.* **2015**, *42*, 727–736. [CrossRef] [PubMed]
36. Gil, P.M.; Gurovich, L.; Schaffer, B.; Alcayaga, J.; Rey, S.; Iturriaga, R. Root to leaf electrical signaling in avocado in response to light and soil water content. *J. Plant Physiol.* **2008**, *165*, 1070–1078. [CrossRef]
37. Zhang, M.; Wu, Y.; Xing, D.; Zhao, K.; Yu, R. Rapid measurement of drought resistance in plants based on electrophysiological properties. *Trans. ASABE* **2015**, *58*, 1441–1446. [CrossRef]
38. Harker, F.R.; Dunlop, J. Electrical impedance studies of nectarines during coolstorage and fruit ripening. *Postharvest Biol. Technol.* **1994**, *4*, 125–134. [CrossRef]
39. Ibba, P.; Falco, A.; Abera, B.D.; Cantarella, G.; Petti, L.; Lugli, P. Bio-impedance and circuit parameters: An analysis for tracking fruit ripening. *Postharvest Biol. Technol.* **2020**, *159*, 110978. [CrossRef]
40. Javed, Q.; Wu, Y.; Xing, D.; Azeem, A.; Ullah, I.; Zaman, M. Re-watering: An effective measure to recover growth and photosynthetic characteristics in salt-stressed *Brassica napus* L. *Chil. J. Agric. Res.* **2017**, *77*, 78–86. [CrossRef]
41. Kertész, Á.; Hlaváčová, Z.; Vozáry, E.; Staroňová, L. Relationship between moisture content and electrical impedance of carrot slices during drying. *Int. Agrophys.* **2015**, *29*, 61–66. [CrossRef]
42. Xing, D.; Chen, X.; Wu, Y.; Zwiazek, J.J. Leaf physiological impedance and elasticity modulus in *Orychophragmus violaceus* seedlings subjected to repeated osmotic stress. *Sci. Hortic.* **2021**, *276*, 109763. [CrossRef]
43. Xing, D.; Mao, R.; Li, Z.; Wu, Y.; Qin, X.; Fu, W. Leaf intracellular water transport rate based on physiological impedance: A possible role of leaf internal retained water in photosynthesis and growth of tomatoes. *Front. Plant Sci.* **2022**, *13*, 845628. [CrossRef]
44. Zhang, C.; Wu, Y.; Su, Y.; Xing, D.; Dai, Y.; Wu, Y.; Fang, L. A plant's electrical parameters indicate its physiological state: A study of intracellular water metabolism. *Plants* **2020**, *9*, 1256. [CrossRef] [PubMed]
45. Zhang, C.; Wu, Y.; Su, Y.; Li, H.; Fang, L.; Xing, D. Plant's electrophysiological properties manifests the composition and nutrient transport characteristics of membrane proteins. *Plant Signal. Behav.* **2021**, *16*, e1918867. [CrossRef]
46. Hajiboland, R.; Yang, X.; Romeld, V. Effects of bicarbonate and high pH on growth of Zn-efficient and Zn-inefficient genotypes of rice, wheat and rye. *Plant Soil.* **2003**, *250*, 349–357. [CrossRef]
47. Haro, R.; Banuelos, M.A.; Quintero, F.J.; Rubio, F.; Rodriguez-Navarro, A. Genetic basis of sodium exclusion and sodium tolerance in yeast, A model for plants. *Physiol. Plant.* **1993**, *89*, 868–874. [CrossRef]
48. Gupta, P.; Seth, C.S. 24-Epibrassinolide Regulates Functional Components of Nitric Oxide Signalling and Antioxidant Defense Pathways to Alleviate Salinity Stress in *Brassica juncea* L. cv. Varuna. *J. Plant Growth Regul.* **2023**, *42*, 183–198. [CrossRef]
49. Shahsavandi, F.; Eshghi, S.; Gharaghani, A.; Ghasemi-Fasaei, R.; Jafarinia, M. Effects of bicarbonate induced iron chlorosis on photosynthesis apparatus in grapevine. *Sci. Hortic.* **2020**, *270*, 109427. [CrossRef]
50. Li, R.; Shi, F.; Fukuda, K.; Yang, Y. Effects of salt and alkali stresses on germination, growth, photosynthesis and ion accumulation in alfalfa (*Medicago sativa* L.). *Soil Sci. Plant Nutr.* **2010**, *56*, 725–733. [CrossRef]
51. Wu, Y. Combined effect of bicarbonate and water in photosynthetic oxygen evolution and carbon neutrality. *Acta Geochim.* **2023**, *42*, 77–88. [CrossRef]
52. Tikhonov, K.; Shevela, D.; Klimov, V.V.; Messinger, J. Quantification of bound bicarbonate in photosystem II. *Photosynthetica* **2017**, *56*, 210–216. [CrossRef]
53. Wu, Y. Is bicarbonate directly used as substrate to participate in photosynthetic oxygen evolution. *Acta Geochim.* **2021**, *40*, 650–658. [CrossRef]
54. Warburg, O.; Krippahl, G. Hill-Reaktionen. *Z. Naturforsch.* **1958**, *13*, 509–514. [CrossRef]
55. Agnihotri, A.; Gupta, P.; Dwivedi, A.; Seth, C.S. Counteractive mechanism (s) of salicylic acid in response to lead toxicity in *Brassica juncea* (L.) Czern. cv. Varuna. *Planta* **2018**, *248*, 49–68. [CrossRef]
56. Singh, D.; Agnihotri, A.; Seth, C.S. Interactive effects of EDTA and oxalic acid on chromium uptake, translocation and photosynthetic attributes in Indian mustard (*Brassica juncea* L. var. Varuna). *Curr. Sci.* **2017**, *112*, 2034–2042. [CrossRef]

57. Kumar, D.; Dhankher, O.P.; Tripathi, R.D.; Seth, C.S. Titanium dioxide nanoparticles potentially regulate the mechanism (s) for photosynthetic attributes, genotoxicity, antioxidants defense machinery, and phytochelatins synthesis in relation to hexavalent chromium toxicity in *Helianthus annuus* L. *J. Hazard. Mater.* **2023**, *454*, 131418. [CrossRef] [PubMed]
58. Giovanna, S.; Francesco, B.; Mario, A.; Federica, C.; Olga, M.; Simona, C. Rapid and positive effect of bicarbonate addition on growth and photosynthetic efficiency of the green microalgae *Chlorella Sorokiniana* (Chlorophyta, Trebouxiophyceae). *Appl. Sci.* **2020**, *10*, 4515. [CrossRef]

agronomy

MDPI

Article

Differential Responses of Nitrate/Ammonium Use to Bicarbonate Supply in Two *Brassicaceae* Species under Simulated Karst Habitat

Antong Xia [1,2] and Yanyou Wu [1,*]

[1] State Key Laboratory of Environmental Geochemistry, Institute of Geochemistry, Chinese Academy of Sciences, Guiyang 550081, China
[2] University of Chinese Academy of Sciences, Beijing 100049, China
* Correspondence: wuyanyou@mail.gyig.ac.cn; Tel.: +86-0851-8439-1746

Abstract: In the karst habitats with nitrate-abundant and ammonium-rare soil, the bicarbonate supply plays a crucial role in both inorganic carbon and nitrogen assimilation in various plant species. Consequently, two carbon sources, carbon dioxide (CO_2) and bicarbonate (HCO_3^-), and two nitrogen sources, namely nitrate (NO_3^-) and ammonium (NH_4^+) are available for plants. However, variations in the absorption and utilization of nitrate, ammonium, and inorganic carbon during bicarbonate supply in different plants are not well-depicted. In this study, we evaluated the nitrate/ammonium use efficiency and their contributions to the total nitrogen assimilation/utilization capacity at different bicarbonate levels using a bidirectional stable nitrogen isotope tracer approach. The inorganic carbon assimilation, such as the photosynthesis, carbon/nitrogen enzymatic activities, carbon/nitrogen content, nitrogen assimilation/utilization capacity, and nitrate/ammonium contributions to plant growth, were also evaluated to decipher the responses of both carbon and nitrogen metabolism to bicarbonate supply in karst habitats. With the increasing bicarbonate level, *Orychophragmus violaceus* (*Ov*) was found to be more available for nitrate to total nitrogen assimilation and utilization than *Bn* (*Brassica napus*). Further, it enhanced the contributions of nitrate and nitrogen accumulation/utilization capacity and inorganic carbon assimilation, increasing photosynthesis, carbon/nitrogen enzymatic activities, and carbon accumulation, and promoted the growth in *Ov*. Though the highest bicarbonate level was conducive to ammonium utilization and water use efficiency in both *Ov* and *Bn*, it inhibited total inorganic carbon and nitrogen assimilation, leading to growth suppression in *Bn* compared to *Ov*. Moreover, considering the optimistic responses of both carbon and nitrogen assimilation to the high bicarbonate supply in nitrate-abundant, as well as ammonium-rare, environments, we conclude that *Ov* was more adaptable to the karst habitats. This study provides a novel approach to elucidate the responses of nitrate/ammonium utilization and inorganic carbon assimilation to bicarbonate. Furthermore, the current study reveals the complex interactions among different carbon–nitrogen metabolism pathways in various plants and their adaptations to karst habitats.

Keywords: adaptation; bidirectional nitrogen isotope; inorganic carbon assimilation; nitrogen utilization; karst habitat

Citation: Xia, A.; Wu, Y. Differential Responses of Nitrate/Ammonium Use to Bicarbonate Supply in Two *Brassicaceae* Species under Simulated Karst Habitat. *Agronomy* **2022**, *12*, 2080. https://doi.org/10.3390/agronomy12092080

Academic Editor: Angel Llamas

Received: 29 June 2022
Accepted: 22 August 2022
Published: 31 August 2022

1. Introduction

High bicarbonate (high pH), abundant nitrite, bare ammonium, and drought are typical characteristics of karst habitats [1]. The widely exposed carbonatite corroded by long-term karstification leads to loose soil structures and poor water storage capacity, which is the major reason causing drought in the karst habitat [2]. Moreover, abundant bicarbonate derived from the dissolved carbonatite promotes soil alkalinity, resulting in high bicarbonate and pH [3,4]. Previously, it has been shown that the bicarbonate content in the karst wet soils of Guizhou, China, is about 5–10 mM or even higher [5]. Thus, the hydrogen ion is rapidly consumed, and ammonium is volatilized in the drought and high bicarbonate (high pH) habitats, leading to a nitrate-abundant and ammonium-rare environment [1,6]. Keeping this in mind, the important role of bicarbonate in karst habitats has been reported in various studies; based on which, the perception of the bicarbonate's role in plant growth has changed universally. Earlier, it was believed that the bicarbonate exerts negative effects, such as the inhibition of photosynthesis, and nitrogen assimilation, leading to the decline in plant growth [7,8]. Although studies have widely reported the variation in inorganic carbon or nitrogen utilization during bicarbonate supplementation among different plant species [4,9], there are few reports on the response of nitrate/ammonium utilization to the bicarbonate supply in plants under karst habitats.

In plants, the inorganic carbon and nitrogen utilization strategy reflect their karst adaptation mechanisms, dominated by both inorganic carbon and nitrogen sources in the environment [10,11]. In the karst habitats, plant growth is affected by two inorganic carbon sources, carbon dioxide and bicarbonate, and two inorganic nitrogen sources, nitrate and ammonium [12]. The positive response to high bicarbonate, which reflected the bicarbonate use capacity (BUC), is one of plant's most essential inorganic carbon metabolism strategies employed for adaptation to karst habitats [3,13]. Compared to the nonadaptable plants, the adaptable species are more efficient in bicarbonate utilization, which not only protects the roots from high bicarbonate and pH environment but also maintains the photosynthesis level to support the normal growth of plants [14,15]. In the karst adaptable plant species, the BUC is attributed to affluent carbonic anhydrase (CA), which facilitates the conversion of bicarbonate into carbon dioxide and water [16]. This process protects plants from insufficient carbon and water conditions and restores photosynthesis in the karst habitats. Wu et al. [3] and Rao et al. [9] confirmed that bicarbonate, which was absorbed by roots, participated in plants' total inorganic carbon assimilation in Camptotheca acuminata. Moreover, karst-adaptable plants are more efficient in BUC than karst-nonadaptable plants to escape their growth inhibitions due to high bicarbonate-containing soil [4]. In this view, it is found that Broussonetia papyrifera (Bp., a typical species of karst-adaptable plant) could utilize more bicarbonate for total carbonate assimilation by enhancing both the glycolytic pathway and pentose phosphate pathway [17] and improving the photosystem electron transfer reaction [18,19] to increase karst drought resistance. Overall, the bicarbonate promotes roots to secrete organic acids, which help in maintaining cellular homeostasis.

Evidence suggests that the efficient utilization of nitrate and ammonium is an important mechanism in plants for karst adaptation [20,21]. It has also been established that karst-adaptable plants exhibit outstanding nitrate utilization efficiency [1], which might attribute to better BUC and total nitrogen assimilation capability, resulting in efficient nitrate reduction, which is promoted by more electronic supplies from excessive bicarbonate in the karst habitats [6]. Altogether these reports provide partial scientific evidence highlighting the role of inorganic carbon or nitrogen metabolisms in plant species adaptable to karst habitats. Interestingly, under high bicarbonate, pH, and drought conditions, intricate interactions between the inorganic carbon and nitrogen metabolism have been observed in plants. However, alteration in the nitrate reduction and ammonium assimilation in response to the bicarbonate supply is not clearly understood to date.

The nitrogen isotope value is performed for the quantification of inorganic nitrogen assimilation in plants [22]. The $\delta^{15}N$ value, ranging between −10‰ and 22‰, represents the inorganic nitrogen assimilation, and the $\delta^{15}N$ value of the leaves reflects the total nitrogen utilization [23]. In complex karst adversities, it is difficult to quantify the responses of nitrate and ammonium to the bicarbonate supply using a single isotope tracer at near-natural abundance levels [24,25]. Therefore, the bidirectional stable nitrogen isotope tracer approach was employed to study the differential contributions of nitrate and ammonium in plants [1]. To solve this difficult problem, two stable nitrogen isotope treatments labeled L and H with a difference in $\delta^{15}N$ values higher than 10‰ were used to calculate the responses of nitrate/ammonium utilization to bicarbonate in two plant species. Further, the plants were transferred to habitats with different bicarbonate levels under the same drought and total inorganic nitrogen supply [21]. Under karst habitats, the bidirectional stable nitrogen isotope treatments were used to quantify the contributions of nitrate and ammonium at differential bicarbonate levels [26]. We observed that the energy and reducing power mechanisms, combined with inorganic carbon assimilation and nitrogen utilization, were dominated by photosynthesis, carbon/nitrogen metabolism enzymes, and growth [27]. Here, we highlighted intricate interactions among different pathways, such as the contributions of nitrate/ammonium, photosynthesis, carbon/nitrogen metabolism enzymes, and plant growth, further revealing the response of nitrate/ammonium utilization to bicarbonate supply in karst habitat-adaptable plants.

Compared with other cruciferous plants, *Orychophragmus violaceus* (*Ov*) has great advantages in adapting well to karst desertification. Moreover, *Ov* is available for widespread cultivation with its abundant fatty acids and fuel properties [13]. In addition, *Brassica napus* (*Bn*) is a typical economic crop, which is widely grown for oil production [16]. In this study, *Ov* and *Bn* were selected as experimental plants. In karst regions, the nutrient content is bare, and the nitrate is dominant compared to ammonium [28]. Under the same nitrogen supply, bicarbonate levels triggered differential nitrate/ammonium utilization and inorganic carbon assimilation efficiencies in *Ov* and *Bn* [12]. We aimed to answer the following questions: (1) Evaluation of the nitrate/ammonium utilization and their contributions to the total inorganic nitrogen assimilation in *Ov* and *Bn* at different bicarbonate levels. (2) Exploring the differential responses of inorganic carbon assimilation to bicarbonate supply between two plant species. (3) Revealing the adaptation mechanisms of karst plants by the interactions among nitrate reduction, ammonium assimilation, and inorganic carbon assimilation at different bicarbonate levels.

2. Materials and Methods

2.1. Plant Materials

The seeds of *Ov* and *Bn* (Zayou NO.59, a variety of *Brassica napus*), selected as experimental material for the study, were obtained, respectively, from the Guizhou Vocational College of Agriculture, Guizhou and Shanxi Agricultural Reclamation Scientific Research Center, Shanxi, China. The experiments were conducted in an artificial greenhouse with a length, width, and height of 10 × 5 × 4 m at the Institute of Geochemistry, Chinese Academy of Sciences (Guiyang, China). A metal halide lamp (HPI-T400 W/645, Philips, The Netherlands) was used as a light source, and the temperature was controlled by air conditioning. The greenhouse environment was maintained as follows: light 500 ± 23 μmol m^{-2} s^{-1}, temperature (day/night): 25/19 °C, constant light time of 12 h per day, and relative humidity range: 55–60%. *Ov* and *Bn* seeds were stirred uniformly with 70% ethanol, sterilized for 1 min, repeatedly rinsed 3–5 times, and soaked for 6–8 h. The cavity trays (twelve-hole size, 19 × 15 × 9.5 cm), the substrate (perlite: vermiculite = 1:3), and modified Hoagland solution were used to grow *Ov* and *Bn* seedlings. In addition, it was reported that karst drought had the properties of drought: high pH and high bicarbonate. Therefore, we provided PEG6000 with 10 g/L solution to simulate a drought habitat, and the pH is kept at 8.30 ± 0.02. The nutrient solution was changed every 3 days, and the seedlings were transplanted for 28 days for the subsequent experiment. The *Ov* and *Bn* seedlings with uniform growth were categorized as 3 plants/pot, 3 pots/group, and 3 groups/treatment for differential bicarbonate treatments for 10 days.

2.2. Bicarbonate Treatments

In this study, the drought treatment induced with 10 g·L^{-1} PEG6000 and the total nitrogen sources (nitrate: ammonium = 14 mM:1 mM) were kept to mimic the karst habitats for *Ov* and *Bn*, along with the supplementation of 1 mM, 5 mM and 15 mM $NaHCO_3$ (Table 1) with the same capacity at 9:00 a.m daily. The soil water content was maintained between 20 and 25% constantly. Two labeled stable nitrogen isotopes were used as the nitrogen source, which were divided into high (H) and low (L) natural ^{15}N abundance in $NaNO_3$ with $\delta^{15}N$ values of 22.67‰ and 8.08‰, respectively. NH_4Cl with a $\delta^{15}N$ value of −2.64‰ was supplied for the ammonium during the plants' growth. The pH of the

Hoagland nutrient solution was adjusted to 8.30 ± 0.05 with 1 mol·L^{-1} KOH to maintain the bicarbonate concentration. The *Ov* and *Bn* under various groups were planted in a growth chamber at 25 °C with a 12-h photo period.

Table 1. Different bicarbonate treatments to *Ov* and *Bn* in simulated karst habitats.

Treatment	Reagent	Substance Content
B1	$NaHCO_3$	1 mM
B3	$NaHCO_3$	5 mM
B5	$NaHCO_3$	15 mM
NO_3^-:NH_4^+	$NaNO_3$ ($\delta^{15}N_H$:22.72‰):NH_4Cl	14 mM:1 mM
NO_3^-:NH_4^+	$NaNO_3$ ($\delta^{15}N_L$:12.7‰): NH_4Cl	14 mM:1 mM
pH	KOH	1 mol·L^{-1}, 8.30 ± 0.05
Drought	PEG6000	10 g·L^{-1}

2.3. Sample Collection for Biomass Estimation

The *Ov* and *Bn* under various treatments were collected in 3 parts: root, stem, and leaf; heated at 108 °C for 30 min; and dried at 70 °C to obtain the organic biomass and total biomass of *Ov* and *Bn*.

2.4. Measurement of Photosynthesis in Plants

The 2nd to 3rd expanded leaves of *Ov* and *Bn* were used for the measurement of photosynthesis from 9:00 to 11:00 a.m. The Li-6400 photosynthesizer (LI-COR, Lincoln, NE, USA) was used to measure the photosynthesis, including the net photosynthetic rate (Pn, μmol/m^2·s^{-1}), stomatal conductance (Cond, mmol H_2O m^{-2} s^{-1}), transpiration rate (Tr, mmol H_2O m^{-2}·s^{-1}), and intercellular CO_2 concentration (Ci, μmol CO_2 mol^{-1} air^{-1}) in *Ov* and *Bn*. The water use efficiency (WUE) was analyzed according to Formula (1). The other parameters were set as temperature 25 °C, CO_2 concentration 400 μmol/mol in the buffered glass bottles, and photosynthetically active radiation intensity 500 μmol/m^2·s^{-1}.

$$\text{WUE}\ (\%) = \text{Pn}/\text{Tr} \tag{1}$$

2.5. Measurement of Carbon and Nitrogen Enzymes

The fresh leaves of *Ov* and *Bn* were collected, frozen in liquid nitrogen, and stored at −20 °C. Enzyme reagent kits (Sangon, Shanghai, China) were used for the quantification of Rubisco (Ribulose bisphosphate carboxylase oxygenase), SS (Sucrose synthetase), NR (nitrate reductase), and glutamate synthase (GOGAT) activities in the leaves.

2.6. Analysis of Leaf Carbon and Nitrogen Content

Dried leaves of *Ov* and *Bn* were used to determine the carbon and nitrogen contents using an elemental analyzer (Vario MACRO cube, Langenselbold, Germany).

2.7. The Total Nitrogen Assimilation/Utilization Capacity

The nitrogen accumulation capacity (NAC) was determined by the quantity of total inorganic nitrogen assimilated in plants. The nitrogen utilization capacity (NUC) according to Formulas (2) and (3) were referred to Wu (2019), who depicted the amount of total inorganic nitrogen used in plants. The atomic mass of N was 14, as well as the relative molecular mass was 62.

$$\text{NAC} = \text{DW} \times \text{N}_{\text{content}};\ (\text{DW: drought weight of plants}) \tag{2}$$

$$\text{NUC}\ (\%) = 100 \times (\text{NAC}/\text{M}_\text{N})/\text{M}(\text{NO}_3^-)\ [\text{M}_\text{N}:14;\ \text{M}(\text{NO}_3^-):62] \tag{3}$$

2.8. Bidirectional Stable Nitrogen Isotope Tracer Method

2.8.1. The Nitrogen Isotope Ratio in Plants

The collected, dried leaves of *Ov* and *Bn* were ground to a fine powder for nitrogen isotope measurement. $\delta^{15}N$ was determined by gas isotope ratio mass spectrometer (MAT 253, Thermo Fisher Scientific, Langenselbold, Germany) according to Formula (4). In the formula, R_{sample} refers to the sample's nitrogen isotope ratio, and Rstandard stands for the nitrogen isotope of N_2 in the air. The reference materials, including IAEA N1, IAEA N2, and IAEA NO_3, were used to calibrate the instrument with a standard deviation of 0.2 % [29]. $\Delta^{15}N$ values refers to the total inorganic nitrogen assimilate capacity of plants. According to Formula (5), $\delta^{15}N$ substrate stands for the nitrogen isotope of samples, and $\delta^{15}N_{product}$ was the reference nitrogen isotope ratio at 8.08‰ under the natural background.

$$\delta^{15}N\ (‰) = (R_{sample}/_{Standard} - 1) \times 1000 \tag{4}$$

$$\delta^{15}N = \delta^{15}N_{product} - \delta^{15}N_{substrate}\ (\delta^{15}N_{product} = 8.08‰) \tag{5}$$

2.8.2. Evaluation of Nitrate and Ammonium Contributions in Plants

The mixed nitrogen sources, including nitrate and ammonium, have been used to determine the nitrogen isotope values in plant species. In this study, two nitrate sources with different nitrogen isotope values (more than 10‰) were supplied for nitrate utilization. The $\delta^{15}N$ value of ammonium was less than those of the nitrate sources. Therefore, the nitrogen isotope value reflected the mixed results of assimilated nitrate and ammonium [30]. Consequently, Formula (6) showed the foliar $\delta^{15}N$ value (δ_T), obtained by measuring nitrogen isotope compos ion. After that, the bidirectional nitrogen isotope method was calibrated to qualify the contributions of nitrate and ammonium. δ_A and δ_B were the $\delta^{15}N$ value derived from nitrate/ammonium assimilation, and the labels of f_A and f_B were the contributions of nitrate/ammonium assimilation.

$$\delta_T = f_A\delta_A + f_B\delta_B = f_A\delta_A + 1 - f_A\delta_B \tag{6}$$

In this study, two kinds of labeled nitrogen isotope ratios, which differed by 10% ($\delta^{15}N$ values, L:12.7–H:22.72‰) were used to qualify the nitrate contribution (f_A) and ammonium contribution (f_B). The two-terminal meta-model was expressed as Formula (7).

$$\delta_{TH} = f_{AH}\delta_{AH} + f_B\delta_B = f_{AH}\,\delta_{AH} + 1 - f_{AH}\delta_B \tag{7}$$

In contrast, only the $\delta^{15}N$ value of leaves changed when the fA and fB were adjusted to the treatment with high-labeled (H) nitrogen isotope groups [8]. According to the above statements, the low-labeled group was expressed in Formula (8).

$$\delta_{TL} = f_{AL}\delta_{AL} + f_B\delta_B = f_{AL}\delta_{AL} + 1 - f_{AL}\text{-}\delta_B \tag{8}$$

In this experiment, the same culture conditions were maintained for both plant species. Therefore, the contributions of nitrate and ammonium were the same, except for the nitrogen isotope value in the high (H) and light (L) treatments [30]. However, the physiological processes, metabolism, and growth were considered consistent under the same total nitrogen level [1]. Consequently, there was a specific formula given in (9) and (10).

$$f_A = f_{AH} = f_A \tag{9}$$

$$1 - f_{AH} = 1 - f_{AL} \tag{10}$$

Based on Formulas (9) and (10), we designed subsequent Formulas (11) and (12).

$$f_A = (\delta_{TH} - \delta_{TL})/(\delta_{AH} - \delta_{AL}) \tag{11}$$

$$f_B = 1 - f_A \quad (12)$$

However, previous studies merely depict the contributions of nitrate and ammonium to total inorganic nitrogen assimilate capacity (NAC) and nitrogen utilization capacity (NUC). The f_A and f_B of plants can be confirmed by bidirectional nitrogen isotope methods with different nitrogen isotope sources. It is also feasible to calculate the contributions of nitrate/ammonium to NAC by Formula (13), (NACA) and Formula (14), (NACB) and the contributions of nitrate/ammonium to NUC by Formula (15), (NUCA) and Formula (16), (NUCB).

$$\text{NACA} = \text{NAC} \times f_A \quad (13)$$

$$\text{NACB} = \text{NAC} \times f_B \quad (14)$$

$$\text{NUCA} = \text{NUC} \times f_A \quad (15)$$

$$\text{NUCB} = \text{NUC} \times f_B \quad (16)$$

2.9. Statistical Analysis

The experimental data were measured by an analysis of variance (ANOVA), and Tukey's test ($p < 0.05$) was performed for the comparison among various experimental treatments. The results were shown as the mean ± standard deviation (SD), and the figures were designed using Origin 2019.

3. Results

3.1. Growth-Related Features of Different Plant Species

The influence of bicarbonate levels on growth was different between *Ov* and *Bn* (Table 2). The *Ov* exhibited less increase in the biomass than the Bn at various bicarbonate supplies ranging from 1 to 15 mM. With the increasing concentration of the bicarbonate, the root, stem, leaves, and total biomass of *Ov* increased significantly, specifically at 5 mM $NaHCO_3$. However, the growth of *Ov* declined at the highest concentration of 15 mM $NaHCO_3$ in the study, especially the stem reduced by 24.86% compared to the lowest bicarbonate level (1 mM $NaHCO_3$). However, the increasing bicarbonate levels persistently inhibited the growth of *Bn*, resulting in a significant decrease in total biomass at the highest bicarbonate concentration (15 mM $NaHCO_3$) used in our study.

Table 2. The growth of *Ov* and *Bn* at various bicarbonate concentrations.

$NaHCO_3^-$, mM	1	5	15	1	5	15
Plant species		*Ov*			*Bn*	
root/mg	96.19 ± 3.09 d	131.97 ± 5.11 b	112.11 ± 2.38 c	315.3 ± 7.98 a	158.99 ± 4.05 b	95.8 ± 3.35 d
stem/mg	125.11 ± 4.5 d	260.48 ± 6.46 c	94.01 ± 4.07 e	464.6 ± 12.69 a	356.47 ± 7.04 b	132.95 ± 4.7 d
leaves/mg	128.11 ± 4.48 e	288.15 ± 4.25 c	144.64 ± 6.27 e	521.85 ± 10.54 a	402.85 ± 9.86 b	218.39 ± 6.98 d
biomass/mg	395.86 ± 15.74 e	742.46 ± 17.75 c	388.59 ± 14.23 e	1443.28 ± 35.21 a	1076.92 ± 26.85 b	525.06 ± 18.69 d

Ov Orychophragmus violaceus and *Bn Brassica napus*. The ratio of nitrate/ammonium within consistent with the total nitrogen concentration was 14 mM:1 mM under various bicarbonate treatment. Each value represents the mean ± SD ($n = 3$), and diverse letters in each value are significantly different by ANOVA ($p > 0.05$).

3.2. The Photosynthetic Efficiency of Plants

The photosynthetic efficiency of *Bn* and *Ov* was affected by different bicarbonate concentrations during various treatments, as depicted in Figure 1. The increase in photosynthesis is attributed to the increasing bicarbonate level in *Ov* as compared to *Bn*. The increasing bicarbonate concentration promoted the photosynthesis in *Ov* with the maximum effect at 5 mM $NaHCO_3$. *Ov* showed the highest Pn during all the bicarbonate concentrations, attributing to the maximum Cond, Ci, and WUE. Therefore, the photosynthesis of *Ov* was decreased at the maximum level of 15 mM $NaHCO_3$, resulting in the decline of Pn, Cond, Tr, and Ci with no significant difference from the lowest level

(1 mM $NaHCO_3$). On the other hand, less photosynthetic efficiency was observed in *Bn* as compared to *Ov* with the increasing concentration of bicarbonate, leading to the maximum decline of Pn, Cond, Ci, and Tr at the highest level (15 mM $NaHCO_3$).

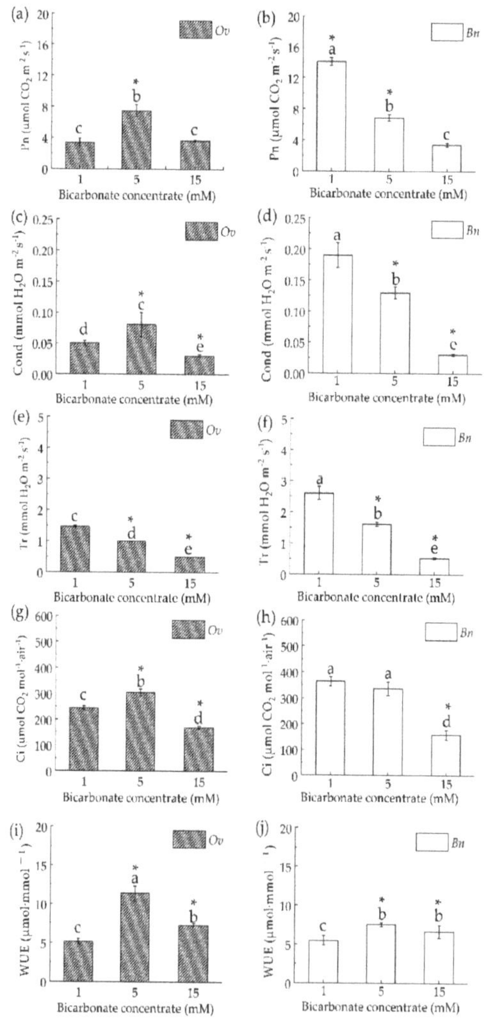

Figure 1. The photosynthesis of *Ov* and *Bn* under different bicarbonate supplies. *Ov*—*Orychophragmus violaceus*, *Bn*—*Brassica napus*, Pn-net—photosynthetic rate, Cond—stomatal conductivity, Tr—transpiration, Ci—intercellular carbon dioxide concentration, and WUE—water use efficiency. (**a**,**b**)—The Pn of *Ov*/*Bn*, (**c**,**d**)—The Cond of *Ov*/*Bn*, (**e**,**f**)—The Tr of *Ov*/*Bn*, (**g**,**h**)—The Ci of *Ov*/*Bn*, (**i**,**j**)—The WUE of *Ov*/*Bn*. The ratio of nitrate/ammonium within consistent total nitrogen concentration was 14 mM:1 mM under various bicarbonate treatments. Each value represents the mean ± SD (n = 3), and diverse letters with "*" in each value are significantly different by ANOVA ($p > 0.05$).

3.3. Responses of Carbon and Nitrogen Metabolizing Enzymes of Two Plant Species

The responses of carbon and nitrogen metabolizing enzymes to different bicarbonate levels were measured between *Ov* and *Bn* (Figure 2). The increasing concentration of bicarbonate (from 1 to 15 mM $NaHCO_3$) promoted the activity of carbon and nitrogen metabolizing enzymes in *Ov* more significantly than *Bn*. At the middle level

(5 mM $NaHCO_3$), the maximum activities of Rubisco, SS, NR, and GOGAT were measured in *Ov*, whereas the inhibited activities of these enzymes were observed in *Bn*. Furthermore, reduction in the activities of Rubisco, SS, NR, and GOGAT was observed less significantly reduced in *Ov* as compared to *Bn* at the highest bicarbonate level (15 mM $NaHCO_3$).

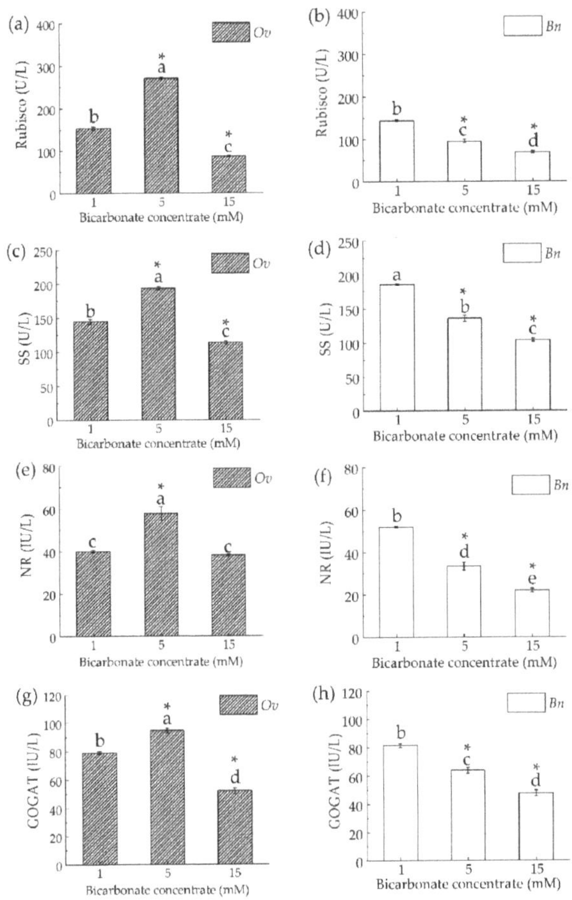

Figure 2. The responses of enzymes involved in carbon and nitrogen metabolism to bicarbonate in *Ov* and *Bn*. *Ov*—*Orychophragmus violaceus* and *Bn*—*Brassica napus*, Rubisco—ribulose bisphosphate carboxylase oxygenase, SS—sucrose synthetase, NR—nitrate reductase, and GOGAT—glutamate synthase. (**a**,**b**)—The Rubisco of *Ov*/*Bn*, (**c**,**d**)—The SS of *Ov*/*Bn*, (**e**,**f**)—The NR of *Ov*/*Bn*, (**g**,**h**)—The GOGAT of *Ov*/*Bn*. The ratio of nitrate/ammonium within consistent total nitrogen concentration was 14 mM:1 mM under various bicarbonate treatments. Each value represents the mean ± SD (n = 3), and diverse letters with "*" in each value are significantly different by ANOVA ($p > 0.05$).

3.4. The Inorganic Carbon and Nitrogen Contents of Leaves in Ov and Bn

With the increasing bicarbonate concentration, the carbon and nitrogen contents in leaves were found to be different between *Ov* and *Bn* (Figure 3). The carbon content of leaves in *Ov* first increased and then declined constantly during treatments from 1 to 15 mM; however, the pattern in *Bn* was subject to *Ov*. Moreover, the carbon content was the maximum in *Ov* at the middle bicarbonate level (5 mM) and significantly higher than that in *Bn*. In addition, the nitrogen content of leaves in *Ov* increased at 5 mM bicarbonate, whereas it declined consonantly in *Bn* and obtained the minimum level at 15 mM. Further, *Ov* exhibited more significantly enhanced NAC and NUC than *Bn* with increasing bicarbonate levels. Consequently, the C/*n* ratio of leaves in *Ov* declined extremely than that in *Bn*.

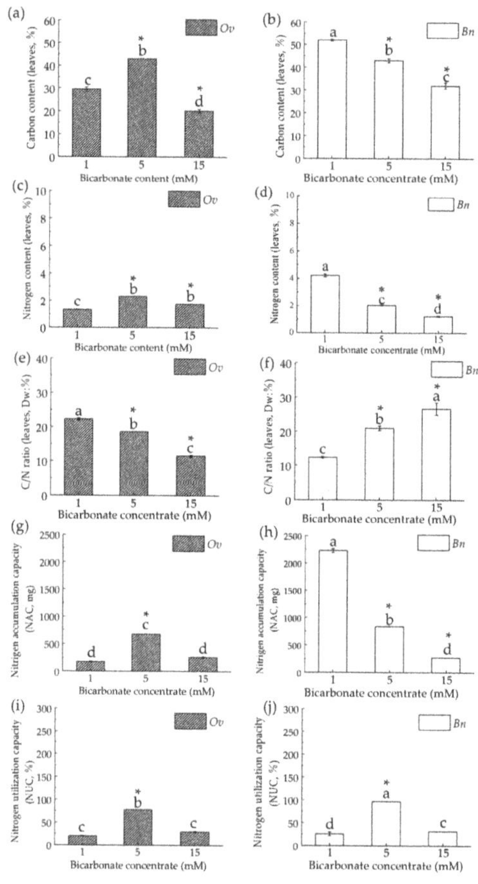

Figure 3. The content of the carbon/nitrogen component in leaves between *Ov* and *Bn* at different bicarbonate levels. *Ov*—*Orychophragmus violaceus* and *Bn*—*Brassica napus*. (**a**,**b**)—The carbon content of *Ov*/*Bn*, (**c**,**d**)—The nitrogen content of *Ov*/*Bn*, (**e**,**f**)—The C/N ratio of *Ov*/*Bn*, (**g**,**h**)—The NAC of *Ov*/*Bn*, (**i**,**j**)—The NUC of *Ov*/*Bn*. The ratio of nitrate/ammonium within consistent total nitrogen concentration was 14 mM:1 mM under various bicarbonate treatments. Each value represents the mean ± SD (n = 3), and diverse letters with "*" in each value are significantly different by ANOVA ($p > 0.05$).

3.5. The Contributions of Nitrate/Ammonium in Two-Plant Species

The contributions of nitrate/ammonium in *Ov* and *Bn* labeled with the high (H) and low (L) treatments were opposite during the treatment with all the concentrations of bicarbonate (Figure 4). The δ^{15}N values of *Ov* and *Bn* leaves were different during all the bicarbonate supplies (Figure 4a,b). In both the H and L groups, the δ^{15}N value of *Ov* first declined and then enhanced with increasing the bicarbonate levels. The minimum and maximum δ^{15}N values of *Ov* were obtained at 5 mM and 15 mM $NaHCO_3$, respectively. However, the δ^{15}N values of *Bn* were increased consistently in all the groups. The differences in the δ^{15}N values between *Ov* and *Bn* were attributed to the distinctive utilization of nitrate (f_A) and ammonium (f_B) in two-plant species (Figure 4c,d), which were determined by different bicarbonate levels. In both *Ov* and *Bn*, the nitrate utilization (f_A) was subjected to the results of δ^{15}N, while the ammonium utilization (f_B) was adjudicative with the increasing bicarbonate supplies. Predictably, the increasing bicarbonate concentration showed more enhancement in *Ov* at the middle bicarbonate level (5 mM $NaHCO_3$), where nitrate was majorly available for inorganic nitrogen assimilation in *Ov*, contributing to the higher nitrate utilization (f_A) as compared to *Bn*. However, when the bicarbonate enhanced consistently exceeded 5 mM $NaHCO_3$, the nitrate utilization was decreased in both *Ov* and *Bn*, leading to lower f_A and higher f_B. At the highest bicarbonate level (15 mM $NaHCO_3$), the ammonium utilization increased more obviously than the nitrate utilization in both *Ov* and *Bn*. Moreover, ammonium could be considered the major source of inorganic nitrate supply in *Bn* at 15 mM $NaHCO_3$, where f_B increased by more than half percent of the lowest bicarbonate level (1 mM $NaHCO_3$) and had the maximum ammonium utilization. Overall, the contribution of ammonium was enhanced to the inorganic nitrogen assimilation in both *Ov* and *Bn* during increasing bicarbonate levels from 1 to 15 mM $NaHCO_3$.

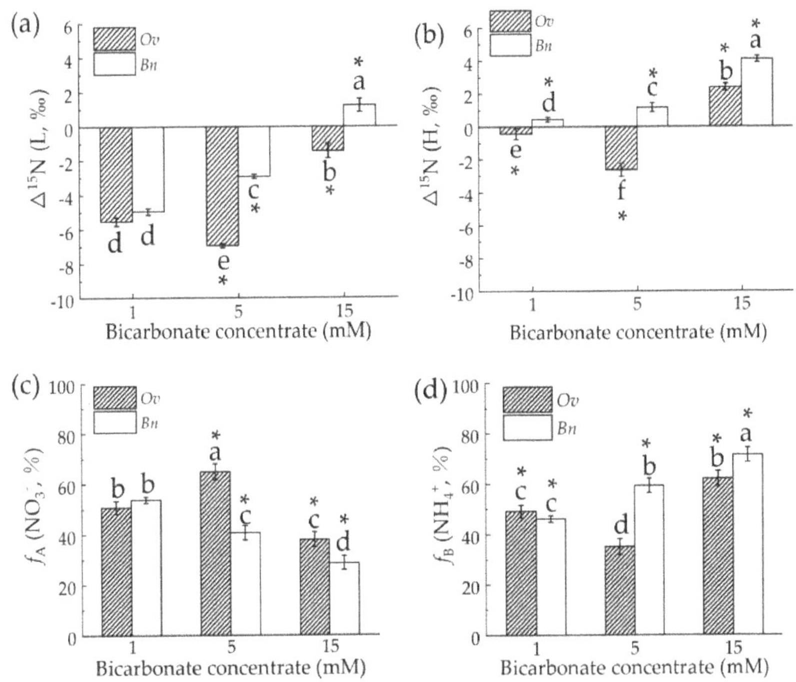

Figure 4. The contributions of nitrate/ammonium of leaves in *Ov* and *Bn* at different bicarbonate levels. *Ov*—*Orychophragmus violaceus*, *Bn*—*Brassica napus*, Δ^{15}N—the δ^{15}N value, f_A—the contribution of nitrate, and f_B—the contribution of ammonium. (**a**,**b**)—the δ^{15}N value of *Ov* and *Bn*, (**c**,**d**)—the contribution of nitrate/ammonium of *Ov* and *Bn*. The ratio of nitrate/ammonium within consistent total nitrogen concentration was 14 mM:1 mM under various bicarbonate treatments. Each value represents the mean ± SD (n = 3), and diverse letters with "*" in each value are significantly different by ANOVA ($p > 0.05$).

3.6. The Contribution of Nitrate/Ammonium to Nitrogen Assimilation and Utilization Capacity

In this study, NACA/NACB and NUCA/NUCB separately represented the contribution of nitrate/ammonium to nitrogen assimilation capacity (NAC) and nitrogen utilization capacity (NUC). As depicted in Figure 5, the increasing bicarbonate levels triggered NAC and NUC in *Ov* more than *Bn*. Compared to the lowest bicarbonate level (1 mM $NaHCO_3$), *Ov* showed enhanced NAC and NUC at 5 mM $NaHCO_3$, utilizing nitrate as the major contributor, leading to the maximum NACA, NACB, NUCA, and NUCB. We observed that the results in *Bn* were different from *Ov*. The increasing bicarbonate level constantly inhibited NAC and NUC in *Bn*, leading to an excessive decline in NACA, NACB, NUCA, and NUCB. The lowest assimilation/utilization in *Bn* was observed at the highest bicarbonate level (15 mM $NaHCO_3$). Altogether, our results suggested that the increasing bicarbonate enhanced more f_A to NAC and NUC in *Ov* than *Bn*.

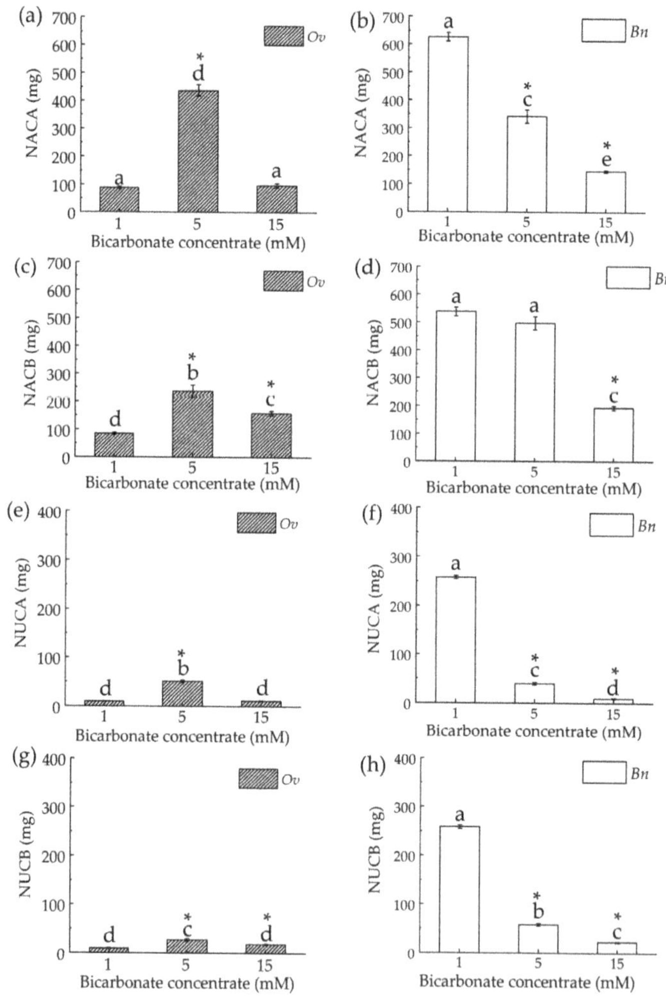

Figure 5. The contribution of nitrate/ammonium to NAC and NUC in *Ov* and *Bn* at different bicarbonate levels. *Ov*—*Orychophragmus violaceus*, *Bn*—*Brassica napus*. NACA—the contribution of nitrate to total nitrogen accumulation capacity, NACB—the contribution of ammonium to total nitrogen accumulation capacity, NUCA—the contribution of nitrate to total nitrogen utilization capacity, and NUCB—the contribution of ammonium to total nitrogen utilization capacity. (**a**,**b**)—the NACA of *Ov*/*Bn*, (**c**,**d**)—the NACB of *Ov* and *Bn*. (**e**,**f**)—the NUCA of *Ov* /*Bn*, (**g**,**h**)—the NUCB of *Ov*/*Bn*. The ratio of nitrate/ammonium within consistent total nitrogen concentration was 14 mM:1 mM under various bicarbonate treatments. Each value represents the mean ± SD (n = 3), and diverse letters with "*" in each value are significantly different by ANOVA ($p > 0.05$).

4. Discussion

4.1. Differential Regulation of Inorganic Carbon Assimilation during Bicarbonate Supplementation in Ov and Bn

In this study, we found that the growth of *Ov* and *Bn* exhibited different patterns in response to the changing bicarbonate supply. With an increase in the bicarbonate concentration, the growth of root, stem, and leaf and total biomass increased at the middle bicarbonate level (5 mM HCO_3^-) in *Ov*, while consistently decreased in *Bn*. Previously, it has been reported that karst-adaptable plants have stronger BUC, which can utilize more bicarbonates for inorganic carbon assimilation [1]. For instance, Camptotheca acuminata could incorporate more bicarbonate into nonstructural carbohydrates (NSC) to alleviate the extreme conditions of karst habitats such as drought, high bicarbonate, and high pH [9]. In addition, at different bicarbonate levels, photosynthesis was promoted in *Ov* compared to *Bn*. Precisely, the photosynthesis in *Ov* obtained the maximum rate at the middle bicarbonate level (5 mM) but consistently decreased in Bn with the minimum photosynthetic rate at 15 mM HCO_3^-. Wu et al. [1] reported that bicarbonate was more readily available for karst-adaptable plants than non-karst plants, which is attributed to the activity of active carbonic anhydrases (CA). CA converts bicarbonate into carbon dioxide and water to alleviate water and carbon source deficits and restore photosynthesis in plants. Wang et al. [13] showed that *Brassica juncea L (Bj)*, a non-karst-adaptable plant, had less bicarbonate use capacity, resulting in a decrease in photosynthesis. Though the Pn, Ci, Cond, and Tr decreased at the highest bicarbonate level more than the middle level, we found that the WUE significantly increased in both *Ov* and *Bn*, indicating that these two plant species have enhanced water use efficiency to adapt to increasing bicarbonate, which is consistent with the results shown by Hang et al. [4] and Gimenez et al. [31].

In addition, we found that change in the bicarbonate concentration was more significantly associated with the activities of carbon metabolizing enzyme in *Ov* compared to *Bn*. With increasing bicarbonate, the Rubisco and SS activities in *Ov* reached to a maximum at the middle bicarbonate level (5 mM) but consistently decreased in *Bn* with the minimum at the highest bicarbonate level (15 mM). It could be due to higher BUC in *Ov* [3,16], which restored the photosynthesis, increasing the carbon metabolizing enzyme activities. Xia et al. [12] also confirmed that *Ov* had more stable carbon metabolic enzyme activities than *Bn* in simulated karst habitats. However, high bicarbonate inhibited the carbon metabolizing enzyme activities in plants [32,33]. In this study, at the highest bicarbonate level (15 mM), Rubisco and SS activities decreased in both *Ov* and *Bn*. However, the growth, photosynthetic capacity, and carbon metabolic enzyme activities in *Ov* decreased less than *Bn*, indicating that bicarbonate imposed slight inhibitions on *Ov*. Further, it highlighted the outstanding adaptability of inorganic carbon assimilation to high bicarbonate supply.

4.2. The Differential Characteristics of Nitrate/Ammonium Utilization at Various Bicarbonate Levels in Two Plant Species

With nitrate-abundant and ammonium-rare soil in karst habitats, the bicarbonate had a pronounced effect on nitrate/ammonium contributions in plants, resulting in the nitrogen isotope value during inorganic nitrogen utilization [34]. It was reported previously that the nitrogen isotope value results from both nitrate and ammonium contributions [35]. In this study, we estimated the contributions of nitrate/ammonium utilization in *Ov* and *Bn* using the bidirectional nitrogen isotope tracer method [21]. The results showed that *Ov* and *Bn* exhibited differential nitrate/ammonium utilization at different bicarbonate levels. Compared to *Bn*, the bicarbonate promoted significant nitrogen utilization in *Ov*, leading to the maximum enhancement of f_A, NAC, and NUC at the middle bicarbonate level (5 mM), which attributed to the higher photosynthetic capacity and growth (Table 1 and Figure 1). It has been shown that increasing photosynthesis promotes the inorganic nitrogen assimilation in plants, increasing both inorganic nitrogen accumulation and utilization capacity [34]. Moreover, the $\Delta^{15}N$ value, which is opposite to the inorganic nitrogen assimilation capacity [35], was reduced to the minimum at 5 mM HCO_3^- (Figure 4). It was shown that karst-induced drought inhibits the stomatal movement of leaves, resulting in a decrease in the carbon dioxide level and water use efficiency [1]. Precisely, it reduced the carbon dioxide assimilation, mediating a decline in the photosynthetic capacity [4]. Conversely, it might inhibit the photo electron transfer by decreasing the water photolysis, which is not conducive to inorganic nitrogen metabolism in plants [3]. Nevertheless, bicarbonate could be used to provide electrons for nitrate reduction, maintaining normal inorganic nitrogen metabolism in plants [6,18]. In this study, 5 mM HCO_3^- enhanced nitrate utilization and NR activity in *Ov*, which was conducive to ammonium assimilation and enzyme formation [27], leading to the enhancement of ammonium utilization, GOGAT activity, and total inorganic nitrogen assimilation. In addition, at the highest level of bicarbonate (15 mM), the nitrate utilization was reduced, but the ammonium utilization was increased in *Ov*, which helped in attaining the NAC and NUC values consistent with those at the minimum bicarbonate level (1 mM). It was observed that ammonium utilization was increased in *Ov* to maintain the nitrogen assimilation capacity at a high bicarbonate level, which conformed to the findings by Lu et al. [6] and Wu et al. [1].

However, with the increasing bicarbonate, we found that the nitrogen assimilation was consistently inhibited in Bn, resulting in the lowest f_A, NACA, and NACB at the maximum bicarbonate level (15 mM). These results were attributed to its weak BUC, since bicarbonate was rarely used, leading to the occurrence of stress on cellular osmotic pressure [36]. Increased osmotic pressure inhibited the absorption of nitrate and ammonium, as well as photosynthesis and growth [6], which is not conducive to inorganic nitrogen assimilation of *Bn*. In addition, we found that, though the utilization of ammonium was increased in *Bn*, NAC and NUC were reduced to a minimum at 15 mM HCO_3^-. It was due to the bare ammonium environment, which was only 1 mM in our study. The abundant nitrate and bare ammonium in karst soil results in the minimal absorption of ammonium by plants [34,35,37]. Additionally, plants rarely receive ammonium for direct assimilation. Thus, nitrate reduction restoring nitrate to ammonium majorly participates in the total inorganic nitrogen assimilation [30]. Consequently, in this experiment, the bare ammonium might be sufficient to enhance total inorganic nitrogen metabolism in *Bn*. Though the bicarbonate supply increased, the ammonium utilization in *Bn* counteracted the negative effects on plant growth, leading to a decline in the total inorganic nitrogen utilization.

4.3. The Differential Responses of Inorganic Carbon and Nitrogen Metabolisms to Bicarbonate Supply in Ov and Bn under Karst Habitats

In the simulated karst habitats with drought, high bicarbonate, abundant nitrate, and bare ammonium soil, we found that the responses of nitrate/ammonium utilization to bicarbonate were contrasting in *Ov* and *Bn* (Figure 6). In this study, the bicarbonate was found to be more conducive to the growth of *Ov* than *Bn*, resulting in the promotion of photosynthesis, enzymatic activities, carbon concentrations, and nitrogen reduction. Bicarbonate concentrations not only affected the contributions of nitrate and ammonium but also impacted the total nitrogen assimilation and utilization capacity in plants [1,6]. However, due to weak adaptability to high bicarbonate and abundant nitrate habitats, *Bn* exhibited more decline in carbon assimilation and nitrogen reduction with increasing bicarbonate, leading to excessive growth inhibitions but an obvious promotion in water use efficiency, as well as ammonium utilization efficiency. Hence, *Ov* exhibited more exceptional responses related to both carbon and nitrogen metabolism to high bicarbonate and nitrate soil than *Bn*, which was primarily attributed to its preferable adaptions to karst habitats.

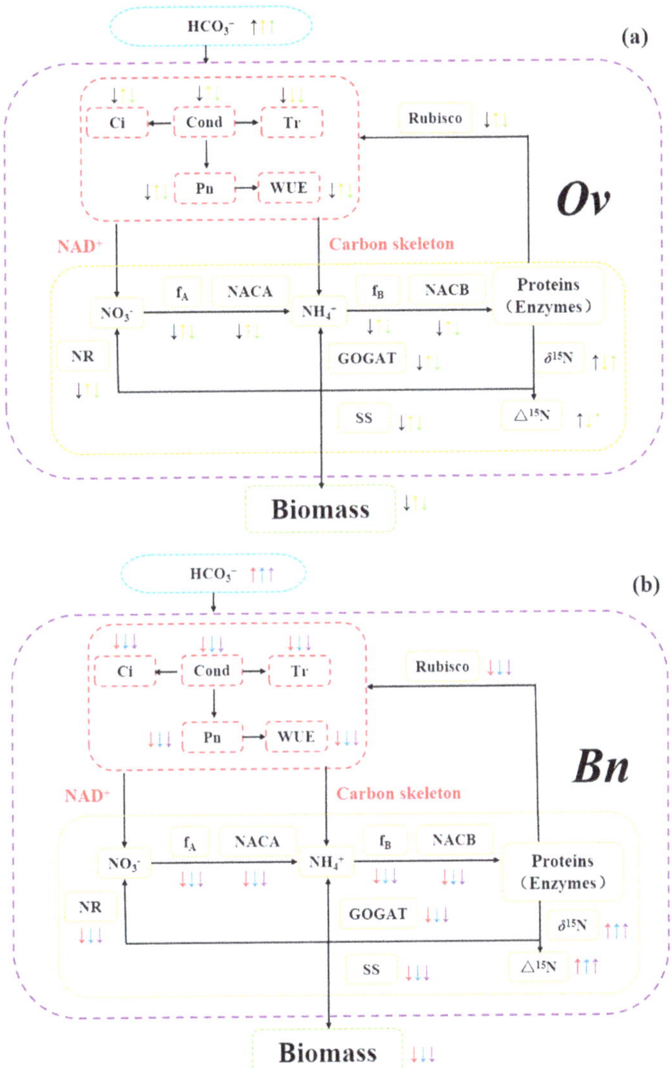

Figure 6. Differential responses of carbon and nitrogen metabolism to bicarbonate in *Ov* and *Bn* under simulated karst habitats. ↑: Increase, ↓: Decrease. (**a**,**b**) represented differential responses of carbon and nitrogen metabolism to bicarbonate in *Ov*/*Bn* under simulated karst habitats. The balck, yellow and green/the red, blue, and purple arrows represent different bicarbonate treatments at 1 mM, 5 mM, and 15 mM in *Ov*/*Bn*.

5. Conclusions

Under simulated karst habitats, the bidirectional stable nitrogen isotope tracing method accurately estimated nitrate and ammonium utilization in plants at different bicarbonate levels. Compared to a non-karst-adaptable plant species (*Bn*), the bicarbonate significantly promoted inorganic carbon assimilation and nitrogen utilization in the karst-adaptable plant (*Ov*), resulting in an increase in the photosynthesis, carbon metabolizing enzyme activities, and biomass. Additionally, the nitrogen enzyme activities, nitrate utilization, and total inorganic nitrogen metabolism were enhanced in *Ov*. Moreover, high bicarbonate promoted ammonium utilization to counteract the decline in the inorganic nitrogen assimilation capacity in *Ov*. Therefore, *Ov* can be considered a widely grown pioneer plant species in karst areas and a valuable resource for oil production and fuel properties, etc.

Author Contributions: A.X. and Y.W. cooperated to complete this article. With the help of correspondence, A.X. determined the experimental outline and analyzed the original data to design this article. Furthermore, Y.W. conceived and funded the research and carefully participated in guiding and revising this article to make it more properly and objectively presented. All authors have read and agreed to the published version of the manuscript.

Funding: This work was supported by the National Natural Science Foundation of China (number U1612441-2) and Support Plan Projects of Science and Technology of Guizhou Province (number (2021)YB453).

Institutional Review Board Statement: Not applicable.

Informed Consent Statement: Not applicable.

Data Availability Statement: Not applicable.

Acknowledgments: We deeply appreciate the essential experimental platform from the Research Center for Environmental Bio-Science and Technology, State Key Laboratory of Environmental Geochemistry, Institute of Geochemistry, Chinese Academy of Sciences. We are also grateful to J.T. and N.A, for their warming hard work, the nitrogen isotope value of these experimental samples could be completely obtained for this article.

Conflicts of Interest: The authors declare no conflict of interest.

Abbreviation:

Abbreviation	Meaning
Bj	*Brassica juncea* L.
Bn	*Brassica napus*
BUC	bicarbonate use capacity
CA	carbonic anhydrase
Ci	intercellular carbon dioxide concentration
Cond	stomatal conductivity
f_A	the contribution of nitrate
f_B	the contribution of ammonium
GOGAT	glutamate synthase
HCO_3^-	bicarbonate
NH_4^+	ammonium
NO_3^-	nitrate
NR	nitrate reductase
NAC	nitrogen accumulation capacity
NUC	nitrogen utilization capacity
NACA	the contribution of nitrate to total nitrogen accumulation capacity
NACB	the contribution of ammonium to total nitrogen accumulation capacity
NUCA	the contribution of nitrate to total nitrogen utilization capacity
NUCB	the contribution of ammonium to total nitrogen utilization capacity
NSC	nonstructural carbohydrate
Ov	*Orychophragmus violaceus*
Pn	photosynthetic rate
Rubisco	ribulose bisphosphate carboxylase oxygenase
SS	sucrose synthetase
Tr	transpiration
WUE	water use efficiency

References

1. Wu, Y.Y.; Xing, D.K.; Hang, H.T.; Zhao, K. *Principles Technology of Determination on Plants' Adaptation to Karst Environment*; Science Press: Beijing, China, 2018; ISBN 978-7-03-059386-3. Available online: https://book.sciencereading.cn/shop/book/Booksimple/show.do?id=B7BC87D1863FC9448E053020B0A0A7668000 (accessed on 1 December 2018).
2. Liu, C.C.; Liu, Y.G.; Guo, K.; Zheng, Y.R.; Li, G.Q.; Yu, L.F.; Yang, R. Influence of drought intensity on the response of six woody karst species subjected to successive cycles of drought and rewatering. *Physiol. Plant.* **2010**, *139*, 39–54. Available online: https://pubmed.ncbi.nlm.nih.gov/20059730 (accessed on 28 June 2022). [CrossRef] [PubMed]
3. Wu, Y.Y.; Xing, D.K. Effect of bicarbonate treatment on photosynthetic assimilation of inorganic carbon in two plant species of *Moraceae*. *Photosynthetica* **2012**, *50*, 587–594. [CrossRef]
4. Hang, H.T.; Wu, Y.Y. Quantification of photosynthetic inorganic carbon utilisation via a bidirectional stable carbon isotope tracer. *Acta Geochim.* **2016**, *35*, 130–137. [CrossRef]
5. Yu, C.; Yang, L.; Leah, N.; Wei, L.; Ling, X.; Hong, S.J. The analysis of leaf traits of Eight ottelia populations and their potential ecosystem functions in karst freshwaters in China. *Front Plant Sci.* **2019**, *9*, 1938. Available online: https://pubmed.ncbi.nlm.nih.gov/30666267 (accessed on 28 June 2022).
6. Ye, L.; Wu, Y.Y.; Zhang, K.Y. Does bicarbonate affect the nitrate utilization and photosynthesis of *Orychophragmus violaceus*. *Acta Geochim.* **2018**, *37*, 875–885. [CrossRef]
7. Wim, F.J.; Jack, J.S. Mechanism of bicarbonate action on photosynthetic electron transport in broken chloroplasts. *Biochim. Biophys. Acta BBA Bioenergetics* **1981**, *2*, 168–174. Available online: https://pubmed.ncbi.nlm.nih.gov/6793066 (accessed on 28 June 2022).
8. Raúl, C.N.; Stéphane, A.; Paul, R. Diurnal nitrate uptake in young tomato (*Lycopersicon esculentum Mill.*) plants: Test of a feedback-based model. *J. Exp. Bot.* **1998**, *321*, 721–730. [CrossRef]
9. Rao, S.; Wu, Y.Y. Root-derived bicarbonate assimilation in response to variable water deficit in *Camptotheca acuminate* seedlings. *Photosynth. Res.* **2017**, *134*, 59–71. Available online: https://pubmed.ncbi.nlm.nih.gov/28623437 (accessed on 28 June 2022). [CrossRef] [PubMed]
10. John, A.R.; Linda, L.H.; Mitchell, A. Global aspects of C/N interactions determining plant–environment interactions. *J. Exp. Bot.* **2004**, *55*, 11–25. Available online: https://pubmed.ncbi.nlm.nih.gov/14645388 (accessed on 28 June 2022).
11. Mitchell, A.; Maule, H.G.; John, A.R.; Mistry, A. Extension growth of impatiens glandulifera at low irradiance: Importance of nitrate and potassium accumulation. *Ann Bot.* **2005**, *95*, 641–648. Available online: https://pubmed.ncbi.nlm.nih.gov/15644384 (accessed on 28 June 2022).
12. Xia, A.; Wu, Y. Joint interactions of carbon and nitrogen metabolism dominated by bicarbonate and nitrogen in *Orychophragmus violaceus* and *Brassica napus* under simulated karst habitats. *BMC Plant Biol.* **2022**, *22*, 264. Available online: https://pubmed.ncbi.nlm.nih.gov/35619072 (accessed on 28 June 2022). [CrossRef] [PubMed]
13. Wang, R.; Wu, Y.; Hang, H.; Liu, Y.; Xie, T.; Zhang, K.; Li, H. *Orychophragmus violaceus* L., a marginal land-based plant for biodiesel feedstock: Heterogeneous catalysis, fuel properties, and potential. *Energy Convers. Manag.* **2014**, *84*, 497–502. [CrossRef]
14. Zhao, K.; Wu, Y.Y. Effects of Zn deficiency and bicarbonate on the growth and photosynthetic characteristics of four plant species. *PLoS ONE* **2017**, *12*, e0169812. Available online: https://pubmed.ncbi.nlm.nih.gov/28076430 (accessed on 28 June 2022). [CrossRef] [PubMed]
15. Ou, Z.; Pang, S.; He, Q.; Peng, Y.; Huang, X.; Shen, W. Effects of vegetation restoration and environmental factors on understory vascular plants in a typical karst ecosystem in southern China. *Sci. Rep.* **2020**, *10*, 12011. Available online: https://pubmed.ncbi.nlm.nih.gov/32694713 (accessed on 28 June 2022). [CrossRef]
16. Wu, Y.Y.; Wu, X.M.; Li, P.P.; Zhao, Y.G.; Li, X.T.; Zhao, X.Z. Comparison of photosynthetic activity of *Orychophragmus violaceus* and *Oil-seed* rape. *Photosynthetica* **2005**, *43*, 299–302. [CrossRef]
17. Yao, K.; Wu, Y.Y. Rhizospheric bicarbonate improves glucose metabolism and stress tolerance of *Broussonetia papyrifera* L. seedlings under simulated drought stress. *Russ. J. Plant. Physiol.* **2021**, *68*, 126–135. Available online: https://link.springer.com/article/10.1134/S1021443721010209?utm_source=xmol&utm_medium=affiliate&utm_content=meta&utm_campaign=DDCN_1_GL01_metadata (accessed on 15 February 2021). [CrossRef]
18. Wu, Y.Y. Is bicarbonate directly used as substrate to participate in photosynthetic oxygen evolution. *Acta Geochim.* **2021**, *40*, 650–658. [CrossRef]
19. Fang, L.; Wu, Y.Y. Bicarbonate uptake experiment show potential karst carbon sinks transformation into carbon sequestration by terrestrial higher plants. *J. Plant Interact.* **2022**, *1*, 419–426. [CrossRef]
20. Wu, Y.Y.; Zhang, K.Y. Sterile dynamic measurement of the in vitro nitrogen use efficiency of plantlets. In *Plant Image Analysis*; CRC Press: Boca Raton, FL, USA, 2014; p. 38. ISBN 9780429072345. [CrossRef]
21. Zhang, K.Y.; Wu, Y.Y. The $\delta^{15}N$ response and nitrate assimilation of *Orychophragmus violaceus* and *Brassica napus* plantlets in vitro during the multiplication stage culturedunder different nitrate concentrations. *Acta Geochim.* **2017**, *36*, 190–197. [CrossRef]
22. Choi, W.-J.; Lee, S.-M.; Ro, H.-M.; Kim, K.-C.; Yoo, S.-H. Natural ^{15}N abundances of maize and soil amended with urea and composted pig manure. *Plant Soil.* **2002**, *245*, 223–232. [CrossRef]
23. Joseph, A.N.; Daniel, M.S.; Paul, J.H. The mechanism of isotope fractionation during algal nitrate assimilation as illuminated by the $^{15}N/^{14}N$ of intracellular nitrate. *J. Phycol.* **2004**, *40*, 517–522. [CrossRef]
24. Reed, B.M.; Wada, S.; DeNoma, J.; Niedz, R.P. Mineral nutrition influences physiological responses of pear in vitro. *In Vitro Cell. Dev. Biol. Plant.* **2013**, *49*, 699–709. [CrossRef]

25. Sukalya, P.; Barbara, M.R. Optimizing shoot culture media for *Rubus germplasm*: The effectsof NH_4^+, NO_3^-, and total nitrogen. *In Vitro Cell. Dev. Biol. Plant.* **2016**, *52*, 265–275.
26. Xie, T.X.; Wu, Y.Y. The biokarst system and its carbon sinks in response to pH changes: A simulation experiment with microalgae. *Geochem.Geophys. Geosyst.* **2017**, *18*, 827–843. [CrossRef]
27. Guo, S.; Zhou, Y.; Shen, Q.; Zhang, F. Effect of ammonium and nitrate nutrition on some physiological processes in higher plants-growth, photosynthesis, photorespiration, and water relations. *Plant Biol.* **2007**, *9*, 21–29. Available online: https://pubmed.ncbi.nlm.nih.gov/17048140 (accessed on 28 June 2022). [CrossRef]
28. Piao, H.C.; Wu, Y.Y.; Hong, Y.T.; Yuan, Z.Y. Soil-released carbon dioxide from microbial biomass carbon in the cultivated soils of karst areas of southwest China. *Biol Fertil Soils.* **2000**, *31*, 422–426. [CrossRef]
29. Salima, Y.; Maria, D.S.; José, L.A. Comparative response of $\delta^{13}C$, $\delta^{18}O$ and $\delta^{15}N$ in durum wheat exposed to salinity at the vegetative and reproductive stages. *Plant Cell Environ.* **2013**, *36*, 1214–1227. Available online: https://pubmed.ncbi.nlm.nih.gov/23240790 (accessed on 28 June 2022).
30. Dev, T.B.; Herbert, J.K. NH_4^+ toxicity in higher plants: A critical review. *J. Plant Physiol.* **2002**, *6*, 567–584. [CrossRef]
31. Gimenez, C.; Mitchell, V.J.; Lawlor, D.W. Regulation of photosynthetic rate of two sunflower hybrids under water stress. *Plant Physiol.* **1992**, *98*, 516–524. [CrossRef]
32. Alhendawi, R.A.; Römheld, V.; Kirkby, E.A.; Marschner, H. Influence of increasing bicarbonate concentrations on plant growth, organic acid accumulation in roots and iron uptake by barley, sorghum, and maize. *J. Plant Nutri.* **1997**, *20*, 1731–1753. [CrossRef]
33. Liu, J.L.; Li, F.R.; Niu, R.X.; Liu, C.A.; Liu, Q.J. Influence of soil salinization on soil animal community in an arid oasis of middle Heihe River basin. *J. App. Ecol.* **2012**, *23*, 1551–1561. Available online: https://pubmed.ncbi.nlm.nih.gov/22937643 (accessed on 28 June 2022).
34. Yoneyama, T.; Matsumaru, T.; Usui, K.; Engelaar, W.M.H.G. Discrimination of nitrogen isotopes during absorption of ammonium and nitrate at different nitrogen concentrations by rice (*Oryza sativa* L.) plants. *Plant Cell Environ.* **2001**, *24*, 133–139. [CrossRef]
35. Waser, N.A.D.; Harrison, P.J.; Nielsen, B.; Calvert, S.E.; Turpin, D.H. Nitrogen isotope fractionation during the uptake and assimilation of nitrate, nitrite, ammonium, and urea by a marine diatom. *Limnol Oceanogr.* **1998**, *43*, 215–224. [CrossRef]
36. Yang, C.W.; Xu, H.H.; Wang, L.L.; Liu, J.; Shi, D.C. Comparative effects of salt-stress and alkali-stress on the growth, photosynthesis, solute accumulation, and ion balance of barley plants. *Photosynthetica* **2009**, *47*, 79–86. [CrossRef]
37. Zhang, K.Y.; Wu, Y.Y.; Hang, H.T. Differential contributions of NO_3^- /NH_4^+ to nitrogen use in response to a variable inorganic nitrogen supply in plantlets of two *Brassicaceae* species in vitro. *Plant Methods* **2019**, *15*, 86. Available online: https://pubmed.ncbi.nlm.nih.gov/31384291 (accessed on 28 June 2022). [CrossRef]

agronomy

Article

Stable Nitrogen Isotopes as an Effective Tool for Estimating the Nitrogen Demand of *Broussonetia papyrifera* (L.) Vent Seedlings under Variable Nitrate Concentrations

Kaiyan Zhang [1,2], Furong Zhang [3], Haitao Li [4], Yue Su [4,*] and Yanyou Wu [2,4,*]

1 School of Karst Science, Guizhou Normal University/State Engineering Technology Institute for Karst Desertification Control, Guiyang 550001, China; kaiyanzhang@126.com
2 State Key Laboratory of Environmental Geochemistry, Institute of Geochemistry, Chinese Academy of Sciences, Guiyang 550081, China
3 School of Agricultural Engineering, Jiangsu University, Zhenjiang 212013, China; furongzhang0218@126.com
4 Engineering Technology Research Center for Protection and Detection of Germplasm Resources of Karst-Adaptable Crops, Guizhou Vocational College of Agriculture, Qingzhen 551400, China; lisea02@126.com
* Correspondence: suyue09136@163.com (Y.S.); wuyanyou@mail.gyig.ac.cn (Y.W.); Tel.: +86-180-8510-5941 (Y.S.); +86-0851-8439-1746 (Y.W.)

Abstract: Poor growth is often observed in artificial young forests due to insufficient inorganic nitrogen in karst soils. However, little is known about the assimilatory demand of the whole plant for nitrate and the partitioning of nitrate assimilation in roots and leaves in woody plants grown in karst habitats. In this study, *Broussonetia papyrifera* (L.) Vent (*B. papyrifera*) seedlings were grown under nearly hydroponic conditions. The isotope mass balance approach was employed to quantify the $\delta^{15}N$ values of the N assimilates in plant organs and in whole plants for *B. papyrifera* seedlings grown at different nitrate concentrations. The $\delta^{15}N$ values of the N assimilates in the whole *B. papyrifera* seedlings showed a rising trend with increasing nitrate concentration. Increasing the supply of nitrate decreased the leaf–root difference in the $\delta^{15}N$ values of the N assimilates for *B. papyrifera* seedlings. Quantifying the $\delta^{15}N$ values of N assimilates in the whole *B. papyrifera* seedlings grown under different nitrate concentrations contributes to estimating the assimilatory demand of the *B. papyrifera* seedlings for nitrate. The leaf–root difference in the $\delta^{15}N$ values of the N assimilates can be used to estimate the partitioning of nitrate assimilation in the roots and leaves.

Keywords: karst; artificial forest; nitrate; isotope mass balance approach; assimilatory demand

Citation: Zhang, K.; Zhang, F.; Li, H.; Su, Y.; Wu, Y. Stable Nitrogen Isotopes as an Effective Tool for Estimating the Nitrogen Demand of *Broussonetia papyrifera* (L.) Vent Seedlings under Variable Nitrate Concentrations. *Agronomy* **2023**, *13*, 1663. https://doi.org/10.3390/agronomy13061663

Academic Editor: Fenliang Fan

Received: 2 June 2023
Revised: 19 June 2023
Accepted: 19 June 2023
Published: 20 June 2023

1. Introduction

Vegetation restoration is the key to controlling karst rocky desertification. However, poor growth is often observed in artificial young forests due to insufficient inorganic nitrogen in karst soils [1,2]. Hence, it is necessary to effectively manage the supply of inorganic nitrogen in artificial forests. The major sources of inorganic nitrogen utilized by plants have been suggested to be nitrate and ammonium [3–5]. However, weakly alkaline soils (pH 7.8–8.4) are often observed in karst rocky desertification areas [6], and high pH values in soils usually lead to ammonia volatilization [7,8]. Hence, a high-nitrate and low-ammonium environment occurs in karst rocky desertification areas [8,9]. As a result, the supply of nitrate may be suitable for artificial young forest growth in such habitats. The uptake and assimilation of nitrate within plants depend on the internal demand and external supply [10,11]. Generally, a balance between internal nitrogen demand and external nitrogen supply is an ideal status for plants and contributes to preventing excess or limited nitrogen supply. However, little is known about the assimilatory demand of the whole plant for nitrate and the partitioning of nitrate assimilation in roots and leaves in woody plants grown in karst rocky desertification areas.

After uptake, nitrate may be assimilated in roots and/or leaves, and partially unassimilated nitrate returns to the medium [10,12]. The lighter N isotope (^{14}N) is favored due to the kinetic process, and as a result, the heavier isotope (^{15}N) is depleted in the product [10,13,14]. Accordingly, the effluxed unassimilated nitrate is enriched in ^{15}N, and the whole-plant $\delta^{15}N$ values are negative relative to the source [15]. The deviations in whole-plant $\delta^{15}N$ values relative to source nitrogen $\delta^{15}N$ values are referred to as the nitrogen isotope fractionation value [16]. Under the condition that nitrate is the sole nitrogen source, the nitrogen isotope discrimination of whole plants depends on nitrate reductase activity and the supply of reductants [17]. Generally, the influxed nitrate will be assimilated to as great a degree as possible with strong nitrate reductase activity and an adequate supply of reductant. Consequently, the amount of effluxed unassimilated nitrate will decrease, which will minimize the observed isotope fractionation for nitrate assimilation by plants. However, when plants experience nitrate reductase activity restriction and/or reductant restriction, the amount of effluxed unassimilated nitrate increases. Then, greater nitrogen isotope fractionation is observed. Hence, the nitrogen isotope fractionation value of whole plants is closely related to the assimilatory demand for nitrate.

Generally, there is preexisting nitrogen in young woody plants [18]. It is difficult to quantify the nitrogen isotope fractionation value of a whole woody plant grown at different nitrate concentrations owing to interference from the $\delta^{15}N$ values of preexisting nitrogen. A simpler, more convenient approach is to quantify the nitrogen isotope fractionation value of N assimilates in whole woody plants grown at different nitrate concentrations to determine the assimilatory demand for nitrate over a greater time scale. Based on the isotope mass balance approach [19], the $\delta^{15}N$ value of N assimilates in whole woody plants can be quantified for woody plants grown at different nitrate concentrations. Moreover, the $\delta^{15}N$ values of N assimilates in plant organs can also be quantified using the isotope mass balance approach [19]. The $\delta^{15}N$ values of newly acquired N assimilates in stems are derived from the mix of $\delta^{15}N$ values of N assimilates in the roots and leaves. Consequently, the proportion of stem nitrogen obtained from the leaves (i.e., $f_{\text{leaf stem}}$) can be estimated using a two end-member isotope mixing model [20,21]. Furthermore, the leaf–root difference in the $\delta^{15}N$ values of N assimilates is closely linked with the partitioning of assimilatory activity between roots and leaves. Hence, quantifying the $\delta^{15}N$ values of N assimilates in plant organs contributes to the estimation of the partitioning of nitrate assimilation in the roots and leaves.

There is increasing interest in using *Broussonetia papyrifera* (L.) Vent (*B. papyrifera*) for ecological reclamation [22]. *B. papyrifera* trees can rapidly colonize abandoned factories and are employed for mining rehabilitation. In addition, this species presents a wide array of potential uses, such as the utilization of its bark for paper production, leaves as a source of forage, roots and fruits for traditional Chinese medicinal purposes, and the entire plant as a bioethanol source. Its rapid growth, strong adaptability to adverse environments, and high economic value make *B. papyrifera* suitable for karst rocky desertification control [23–26]. In the present study, *B. papyrifera* seedlings were subjected to different nitrate regimes. The effects of different nitrate concentrations on the growth, photosynthesis, chlorophyll fluorescence, nitrate reductase activity in leaves and roots, nitrogen content of plant organs, and $\delta^{15}N$ values of plant organs of *B. papyrifera* seedlings were investigated. The following were our main aims: (1) to estimate the assimilatory demand of whole *B. papyrifera* seedlings for nitrate over a greater time scale and (2) to estimate the partitioning of nitrate assimilation in the roots and leaves for *B. papyrifera* seedlings grown at different nitrate concentrations.

2. Materials and Methods

2.1. Plant Material and Experimental Treatments

Seeds of *B. papyrifera* were germinated in 12 drainage-hole-containing double-layer basin with a mixture of perlite and vermiculite (1:1 *v*/*v*) for 2 weeks at a temperature of 26/20 °C in the light/dark and 50–55% relative humidity. The lower basin contained a certain amount of water to keep the mixture moist. Seedlings of *B. papyrifera* were then incubated under a 12-h photoperiod, with 500 ± 20 μmol m^{-2} s^{-1} of photosynthetic photon flux density (PPFD). The lower basin contained adequate 1/8 strength Hoagland nutrient solution [27], and the solution was completely replaced every 3 days. After 6 weeks, vigorous seedlings were transplanted to pots (height of 8.5 cm, bottom diameter of 9 cm, and 12 holes at the bottom, which were 0.9 cm in diameter). Two layers of nylon mesh were placed inside the pot, and then a mixture of perlite and vermiculite (1:1 v/v) was added to the pot to fix the roots of the seedling. Each pot contained only one seedling. Six pots were placed in a tray that contained adequate 1/4 strength Hoagland nutrient solution [27]. The seedlings in the pots grew well without extra aeration. The tray (not including the pot) was covered with aluminum foil to prevent algal growth from light infiltration into the solution. The solution in the tray was completely replaced every 3 days, and the surface of the pot and the tray were cleaned to avoid algal contamination. After 3 weeks of growth, the nutrient solution was replaced by a modified Hoagland solution containing 1 mM $MgSO_4{\cdot}7H_2O$, 0.125 mM KH_2PO4, 2.5 mM KCl, 4 mM $CaCl_2$, 0.1875 mM K_2SO_4, 50 μM Fe(Na)EDTA, 25 μM H_3BO_3, 2 μM $MnSO_4{\cdot}1H_2O$, 2 μM $ZnSO_4{\cdot}7H_2O$, 0.1 μM $CuSO_4$, 0.04 μM $CoCl_2{\cdot}6H_2O$, and 0.1 μM $Na_2MoO_4{\cdot}2H_2O$ at a pH of 7.3 ± 0.1. $NaNO_3$, with a $\delta^{15}N$ of 22.35‰, was employed as the sole nitrogen source. The nitrate concentrations in the three treatments were set at 0.5 mM, 2 mM, and 8 mM. Each treatment contained three replicates, and each seedling was treated as a replicate. The seedlings in all treatments achieved the same growth status. Each pot was placed in a tray that contained 500 mL modified Hoagland solution. The small tray (not including the pot) was covered with aluminum foil to prevent algal growth from light infiltration into the solution. The modified Hoagland solution in the small tray was completely replaced every other day, and the surface of the pot and the tray were cleaned to avoid algal contamination. The treatments lasted for 20 days.

2.2. Measurements of Growth

After 20 days of culture, all *B. papyrifera* seedlings were harvested and divided into leaves, stems, and roots. The dry weights of the leaves, stems, and roots were determined after oven drying to constant mass at 80 °C. To obtain the dry weight of the leaves, stems, and roots of the *B. papyrifera* seedlings at the start of the experiment, three *B. papyrifera* seedlings with the same growth status were selected, and the corresponding dry weights were measured. The average dry weight of the leaves, stems, and roots of the three *B. papyrifera* seedlings at the start of the experiment was approximately equal to the initial dry weight of the leaves, stems, and roots of the *B. papyrifera* seedlings in this study [see Table A1]. Accordingly, the average nitrogen contents of the leaves, stems, and roots of the three *B. papyrifera* seedlings at the start of the experiment were approximately equal to the initial nitrogen contents of the leaves, stems, and roots of the *B. papyrifera* seedlings in this study [see Table A1]. In addition, the average $\delta^{15}N$ values of the leaves, stems, and roots of the three *B. papyrifera* seedlings at the start of the experiment were approximately equal to the initial $\delta^{15}N$ values of the leaves, stems, and roots of the *B. papyrifera* seedlings in this study [see Table A1].

2.3. Measurement of Chlorophyll Content and Gas Exchange

At the final harvest, the chlorophyll (Chl) content in the second fully expanded leaf was determined using the chlorophyll meter SPAD-502Plus (Konica Minolta, Tokyo, Japan). The gas exchange measurements were performed with a portable photosynthesis system LI-6800 (LI-COR, Lincoln, NE, USA). The photosynthetically available radiation, leaf temperature, relative humidity, and CO_2 concentration during the measurements were 500 μmol m^{-2} s^{-1}, 27.0 °C, 55%, and 400 μmol m^{-2} s^{-1}, respectively. The net photosynthetic rate (Pn), stomatal conductance (Gs), transpiration rate (Tr), and intercellular CO_2 concentration (Ci) of the second fully expanded leaves were measured in the fluorescence leaf chamber using 6800-01A from 09:00 to 11:00.

2.4. Chlorophyll Fluorescence Measurements

Leaves were dark-adapted for 30 min to ensure complete relaxation of all reaction centers before the measurements. As mentioned earlier, the second fully expanded leaves were selected for Chl fluorescence measurements with a portable photosynthesis system LI-6800 (LI-COR, Lincoln, NE, USA). The initial (F_o) and maximum (F_m) Chl fluorescence were measured, and then, the maximum photochemical efficiency (F_v/F_m) was calculated. The maximum fluorescence (F_m') in the light-adapted state, basic fluorescence after induction (F_o'), and fluorescence yield in the steady state (F_s) were simultaneously determined while determining the Pn. The F_v/F_m, the actual photochemical efficiency of PSII (Φ_p), the photochemical quenching coefficient (q_P), the nonphotochemical quenching coefficient (q_N), and the electron transport rate (ETR) were calculated according to the following formulae:

$$F_v/F_m = (F_m - F_o)/F_m \tag{1}$$

$$\Phi_p = (F_{m'} - F_s)/F_{m'} \tag{2}$$

$$q_p = (F_{m'} - F_s)/(F_{m'} - F_{o'}) \tag{3}$$

$$q_N = (F_m - F_{m'})/(F_m - F_{o'}) \tag{4}$$

$$ETR = PPFD \times \Phi_p \times 0.85 \times 0.5 \tag{5}$$

2.5. Analysis of Elements and Determination of $\delta^{15}N$ in Plants

The nitrogen contents of the dried leaves, stems, and roots were determined using an elemental analyzer (vario MACRO cube, Langenselbold, Germany). The $\delta^{15}N$ values of the leaves, stems, and roots were measured using a gas isotope ratio mass spectrometer (MAT-253, Thermo Fisher Scientific, Langenselbold, Germany). The $\delta^{15}N$ values were calculated according to the following equation:

$$\delta^{15}N(‰) = (R_{sample}/R_{standard} - 1) \times 1000 \tag{6}$$

where R_{sample} refers to the nitrogen isotope ratio of the plant material and $R_{standard}$ refers to the isotope ratio of a known standard (N_2 in air). IAEA N_1, IAEA N_2, and IAEA NO_3 reference materials were used to calibrate the instrument to reach a precision of 0.2‰ [28].

2.6. Photosynthetic Nitrogen Use Efficiency

After determining the nitrogen contents and net photosynthetic rate of leaves, the photosynthetic nitrogen use efficiency (PNUE) was obtained by dividing the value of Pn by the N content of the leaf [29,30].

2.7. Nitrate Reductase Activity Determination

The nitrate reductase activity (NRA) in the leaves and roots was tested using a Micro Nitrate Reductase Assay Kit (Solarbio, Beijing, China). The NRA was expressed as U/g FW.

2.8. The $\delta^{15}N$ of the Whole Plant

Using an isotope mass balance approach [19], the integrated $\delta^{15}N$ values of the whole plant consisting of leaves, stems, and roots were calculated as follows:

$$\delta^{15}N_{whole\text{-}plant}(‰) = (m_{leaf} \times \delta^{15}N_{leaf} + m_{stem} \times \delta^{15}N_{stem} + m_{root} \times \delta^{15}N_{root})/(m_{leaf} + m_{stem} + m_{root}) \tag{7}$$

where m_{leaf}, m_{stem}, and m_{root} are the total N accumulation (g) of the leaves, stems, and roots, respectively. $\delta^{15}N_{leaf}$, $\delta^{15}N_{stem}$, and $\delta^{15}N_{root}$ represent the $\delta^{15}N$ values of the leaves, stems, and roots, respectively.

2.9. The $\delta^{15}N$ of N Assimilates in the Whole Plant

Based on the isotope mass balance approach [19], we were able to calculate the $\delta^{15}N$ of the N assimilates in the whole plant when the $\delta^{15}N$ values of the initial and final whole plant (i.e., integrated $\delta^{15}N$ values of the whole plant at the beginning and harvest stages) were calculated. The $\delta^{15}N$ of the N assimilates in the whole plant ($\delta^{15}N_{assimilates}$) was calculated using the following equations:

$$\begin{aligned}\delta^{15}N_{assimilates}(‰) &= (m \times \delta^{15}N_{whole\text{-}plant1} - m_0 \times \delta^{15}N_{whole\text{-}plant0})/(m - m_0) \\ &= (m_l \times \delta^{15}N_l + m_s \times \delta^{15}N_s + m_r \times \delta^{15}N_r - m_{l0} \times \delta^{15}N_{l0} - m_{s0} \times \delta^{15}N_{s0} - m_{r0} \times \delta^{15}N_{r0})/(m - m_0)\end{aligned} \tag{8}$$

where $\delta^{15}N_{whole\text{-}plant1}$ and $\delta^{15}N_{whole\text{-}plant0}$ represent the $\delta^{15}N$ values of the final and initial whole plant, respectively. m and m_0 are the total N accumulation (g) of the final and initial whole plant, respectively. $\delta^{15}N_l$, $\delta^{15}N_s$, and $\delta^{15}N_r$ represent the $\delta^{15}N$ values of the leaves, stems, and roots at the final harvest, respectively. m_l, m_s, and m_r are the total N accumulation (g) of the leaves, stems, and roots at the final harvest, respectively. $\delta^{15}N_{l0}$, $\delta^{15}N_{s0}$, and $\delta^{15}N_{r0}$ represent the $\delta^{15}N$ values of the leaves, stems, and roots at the start of the experiment, respectively. m_{l0}, m_{s0}, and m_{r0} are the total N accumulation (g) of the leaves, stems, and roots at the start of the experiment, respectively. The standard error (SE) of the $\delta^{15}N_{assimilates}$ was achieved using the error propagation formula [31].

2.10. The $\delta^{15}N$ of N Assimilates in Plant Organs

For woody plant species, we treat the plant as having three major organs. We assume that the major sites of nitrogen assimilation are only leaves and roots. The $\delta^{15}N$ of the N assimilates in the stem can be predicted using a two end-member isotope mixing model using the $\delta^{15}N$ of the N assimilates in the leaves and roots as end members depending on the source for that stem nitrogen [20,21]:

$$\delta^{15}N_{stem\text{-}assimilates}(‰) = (\delta^{15}N_{leaf\text{-}assimilates} \times f_{leaf\ stem}) + (\delta^{15}N_{root\text{-}assimilates} \times f_{root\ stem}) \tag{9}$$

Equation (9) was rearranged to yield the fraction of stem N that is from the roots ($f_{root\ stem}$) or the leaves ($f_{leaf\ stem}$). For the fraction from the leaves, note that $f_{root\ stem} = 1 - f_{leaf\ stem}$:

$$f_{leaf\ stem} = \frac{(\delta^{15}N_{stem\text{-}assimilates} - \delta^{15}N_{root\text{-}assimilates})}{(\delta^{15}N_{leaf\text{-}assimilates} - \delta^{15}N_{root\text{-}assimilates})} \tag{10}$$

where $\delta^{15}N_{leaf\text{-}assimilates}$, $\delta^{15}N_{stem\text{-}assimilates}$, and $\delta^{15}N_{root\text{-}assimilates}$ can be calculated using an isotope mass balance approach. The mass balance equations for $\delta^{15}N_{leaf\text{-}assimilates}$, $\delta^{15}N_{stem\text{-}assimilates}$, and $\delta^{15}N_{root\text{-}assimilates}$ are as follows:

$$\delta^{15}N_{leaf\text{-}assimilates}(‰) = (m_l \times \delta^{15}N_l - m_{l0} \times \delta^{15}N_{l0})/(m_l - m_{l0}) \tag{11}$$

$$\delta^{15}N_{stem\text{-}assimilates}(‰) = (m_s \times \delta^{15}N_s - m_{s0} \times \delta^{15}N_{s0})/(m_s - m_{s0}) \tag{12}$$

$$\delta^{15}N_{root\text{-}assimilates}(‰) = (m_r \times \delta^{15}N_r - m_{r0} \times \delta^{15}N_{r0})/(m_r - m_{r0}) \tag{13}$$

where $\delta^{15}N_{leaf\text{-}assimilates}$ and $\delta^{15}N_{root\text{-}assimilates}$ represent the $\delta^{15}N$ values of N assimilates in leaves and roots, respectively. $\delta^{15}N_{stem\text{-}assimilates}$ represents the $\delta^{15}N$ values of newly acquired N assimilates in stems. The standard error (SE) of the $f_{leaf\ stem}$, $\delta^{15}N_{leaf\text{-}assimilates}$, $\delta^{15}N_{stem\text{-}assimilates}$, and $\delta^{15}N_{root\text{-}assimilates}$ was determined using the error propagation formula [31].

2.11. Statistical Analysis

The data were subjected to analysis of variance (ANOVA). The means of the different groups were compared via Tukey's test ($p < 0.05$). The data are shown as the mean ± standard deviation (SE). All analyses were conducted using Data Processing System (DPS) software 7.05 (Hangzhou Ruifeng Information Technology Co., Ltd., Hangzhou, China).

3. Results

3.1. Growth

The nitrate concentration had a significant effect on the growth of *B. papyrifera* seedlings. As shown in Table 1, the dry weight of the *B. papyrifera* seedlings increased significantly with increasing nitrate supply. However, compared to the dry weight of the leaves, stems, and roots at a 2 mM nitrate concentration, the excessive nitrate supply (8 mM) did not lead to a significant increase in dry biomass accumulation. In addition, the shoot length of *B. papyrifera* seedlings did not show a significant increase when the nitrate concentration increased from 2 to 8 mM. These results indicated that excessive nitrate supply did not significantly promote the growth of *B. papyrifera* seedlings.

Table 1. The growth parameters of *Broussonetia papyrifera* (L.) Vent seedlings under the three nitrate regimes.

Parameters	Nitrate Concentration (mM)		
	0.5	2	8
Dry weight (g/plant)	1.917 ± 0.078 b	2.814 ± 0.172 a	3.035 ± 0.112 a
Leaf DW (g)	0.771 ± 0.070 b	1.224 ± 0.111 ab	1.460 ± 0.132 a
Stem DW (g)	0.464 ± 0.037 b	0.622 ± 0.017 ab	0.653 ± 0.057 a
Root DW (g)	0.683 ± 0.071 a	0.967 ± 0.102 a	0.921 ± 0.075 a
Shoot length (cm)	12.8 ± 0.6 b	16.1 ± 0.9 ab	17.8 ± 1.2 a

Each value represents the mean ± SE (n = 3). Values with the same letter in each line are not significantly different based on Tukey's test ($p > 0.05$).

3.2. Photosynthesis, SPAD, and Chl Fluorescence

We monitored the SPAD values and gas exchange parameters to determine the effects of different nitrate levels. With increasing nitrate supply, SPAD, Pn, Gs, and Tr showed a significant increasing trend (Table 2). The SPAD values and gas exchange parameters of *B. papyrifera* seedlings showed a positive response to nitrate concentration.

Table 2. The photosynthetic parameters, chlorophyll fluorescence, and SPAD of *Broussonetia papyrifera* (L.) Vent seedlings under the three nitrate regimes.

Parameters	Nitrate Concentration (mM)		
	0.5	2	8
Pn (μmol m^{-2} s^{-1})	4.714 ± 0.244 c	11.005 ± 0.189 b	14.083 ± 0.815 a
Gs (mol m^{-2} s^{-1})	0.059 ± 0.012 b	0.115 ± 0.002 ab	0.169 ± 0.022 a
Tr (mmol m^{-2} s^{-1})	1.067 ± 0.199 b	2.037 ± 0.034 ab	2.862 ± 0.334 a
Fv/Fm	0.740 ± 0.010 b	0.801 ± 0.005 a	0.819 ± 0.003 a
Φp	0.287 ± 0.009 c	0.509 ± 0.011 b	0.623 ± 0.007 a
qP	0.641 ± 0.012 c	0.802 ± 0.001 b	0.866 ± 0.005 a
qN	0.788 ± 0.015 a	0.586 ± 0.024 b	0.379 ± 0.025 c
ETR	60.477 ± 1.939 c	107.343 ± 2.305 b	131.276 ± 1.549 a
SPAD	40.533 ± 1.408 c	47.367 ± 1.049 b	57.200 ± 0.436 a

Each value represents the mean ± SE (n = 3). Values with the same letter in each line are not significantly different based on Tukey's test ($p > 0.05$).

Chl fluorescence was further investigated to understand the internal causes of the effects of different nitrate levels on photosynthesis. The *B. papyrifera* seedlings grown at the lowest nitrate concentration had a significantly lower F_v/F_m, Φ_p, qP, and ETR, while the qN was the highest (Table 2). These results suggested that the activity of PSII was significantly affected by the supply of nitrate. Because of the N deficiency (0.5 mM nitrate), the *B. papyrifera* seedlings seemed to suffer from photoinhibition.

3.3. Nitrogen Content in the Leaves, Stems, and Roots of B. papyrifera Seedlings

Increasing the nitrate supply significantly promoted nitrogen assimilation in the *B. papyrifera* seedlings. As shown in Figure 1, the nitrogen content in the leaves, stems, and roots showed a significant rising trend with increasing nitrate concentration. In addition, the nitrogen content of leaves was markedly higher than that of roots and stems in all treatments.

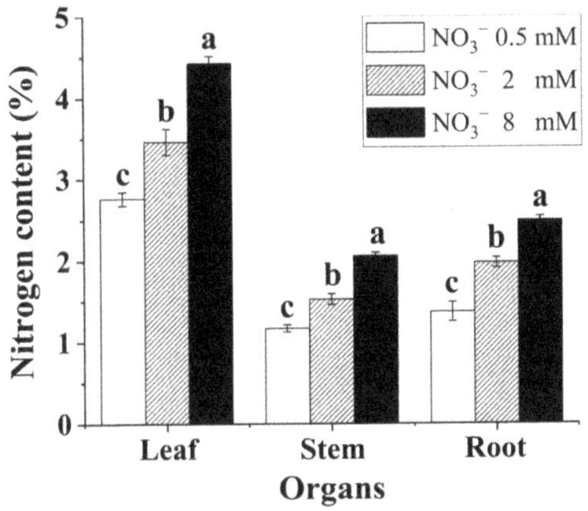

Figure 1. Nitrogen content of the *Broussonetia papyrifera* (L.) Vent seedlings under the three nitrate regimes. The mean ± SE (n = 3) followed by different letters in the same legend differ significantly (Tukey's test, $p < 0.05$).

3.4. Photosynthetic Nitrogen-Use Efficiency (PNUE)

The PNUE of the *B. papyrifera* seedlings only showed significant differences at the lowest nitrate concentration. As shown in Figure 2, increasing the nitrate supply did not significantly affect the PNUE of the *B. papyrifera* seedlings when the nitrate concentration was in the range of 2 to 8 mM.

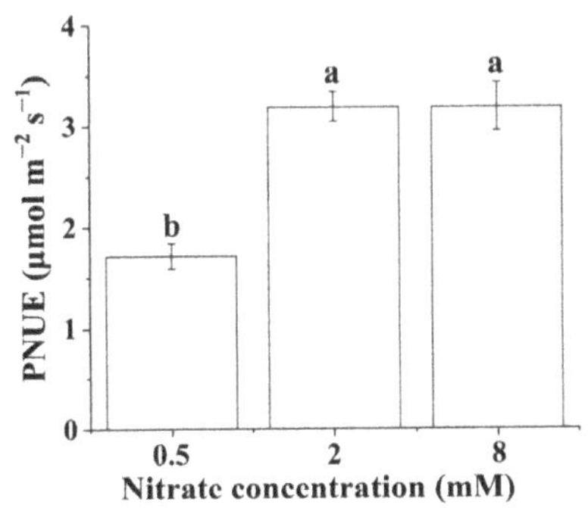

Figure 2. Photosynthetic nitrogen-use efficiency of *Broussonetia papyrifera* (L.) Vent seedlings under the three nitrate regimes. The mean ± SE (n = 3) followed by different letters in the same legend differ significantly (Tukey's test, $p < 0.05$).

3.5. NRA in the Leaves and Roots

Nitrate supply had a significant effect on the NRA in the leaves and roots (Figure 3). Increasing the nitrate concentration contributed to enhancing the NRA in the leaves and roots. Generally, the NRA in the leaves was markedly higher than that in the roots for the *B. papyrifera* seedlings.

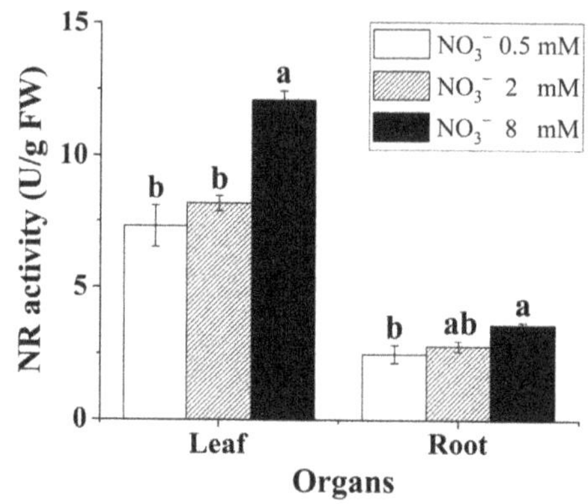

Figure 3. NRA in the leaves and roots of *Broussonetia papyrifera* (L.) Vent seedlings under the three nitrate regimes. The mean ± SE (n = 3) followed by different letters in the same legend differ significantly (Tukey's test, $p < 0.05$).

3.6. *Nitrogen Isotope Composition in the Leaves, Stems, and Roots of the B. papyrifera Seedlings*

There were differences in organ-level $\delta^{15}N$ for the *B. papyrifera* seedlings grown under the three nitrate regimes (Figure 4). Generally, the leaves were consistently enriched in ^{15}N relative to the roots. In addition, the $\delta^{15}N$ in the leaves, stems, and roots showed a significant rising trend with increasing nitrate concentration. Increasing the supply of nitrate contributed to enriching ^{15}N in the *B. papyrifera* seedlings.

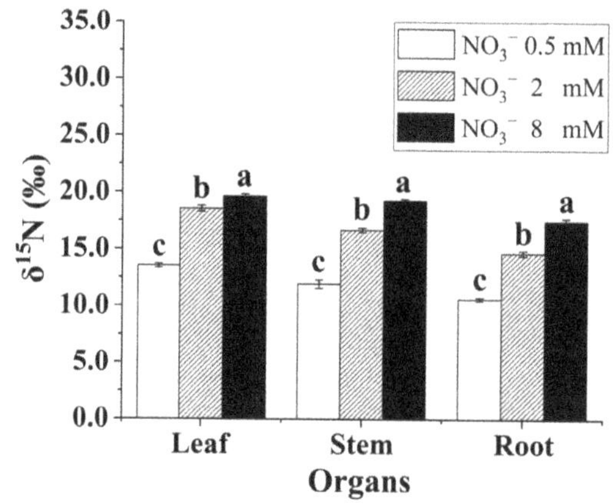

Figure 4. The $\delta^{15}N$ in leaves, stems, and roots of the *Broussonetia papyrifera* (L.) Vent seedlings under the three nitrate regimes. The mean ± SE (n = 3) followed by different letters in the same legend differ significantly (Tukey's test, $p < 0.05$).

3.7. *The $\delta^{15}N$ Values of N Assimilates in the Whole Plant*

The $\delta^{15}N$ values of the N assimilates in the whole *B. papyrifera* seedlings were negative compared to the initial source $\delta^{15}N$ under the three nitrate regimes (Figure 5), which suggested that isotope discrimination occurred during the process of nitrate assimilation. Increasing the nitrate concentration contributed to enriching ^{15}N in the N assimilates in the whole *B. papyrifera* seedlings. The $\delta^{15}N$ values of the N assimilates in the whole *B. papyrifera* seedlings at the lowest nitrate concentration were distinctly lower than that at other nitrate concentrations.

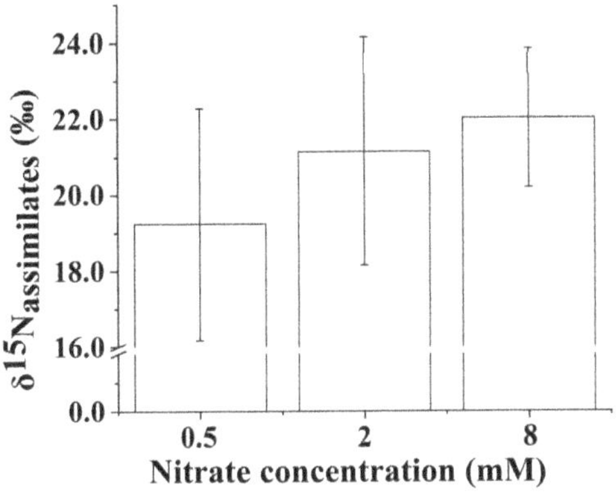

Figure 5. The $\delta^{15}N$ values of the N assimilates in the whole *Broussonetia papyrifera* (L.) Vent seedlings under the three nitrate regimes. The error bars were calculated using the error propagation formula.

3.8. Proportion of Stem N Derived from the Leaves

The proportion of stem N derived from the leaves showed a linear increase with nitrate concentration for the *B. papyrifera* seedlings (Figure 6). As shown in Figure 6, increasing the supply of nitrate contributed to the translocation of the N assimilates from the leaves to the stems. We observed that approximately 60% of stem N was derived from the translocation of N assimilates from the leaves when the nitrate concentration reached 8 mM.

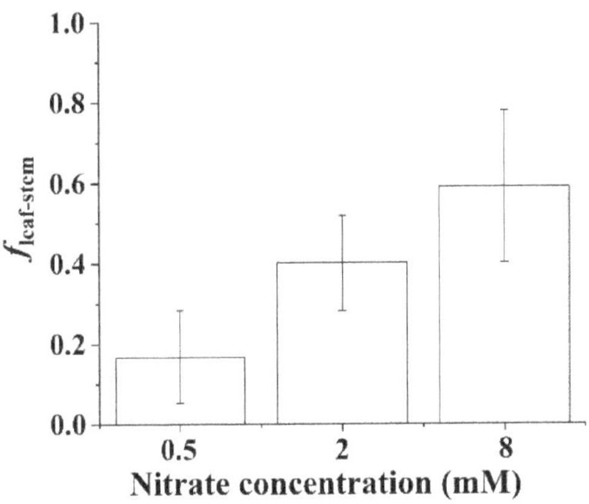

Figure 6. Proportion of stem N derived from the leaves of the *Broussonetia papyrifera* (L.) Vent seedlings under the three nitrate regimes. The error bars were calculated using the error propagation formula.

3.9. The $\delta^{15}N$ of N Assimilates in Plant Organs

The $\delta^{15}N$ values of the N assimilates in the leaves, stems, and roots of the *B. papyrifera* seedlings depended on the nitrate supply. As shown in Figure 7, the $\delta^{15}N$ values of the N assimilates in the leaves showed a decreasing trend with increasing nitrate concentration. However, increasing the supply contributed to enriching ^{15}N in the N assimilates of the stems and roots of the *B. papyrifera* seedlings. In addition, the difference between the $\delta^{15}N$ values of the N assimilates in the leaves and roots decreased gradually with increasing nitrate concentration.

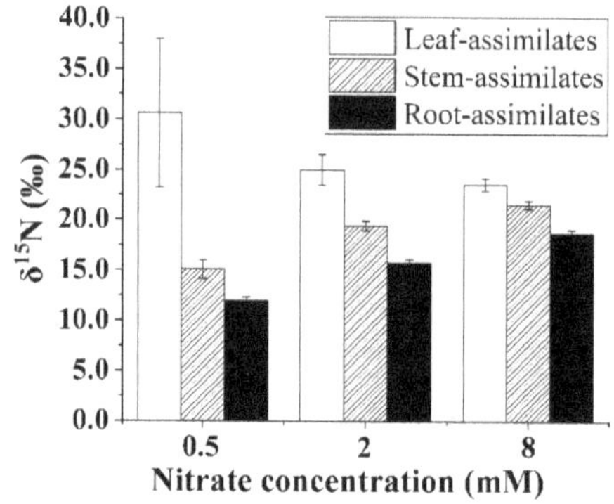

Figure 7. The $\delta^{15}N$ values of the N assimilates in the *Broussonetia papyrifera* (L.) Vent organs under the three nitrate regimes. The error bars were calculated using the error propagation formula.

4. Discussion

Plant $\delta^{15}N$ is a physiological indicator of N demand and fractionation that reflects changes in metabolic N fluxes and/or environmental effects [10,11]. Generally, variation in N supply can affect organ-level nitrogen isotope composition [13,32]. As shown in Figure 4, there was variation in organ-level $\delta^{15}N$ for the *B. papyrifera* seedlings grown under the three nitrate regimes. Enrichment of leaf $\delta^{15}N$ relative to root $\delta^{15}N$ for *B. papyrifera* grown at all nitrate concentrations indicated that some unassimilated inorganic nitrogen in the roots translocated to the shoot through the xylem [33]. Increasing the supply of nitrate enhanced the enrichment of ^{15}N in the leaves, stems, and roots of the *B. papyrifera* seedlings, which might be attributed to an increased capacity to assimilate nitrate in the leaves and/or roots [33]. Increasing the supply of nitrate also led to a significant increase in the nitrogen content of the leaves, stems, and roots (Figure 1), which suggested that increasing the nitrate concentration contributed to improving the nitrogen assimilation ability of *B. papyrifera* seedlings. Hence, there might be a positive response between the enrichment of ^{15}N and the nitrogen assimilation ability of *B. papyrifera* seedlings.

In the present study, the roots of *B. papyrifera* seedlings were always grown in the solution, which contributed to minimizing localized ^{15}N enrichment of the solution around the roots. The uptake of nitrogen by *B. papyrifera* seedlings caused the nitrate concentration in the tray to decrease, which resulted in an ^{15}N enrichment of the residual nitrate over time. Generally, the degree of ^{15}N enrichment in the residual nitrate depended on the change in nitrate concentration in the solution; namely, the more the nitrate concentration in the solution decreased, the greater the residual nitrate enriched ^{15}N [34]. After 20 days of culture, the *B. papyrifera* seedlings grown in three nitrate regimes (0.5 mM, 2 mM, and 8 mM) accumulated 0.0157 ± 0.0021 g N (n = 3, SE), 0.0510 ± 0.0056 g N (n = 3, SE), and 0.0807 ± 0.0048 g N (n = 3, SE), respectively. The total nitrogen supply of the three nitrate regimes (0.5 mM, 2 mM, and 8 mM) was 0.035 g N, 0.140 g N, and 0.560 g N, respectively. Hence, the $\delta^{15}N$ value of the residual nitrate at low nitrate concentrations would be greater than that at high nitrate concentrations. Based on the isotope mass balance approach [19], the $\delta^{15}N$ values of the N assimilates in the whole *B. papyrifera* seedlings grown at three nitrate regimes could be quantified. Because the actual $\delta^{15}N$ value of the source (i.e., the residual nitrate in the solution) could not be known, it was very difficult to precisely calculate the nitrogen isotope discrimination for the N assimilates in the whole *B. papyrifera* seedlings. However, the $\delta^{15}N$ values of the N assimilates in the whole *B. papyrifera* seedlings could still be used to indirectly indicate the degree of nitrogen isotope discrimination. The $\delta^{15}N$ values of the N assimilates in the whole *B. papyrifera* seedlings gradually increased with increasing nitrate concentration (Figure 5). The actual $\delta^{15}N$ value of the source (i.e., the residual nitrate in the solution) at the three nitrate concentrations was above 22.35‰ (the $\delta^{15}N$ value of the initial source). Meanwhile, the actual $\delta^{15}N$ value of the source (i.e., the residual nitrate in the solution) gradually decreased with increasing nitrate concentration. Hence, we concluded that increasing the nitrate concentration reduced the nitrogen isotope discrimination for *B. papyrifera* seedlings when the nitrate concentration was in the range of 0.5 to 8 mM [16].

In general, nitrogen isotope discrimination is dependent on the relationship between nitrogen supply and nitrogen demand [11]. Increased nitrogen isotope discrimination usually corresponds to a reduced assimilatory demand for nitrogen [35], while the enhanced assimilatory demand for nitrogen will reduce the efflux and result in decreased nitrogen isotope discrimination [34]. As a result, we speculated that the decrease in nitrogen isotope discrimination value (i.e., the increase in the δ^{15}N value of the N assimilates in the whole *B. papyrifera* seedlings) was caused by enhanced assimilatory demand for nitrate in this study. Generally, increasing the supply of inorganic nitrogen leads to an increase in the nitrogen isotope discrimination value [13]. However, nitrogen isotope discrimination is not only dependent on the external inorganic nitrogen concentration. Nitrogen isotope discrimination was observed only when the demand for nitrogen was lower than the nitrogen supply [13]. Buschhaus [34] also found that increasing the supply of inorganic nitrogen resulted in a decrease in nitrogen isotope discrimination, which was attributed to the increased nitrogen demand resulting from stimulating growth. Hence, the assimilatory demand for nitrogen could be estimated using the δ^{15}N value of the N assimilates in the whole plant. In this study, increasing the supply of nitrate increased the δ^{15}N values of the N assimilates in the whole *B. papyrifera* seedlings when the nitrate concentration was in the range of 0.5 to 8 mM, which suggested that a relatively high external nitrate supply (8 mM) did not exceed the demand for nitrogen. In addition, the growth potential of the *B. papyrifera* seedlings also reflected that increasing the supply of nitrate contributed to enhancing the assimilatory demand for nitrate. As shown in Table 2, the Fv/Fm values of *B. papyrifera* seedlings were below 0.75 when the nitrate concentration was only 0.5 mM, which suggested that low N stress caused photoinhibition [36,37]. Photoinhibition usually damages the photosynthetic structure, thus significantly reducing the electron transport rate [37]. As a result, poor photosynthetic capacity was observed for *B. papyrifera* seedlings grown at 0.5 mM nitrate. Accordingly, the assimilatory demand for nitrate would be very low for *B. papyrifera* seedlings grown at 0.5 mM nitrate. As a whole, increasing the supply of nitrate could significantly enhance the photosynthetic capacity of *B. papyrifera* seedlings (Table 2). The enhanced photosynthetic capacity promoted the growth of *B. papyrifera* seedlings, which suggested that the demand for nitrate must have increased. Furthermore, increasing the nitrate concentration significantly elevated the nitrogen content of leaves, stems, and roots of *B. papyrifera* seedlings, which suggested that increasing the supply of nitrate enhanced the assimilatory demand for nitrate for *B. papyrifera* seedlings. Hence, quantifying the δ^{15}N value of the N assimilates in the whole plant could estimate the assimilatory demand of the whole plant for nitrate over a greater time scale, which contributed to preventing the waste and insufficiency of the nitrate supply.

Given that effluxed nitrogen is enriched in ^{15}N [15], the increased δ^{15}N values of the N assimilates in the whole *B. papyrifera* seedlings indicated that the net efflux of nitrate in the roots decreased. Correspondingly, increasing the supply of nitrate enhanced the assimilation of nitrate in the *B. papyrifera* seedlings. Leaves and roots are thought to be the major sites of nitrogen assimilation [11]. Hence, the efflux of nitrate in the roots was closely related to the assimilation of nitrate in the leaves and roots. Generally, nitrate reductase is required for the assimilation of nitrate, NRA is inducible with available nitrate [38,39], and increasing the supply of nitrate contributes to enhancing the NRA [39,40]. In this study, the NRA in the leaves and roots showed a positive response to the nitrate concentration (Figure 3). However, the NRA was not the only factor that affected nitrogen isotopic fractionation. As shown in Figure 3, when the nitrate concentration was in the range of 0.5 to 2 mM, no significant difference was observed in the NRA in the roots, and the NRA in the leaves did not show a significant change. The δ^{15}N values of the N assimilates in the whole *B. papyrifera* seedlings showed a marked increase when the nitrate concentration increased from 0.5 to 2 mM, which suggested that the increased assimilation of nitrate might benefit from an adequate supply of reductant [17]. As shown in Figure 2, the PNUE level was significantly higher in the 2 mM nitrate concentration than in the 0.5 mM nitrate concentration. The high PNUE level was accompanied by an adequate supply of reductant. The PNUE level did not show a significant increase when the nitrate concentration increased from 2 to 8 mM. Hence, the increased δ^{15}N values of the N assimilates in the whole *B. papyrifera* seedlings might be attributed to the significantly enhanced NRA in the leaves when the nitrate concentration increased from 2 to 8 mM (Figure 3). Overall, the degree of nitrogen isotopic fractionation observed for nitrate assimilation depended on the NRA and the supply of reductant for the young, rapidly growing *B. papyrifera* seedlings.

Generally, the leaves had higher δ^{15}N values than the roots [15,18,33], which was the expected pattern because residual unassimilated nitrate (enriched in ^{15}N) is transported from roots to shoots. Accordingly, the δ^{15}N values of the N assimilates in the leaves were higher than those in the roots for the *B. papyrifera* seedlings (Figure 7). As shown in Figure 7, increasing the nitrate concentration decreased the leaf–root difference in the δ^{15}N values of the N assimilates, which might have been related to the partitioning of assimilatory activity between the roots and leaves. Although the NRA in the leaves was considerably greater than that in the roots (Figure 3), the location of nitrate assimilation appeared to depend on the internal demand of the *B. papyrifera* seedlings as well as the external nitrate concentration [33]. Under nonsaturating nitrate conditions, substantial nitrate will be assimilated by root nitrate reductase, and less nitrate may be assimilated in leaves [10]. A previous study also showed that assimilation is weighted more toward roots, particularly at concentrations below 1 mM nitrate [41]. Low assimilatory demand for nitrate was evidenced by the δ^{15}N value of the N assimilates in the whole *B. papyrifera* seedlings grown at 0.5 mM nitrate. Hence, only a smaller proportion of nitrate was assimilated in the leaves of the *B. papyrifera* seedlings grown at 0.5 mM nitrate. As a result, distinct leaf–root differences in the δ^{15}N values of the N assimilates were observed for the *B. papyrifera* seedlings grown at 0.5 mM nitrate (Figure 7). With increasing nitrate concentration, the assimilatory demand for nitrate in the roots reached saturation, and then, substantial nitrate was assimilated in the leaves, which led to a decrease in the δ^{15}N values of the N assimilates in the leaves. Consequently, a gradually decreased leaf–root difference in the δ^{15}N values of the N assimilates was observed for *B. papyrifera* seedlings when the nitrate concentration increased from 2 to 8 mM (Figure 7). Hence, the leaf–root difference in the δ^{15}N values of the N assimilates could be used to estimate the partitioning of nitrate assimilation in the roots and leaves. The newly acquired organic nitrogen in the stems is translocated from the leaves and roots [11,15]. Hence, when the δ^{15}N values of the N assimilates in the leaves, stems, and roots are quantified using the isotope mass balance approach [19], the proportion of stem N derived from the leaves (i.e., $f_{\text{leaf stem}}$) can be calculated using Equation (10). As shown in Figure 6, the proportion of stem nitrogen obtained from the leaves showed an obviously rising trend with increasing nitrate concentration, which suggested that increasing the supply of nitrate promoted the translocation of the N assimilates from the leaves to the stems. The increased translocation of the N assimilates from the leaves to the stems might imply that the nitrate assimilation was weighted more toward leaves with increasing nitrate concentration. Hence, quantifying the δ^{15}N values of the N assimilates in the plant organs not only contributes to estimating the partitioning of nitrate assimilation in roots and leaves but also provides an alternate way to calculate the proportion of stem N derived from the leaves.

5. Conclusions

Based on the isotope mass balance approach, the δ^{15}N values of N assimilates in plant organs and in whole plants can be quantified for *B. papyrifera* seedlings grown at different nitrate concentrations. The δ^{15}N values of N assimilates in whole *B. papyrifera* seedlings can be used to estimate the assimilatory demand of whole *B. papyrifera* seedlings for nitrate over a greater time scale. Increasing the supply of nitrate contributes to enhancing the assimilatory demand of the whole *B. papyrifera* seedlings for nitrate when the nitrate concentration was in the range of 0.5 to 8 mM. However, our results suggested that a concentration of 0.5 mM nitrate was insufficient to maintain the health of *B. papyrifera* seedlings. The optimal nitrate supply for *B. papyrifera* seedlings varies depending on the intended objective. To achieve ecological restoration, a 2 mM nitrate supply is recommended, whereas a nitrate supply of 8 mM is optimal for achieving a leaf forage with high protein content. Hence, quantifying the $\delta^{15}N$ values of N assimilates in whole *B. papyrifera* seedlings grown under different nitrate concentrations contributes to preventing the waste and insufficiency of the nitrate supply, which provides a theoretical basis for effective inorganic nitrogen management in *B. papyrifera* seedlings grown in karst regions. The partitioning of nitrate assimilation in the roots and leaves can be estimated using the leaf–root difference in the $\delta^{15}N$ values of the N assimilates for *B. papyrifera* seedlings grown under different nitrate concentrations. Increasing the nitrate supply contributes to increasing the partitioning of nitrate assimilation in leaves relative to roots.

Author Contributions: K.Z. and Y.W. conceived and designed the experiment. F.Z. performed most of the experiment. H.L. performed some of the experiment. K.Z. and F.Z. analyzed the data. K.Z. wrote the manuscript. Y.W. and Y.S. reviewed and edited the manuscript. All authors have read and agreed to the published version of the manuscript.

Funding: This work was supported by Guizhou Provincial Science and Technology Foundation ([2020]1Y172) and the National Natural Science Foundation of China (32001101).

Data Availability Statement: All data generated or analyzed during this study are included in this published article and Table A1.

Acknowledgments: The authors would like to thank the technical staff at the State Key Laboratory of Environmental Geochemistry, Institute of Geochemistry, Chinese Academy of Sciences, for technical assistance during the measurement of the $\delta^{15}N$, in particular J.T.

Conflicts of Interest: The authors declare no conflict of interest.

Appendix A

Table A1. The initial biomass, nitrogen content and $\delta^{15}N$ of the *Broussonetia papyrifera* (L.) Vent seedlings at the start of the experiment.

Parameters	Plant Organs		
	Leaves	Stems	Roots
Dry weight (g)	0.348 ± 0.031	0.075 ± 0.006	0.070 ± 0.011
Nitrogen content (%)	4.53 ± 0.02	2.81 ± 0.01	3.15 ± 0.01
$\delta^{15}N$ (‰)	7.51 ± 0.09	6.97 ± 0.04	6.46 ± 0.02

Note: Each value represents the mean ± SE (n = 3).

References

1. Garousi, F.; Shan, Z.J.; Ni, K.; Yang, H.; Shan, J.; Cao, J.H.; Jiang, Z.C.; Yang, J.L.; Zhu, T.B.; Müller, C. Decreased inorganic N supply capacity and turnover in calcareous soil under degraded rubber plantation in the tropical karst region. *Geoderma* **2021**, *381*, 114754. [CrossRef]
2. Tang, Y.Q.; Tian, J.; Li, X.Z.; Yao, M.J.; Wang, S.Q.; Kuzyakov, Y.; Dungait, J.A.J. Higher free-living N_2 fixation at rock-soil interfaces than topsoils during vegetation recovery in karst soils. *Soil Biol. Biochem.* **2021**, *159*, 108286. [CrossRef]
3. Cui, J.; Yu, C.; Qiao, N.; Xu, X.; Tian, Y.; Ouyang, H. Plant preference for NH_4^+ versus NO_3^- at different growth stages in an alpine agroecosystem. *Field Crop. Res.* **2017**, *201*, 192–199. [CrossRef]
4. Tho, B.T.; Lambertini, C.; Eller, F.; Brix, H.; Sorrell, B.K. Ammonium and nitrate are both suitable inorganic nitrogen forms for the highly productive wetland grass *Arundo donax*, a candidate species for wetland paludiculture. *Ecol. Eng.* **2017**, *105*, 379–386. [CrossRef]
5. Hessini, K.; Issaoui, K.; Ferchichi, S.; Abdelly, C.; Siddique, K.; Cruz, C. Interactive effects of salinity and nitrogen forms on plant growth, photosynthesis and osmotic adjustment in maize. *Plant Physiol. Biochem.* **2019**, *139*, 171–178. [CrossRef] [PubMed]
6. Hui, N.; Sun, N.X.; Du, H.M.; Umair, M.; Kang, H.Z.; Liu, X.X.; Romantschuk, M.; Liu, C.J. Karst rocky desertification does not erode ectomycorrhizal fungal species richness but alters microbial community structure. *Plant Soil* **2019**, *445*, 383–396. [CrossRef]

7. Wu, Q.X.; Han, G.L.; Tao, F.X.; Yang, T. Chemical composition of rainwater in a karstic agricultural area, Southwest China: The impact of urbanization. *Atmos. Res.* **2012**, *111*, 71–78. [CrossRef]
8. Xia, A.T.; Wu, Y.Y. Joint interactions of carbon and nitrogen metabolism dominated by bicarbonate and nitrogen in *Orychophragmus violaceus* and *Brassica napus* under simulated karst habitats. *BMC Plant Biol.* **2022**, *22*, 264. [CrossRef]
9. Wu, Y.Y.; Wu, Y.S. The increase in the karstification-photosynthesis coupled carbon sink and its implication for carbon neutrality. *Agronomy* **2022**, *12*, 2147. [CrossRef]
10. Kalcsits, L.A.; Guy, R.D. Whole-plant and organ-level nitrogen isotope discrimination indicates modification of partitioning of assimilation, fluxes and allocation of nitrogen in knockout lines of *Arabidopsis thaliana*. *Physiol. Plant.* **2013**, *149*, 249–259. [CrossRef]
11. Kalcsits, L.A.; Buschhaus, H.A.; Guy, R.D. Nitrogen isotope discrimination as an integrated measure of nitrogen fluxes, assimilation and allocation in plants. *Physiol. Plant.* **2014**, *151*, 293–304. [CrossRef]
12. Hu, Y.; Guy, R.D.; Soolanayakanahally, R.Y. Nitrogen isotope discrimination in open-pollinated and hybrid canola suggests indirect selection for enhanced ammonium utilization. *Front. Plant Sci.* **2022**, *13*, 1024080. [CrossRef] [PubMed]
13. Evans, R.D. Physiological mechanisms influencing plant nitrogen isotope composition. *Trends Plant Sci.* **2001**, *6*, 121–126. [CrossRef] [PubMed]
14. Robinson, D. $\delta^{15}N$ as an integrator of the nitrogen cycle. *Trends Ecol. Evol.* **2001**, *16*, 153–162. [CrossRef]
15. Hu, Y.; Guy, R.D. Isotopic composition and concentration of total nitrogen and nitrate in xylem sap under near steady-state hydroponics. *Plant Cell Environ.* **2020**, *43*, 2112–2123. [CrossRef] [PubMed]
16. Evans, R.D.; Bloom, A.J.; Sukrapanna, S.S.; Ehleringer, J.R. Nitrogen isotope composition of tomato (Lycopersicon esculentum Mill. cv. T-5) grown under ammonium or nitrate nutrition. *Plant Cell Environ.* **1996**, *19*, 1317–1323. [CrossRef]
17. Mariotti, A.; Mariotti, F.; Champigny, M.L.; Amarger, N.; Moyse, A. Nitrogen isotope fractionation associated with nitrate reductase activity and uptake of NO_3^- by pearl millet. *Plant Physiol.* **1982**, *69*, 880–884. [CrossRef]
18. Kalcsits, L.A.; Guy, R.D. Quantifying remobilization of pre-existing nitrogen from cuttings to new growth of woody plants using ^{15}N at natural abundance. *Plant Methods* **2013**, *9*, 27. [CrossRef]
19. Hayes, J.M. *An Introduction to Isotopic Calculations*; Woods Hole Oceanographic Institution: Woods Hole, MA, USA, 2004; pp. 1–10. Available online: https://www.whoi.edu/cms/files/jhayes/2005/9/IsoCalcs30Sept04_5184.pdf (accessed on 25 May 2023).
20. Zhang, K.Y.; Wu, Y.Y.; Hang, H.T. Differential contributions of NO_3^-/NH_4^+ to nitrogen use in response to a variable inorganic nitrogen supply in plantlets of two Brassicaceae species in vitro. *Plant Methods* **2019**, *15*, 86. [CrossRef]
21. Zhang, K.Y.; Wu, Y.Y.; Su, Y.; Li, H.T. Implication of quantifying nitrate utilization and CO_2 assimilation of *Brassica napus* plantlets in vitro under variable ammonium/nitrate ratios. *BMC Plant Biol.* **2022**, *22*, 392. [CrossRef]
22. Ni, J.W.; Su, S.; Li, H.; Geng, Y.H.; Zhou, H.J.; Feng, Y.Z.; Xu, X.Q. Distinct physiological and transcriptional responses of leaves of paper mulberry (*Broussonetia kazinoki* × *B. papyrifera*) under different nitrogen supply levels. *Tree Physiol.* **2020**, *40*, 667–682. [CrossRef]
23. Li, M.; Li, Y.; Li, H.; Wu, G. Overexpression of atnhx5 improves tolerance to both salt and drought stress in *Broussonetia papyrifera* (L.) vent. *Tree Physiol.* **2011**, *31*, 349–357. [CrossRef] [PubMed]
24. Mei, R.; Wang, Y.; Du, G.; Liu, G.; Zhang, L.; Cheng, Y. Antioxidant lignans from the fruits of *Broussonetia papyrifera*. *J. Nat. Prod.* **2009**, *72*, 621. [CrossRef] [PubMed]
25. Yao, L.; Yang, H.; Yoo, C.G.; Meng, X.; Li, M.; Pu, Y.; Ragauskas, A.J.; Sykes, R.W. Adsorption of cellobiohydrolases I onto lignin fractions from dilute acid pretreated *Broussonetia papyrifera*. *Bioresour. Technol.* **2017**, *244*, 957–962. [CrossRef] [PubMed]
26. Peng, X.J.; Liu, H.; Chen, P.L.; Tang, F.; Hu, Y.M.; Wang, F.F.; Pi, Z.; Zhao, M.L.; Chen, N.Z.; Chen, H.; et al. A chromosome-scale genome assembly of paper mulberry (*Broussonetia papyrifera*) provides new insights into its forage and papermaking usage. *Mol. Plant* **2019**, *12*, 661–677. [CrossRef] [PubMed]
27. Hoagland, D.R.; Arnon, D.I. The water-culture method for growing plants without soil. *Circ. Calif. Agric. Exp. Stn.* **1950**, *347*, 32. Available online: https://www.researchgate.net/file.PostFileLoader.html?id=54aefd7ed4c118b6358b45db&assetKey=AS%3A273668901408776%401442259158553 (accessed on 25 May 2023).
28. Yousfi, S.; Serret, M.D.; Araus, J.L. Comparative response of $\delta^{13}C$, $\delta^{18}O$ and $\delta^{15}N$ in durum wheat exposed to salinity at the vegetative and reproductive stages. *Plant Cell. Environ.* **2013**, *36*, 1214–1227. [CrossRef]
29. Poorter, H.; Evans, J.R. Photosynthetic nitrogen-use efficiency of species that differ inherently in specific leaf area. *Oecologia* **1998**, *116*, 26–37. [CrossRef]
30. Suárez, J.C.; Polanía, J.A.; Anzola, J.A.; Contreras, A.T.; Méndez, D.L.; Vanegas, J.I.; Noriega, J.E.; Rodríguez, L.; Urban, M.O.; Beebe, S.; et al. Influence of nitrogen supply on gas exchange, chlorophyll fluorescence and grain yield of breeding lines of common bean evaluated in the Amazon region of Colombia. *Acta Physiol. Plant.* **2021**, *43*, 66. [CrossRef]
31. Taylor, J. *Introduction to Error Analysis, the Study of Uncertainties in Physical Measurements*; University Science Books: Sausalito, CA, USA, 1997; Available online: https://ui.adsabs.harvard.edu/abs/1997ieas.book.....T/abstract (accessed on 25 May 2023).
32. Pritchard, E.S.; Guy, R.D. Nitrogen isotope discrimination in white spruce fed with low concentrations of ammonium and nitrate. *Trees* **2005**, *19*, 89–98. [CrossRef]
33. Kalcsits, L.A.; Min, X.; Guy, R.D. Interspecific variation in leaf-root differences in $\delta^{15}N$ among three tree species grown with either nitrate or ammonium. *Trees* **2015**, *29*, 1069–1078. [CrossRef]

34. Buschhaus, H.A. ^{15}N Discrimination as an Indicator of Nitrogen Dynamics in Populus Trichocarpa. Master's Thesis, University of British Columbia, Vancouver, BC, Canada, 2007; p. 70. [CrossRef]
35. Yousfi, S.; Serret, M.D.; Araus, J.L. Shoot $\delta^{15}N$ gives a better indication than ion concentration or $\delta^{13}C$ of genotypic differences in the response of durum wheat to salinity. *Funct. Plant Biol.* **2009**, *36*, 144–155. [CrossRef] [PubMed]
36. Ogaya, R.; Penuelas, J.; Asensio, D.; Llusià, J. Chlorophyll fluorescence responses to temperature and water availability in two co-dominant Mediterranean shrub and tree species in a long-term field experiment simulating climate change. *Environ. Exp. Bot.* **2011**, *71*, 123–127. [CrossRef]
37. Wu, Y.W.; Li, R.; Chen, W.; Liu, X.L.; Kong, F.L.; Ke, Y.P.; Shi, H.C.; Yuan, J.C. Effect of low-nitrogen stress on photosynthesis and chlorophyll fluorescence characteristics of maize cultivars with different low-nitrogen tolerances. *J. Integr. Agric.* **2019**, *18*, 1246–1256. [CrossRef]
38. Campbell, W.H. Nitrate reductase structure, function and regulation: Bridging the gap between biochemistry and physiology. *Annu. Rev. Plant Biol.* **1999**, *50*, 277–303. [CrossRef] [PubMed]
39. Kaiser, W.M.; Huber, S.C. Post-translational regulation of nitrate reductase: Mechanism, physiological relevance and environmental triggers. *J. Exp. Bot.* **2001**, *52*, 1981–1989. [CrossRef]
40. Black, B.L.; Fuchigami, L.H.; Coleman, G.D. Partitioning of nitrate assimilation among leaves, stems and roots of poplar. *Tree Physiol.* **2002**, *22*, 717–724. [CrossRef]
41. Andrews, M.; Morton, J.D.; Lieffering, M.; Bisset, L. The partitioning of nitrate assimilation between root and shoot of a range of temperate cereals and pasture grasses. *Ann. Bot.* **1992**, *70*, 271–276. [CrossRef]

agronomy

MDPI

Article

Diurnal Variation in Transport and Use of Intracellular Leaf Water and Related Photosynthesis in Three Karst Plants

Xiaojie Qin [1], Deke Xing [1,*], Yanyou Wu [2], Weixu Wang [1], Meiqing Li [1] and Kashif Solangi [1]

[1] School of Agricultural Engineering, Jiangsu University, Zhenjiang 212013, China
[2] State Key Laboratory of Environmental Geochemistry, Institute of Geochemistry, Chinese Academy of Sciences, Guiyang 550081, China
* Correspondence: xingdeke@ujs.edu.cn; Tel.: +86-511-88797338

Abstract: Except for transpired water, the intracellular water stored in leaves accounts for only 1–3% of the water absorbed by roots. Understanding water transport and use, as well as the related photosynthetic response, helps with determining plant water status and improving the revegetation efficiency in fragile karst habitats. In this study, we conducted experiments on 8 year old naturally growing plants of *Coriaria nepalensis* Wall., *Broussonetia papyrifera* (L.) Vent., and *Elaeocarpus decipiens* Hemsl. in karst areas. We determined the diurnal variations in leaf electrophysiology, water potential, gas exchange, and chlorophyll fluorescence parameters. The results indicated that *C. nepalensis* plants maintained a high photosynthetic rate, with a high root water uptake ability and leaf intracellular water-holding capacity (LIWHC). The stomata quickly closed to conserve water within cells and protect the photosynthetic structure. *B. papyrifera* maintained stable intracellular water transport rate (LIWTR), and the photosynthetic efficiency was increased with increasing intracellular water-use efficiency (LIWUE). *B. papyrifera* also maintained its photosynthesis by efficiently using the transpired water when the LIWHC was increased. The inter- and intracellular water in the leaves of *E. decipiens* remained stable, which could be attributed to the leathery leaves and its high water-holding capacity. The photosynthesis of *E. decipiens* was low and stable. Compared with the high photosynthesis, high transpiration, and low instantaneous water-use efficiency (WUE_i) pattern in *C. nepalensis* plants, *E. decipiens* plants exhibited low photosynthesis, low transpiration, and low WUE_i, whereas *B. papyrifera* plants presented high photosynthesis, low transpiration, and high WUE_i. Plants in karst regions change their transport and use of intracellular leaf water to regulate the photosynthetic performance, which differs among different plant species.

Keywords: electrophysiology; stomatal conductance; water-use efficiency; chlorophyll fluorescence; adaptability

Citation: Qin, X.; Xing, D.; Wu, Y.; Wang, W.; Li, M.; Solangi, K. Diurnal Variation in Transport and Use of Intracellular Leaf Water and Related Photosynthesis in Three Karst Plants. *Agronomy* **2022**, *12*, 2758. https://doi.org/10.3390/agronomy12112758

Academic Editor: Begoña Blasco

Received: 18 October 2022
Accepted: 4 November 2022
Published: 6 November 2022

1. Introduction

The karst landscape in Guizhou province, China, is one of the largest continuous landscapes in southwestern China [1]. This karst landscape is located in a subtropical monsoon climate zone and has an annual average temperature of 14 °C and precipitation of 1300 mm [2]. In recent years, rocky karst desertification is becoming increasingly serious due to frequent human activities, resulting in a shallow and discontinuous soil layer, which has gradually decreased the vegetation cover and soil water retention capacity [3]. The vegetation in these regions is mainly composed of uneven grasses and shrubs [4,5], which has produced an uneven surface hydrologic permeability distribution. Despite receiving enough precipitation, the plants in these regions usually suffer from temporary water stress. Most importantly, the droughts within karst regions are highly heterogeneous [6–8]. Therefore, plant adaptability to the heterogeneous drought environments must be studied, allowing the selected plant species to be matched with fragile karst habitats and the vegetation restoration efficiency to be increased.

Water is crucial for plant growth and development, but water-use strategies differ among plant species [9,10]. Stomata are the first line of defense against drought stress [11–13] and play an important role in the water transport within plants by regulating the apertures [14]. With opened stomata apertures, a high transpiration rate (E) is accompanied by strong photosynthetic CO_2 assimilation. However, a high E decreases the leaf water potential (Ψ_L) and can damage the mesophyll, thus depressing photosynthesis [12]. Stomatal closure can reduce water loss and prevent irreversible damage caused by water deficits to the photosynthetic structures [15]. Photosynthetic performance under water stress can be used for investigating the water-use traits in plants [16]. Photosynthesis is a dynamic reaction process. The diurnal variations in photosynthetic parameters can indicate the photosynthetic capacity of a plant [17]. Water stress also influences the chlorophyll fluorescence parameters, which are closely related to each reaction process in photosynthesis [18]. Chlorophyll fluorescence parameters can reflect the characteristics of plant internality [19]. Photochemical quenching (qP) represents the openness of the photosystem II (PSII) reaction center [20,21], whereas nonphotochemical quenching (NPQ) is a plant self-protective mechanism against excess light energy. Under water stress conditions, the daily average minimal fluorescence (F_o) increases, while the maximum fluorescence yield (F_m) and primary light energy conversion efficiency (F_v/F_m) of PSII decrease [22]. Therefore, the chlorophyll fluorescence can be used to nondestructively determine the internal reaction process of the photosynthetic structure and evaluate the response of a plant to adversity [20].

Most (~97%) of the water absorbed by roots is dissipated through transpiration, while only a small part (1~3%) is stored within leaf cells, which directly determines the photosynthesis, growth, and other metabolic processes of the plant [14]. Photosynthesis is directly correlated with intracellular water transport and use. When plants are subjected to water deficits, changes in leaf intracellular water alter the concentrations of cell sap and electrolytes (i.e., ions, ion groups, and electric dipoles in cells) [23]. The cell membrane, having strictly selective permeability, also influences the concentration of intracellular electrolytes, which mainly exist in the vacuoles and cytoplasm [24]. The cytoplasm contains numerous organelles with specific membranes, and electrical features vary across the organelles, vacuoles, and cytoplasm, which occupy most of intracellular space and can be regarded as resistors, whereas the plasma membrane shows capacitance. The electrolyte solutions on the two sides of a cell membrane form a specific conductive state. A mesophyll cell can be modeled as a concentric sphere capacitor due to the abovementioned special composition and structure [24,25]. The water metabolism in leaves alters the electrolyte concentration and changes the corresponding electrophysiological parameters. Therefore, electrophysiology is increasingly being used to diagnose the status of plant water [26,27]. Electrophysiological parameters, such as physiological capacitance and impedance [23,24,27], are related to the change in solute concentration within leaf cells; they can reflect the transport of intracellular dielectric substances and be used to detect the dynamic characteristics of intracellular water. The findings of studies on the transport and use of intracellular water stored in the leaves, rather than the transpired water, can help to accurately determine the water status of plants.

Coriaria nepalensis Wall. is a deciduous shrub with strong reproductive capacity and papery or thin leathery leaves [28,29]. It is widely distributed in the Shanxi, Sichuan, and Gansu provinces in China. *C. nepalensis* has a strong adaptability and can resist drought and barren environments. This type of plant grows well in neutral alkali soil. *Broussonetia papyrifera* (L.) Vent. is a deciduous tree with a shallow root system, wide lateral root distribution, and papery leaves [30]. *B. papyrifera* also has a strong adaptability and can resist to drought and barren environments, thus being widely distributed. *Elaeocarpus decipiens* Hemsl. is an evergreen tree that is slightly shade-tolerant, with a developed root system, strong germination, and leathery leaves [31]; it has high ornamental value and is suitable for planting in the mining areas due to its strong tolerance to SO_2 [32]. These three plant species all grow well in karst areas, but the water transport and use traits of these plants remain unknown. As such, in this study, we selected *C. nepalensis*, *B. papyrifera*, and *E. decipiens* as the experimental object, and we measured the diurnal variations in electrophysiology, Ψ_L, gas exchange, and chlorophyll fluorescence parameters of these plants. We calculated and analyzed the dynamic characteristics of leaf intracellular water transport as a function of the electrophysiological parameters. Our aims in this study were to (i) analyze the relationship between leaf intracellular water and photosynthesis, (ii) explore the diurnal variation in transport and use of intracellular leaf water in different plants, and (iii) compare the adaptabilities of the three plant species to a karst environment. The results provide a basis for selecting appropriate plant species matched to the heterogeneous karst environment during the vegetation restoration in fragile karst habitats.

2. Materials and Methods

2.1. Study Area

We conducted this study at the National Observation Station of Karst Ecosystem (Puding Station) in Puding County, Guizhou province, southwest China (26°22′07″ N, 105°45′06″ E). This experimental site was 1158 m above sea level with a humid subtropical monsoon climate, and an average annual temperature of 15.1 °C, with a maximum of 34.4 °C in summer and a minimum of –11.0 °C in winter. The average annual precipitation was 1397 mm·year^{-1}. The vegetation types were mainly degraded rattan shrub and secondary evergreen and deciduous broadleaved forest. The soil type was mainly lime soil with a loam texture. Hang et al. (2018) reported that the soils in this site were characterized by high calcium, phosphorus deficiency, high bicarbonate, and alkaline properties [33]. We conducted the experiment in July 2021, which was a period with a relatively high temperature and water deficiency.

According to the planning and construction of the National Observation Station and the branch growth of the selected plants, we selected the leaves of approximately 8 year old *C.nepalensis*, *B. papyrifera*, and *E. decipiens* plants that were naturally growing around Puding station as the experimental subject. We used five leaves from five different randomly selected plants for each parameter determination in each plant species. The abovementioned plant species differed in distribution, adaptive habitats, and biological traits (Table 1). We recorded the measurements on the fourth or fifth fully expanded leaves that were growing well and uniformly. We measured the parameters every 2 h from 8:00 a.m. to 6:00 p.m. to study their diurnal variations.

Table 1. Comparison between *C.nepalensis, B. papyrifera, and E. decipiens.*

Plant Species	Distribution	Adaptive Habitats	Biological Traits
C. nepalensis	Yunnan, Guizhou, Sichuan, Hubei, Shanxi, Gansu, Xizang provinces in China; India; Nepal	Drought, low nutrition, neutral alkaline, heliophilous, low temperature	Shrub, fast-growing, leathery leaf, shallow root system with many horizontal and oblique roots
B. papyrifera	Yellow, Yangtze and Pearl River Basins in China; Vietnam; Japan; India; Malaysia; Thailand; Burma	Drought, low nutrition, waterlog, heliophilous, acidic and neutral soil, chimney, air pollution, limestone	Tree, fast-growing, papery tomentose leaf, shallow root system with wide lateral root distribution
E. decipiens	Guangxi, Guangdong, Guizhou, Jiangxi, Fujian, Zhejiang, Yunnan, Hunan, Taiwan provinces in China; Vietnam; Japan	Slightly shade-tolerant, liked warmth and humid, acid soil, strong tolerance to SO_2	Tree, fast-growing, leathery leaf, developed deep root system

2.2. Determination of Electrophysiological Parameters and Leaf Water Potential

Electrophysiological parameters are easily affected by environmental stimuli. Therefore, we removed the plant leaves from the branches and immediately immersed them in water for 30 min. As such, we ensured that the cells of the leaves reached a standard and uniform state, which helped to minimize the influence of environmental stimuli on the values of the electrophysiological parameters. Then, we dried the leaf surfaces with tissues, and we clamped the leaves with a custom-made parallel-plate capacitor [24]. We only used the rehydrated leaves for determining the electrophysiological parameters. We recorded the variations in the electrophysiological parameters (i.e., physiological capacitance, resistance, and impedance) with increasing gripping force using an LCR tester (*Model 3532-50, Hioki*, Nagano, Japan). We set the gripping forces to 2.1, 4.1, 6.1, and 8.1 N. The leaves were not to be damaged by these gripping forces, as they were not strong enough. We first placed the leaf between two parallel electrode plates with a diameter of 7 cm, and we applied different gripping forces by adding the same quality iron blocks. We selected three sites on each leaf for recording the electrophysiological parameters at each gripping force, and we calculated the average value of each parameter. We recorded the measurement on five leaves from five different randomly selected plants for each plant species.

We established the coupling models of gripping force and electrophysiological parameters according to the Nernst equation and the law of energy conservation, respectively. We then calculated the leaf intracellular water transport rate (LIWTR) [24], water-holding capacity, and water-use efficiency [23]. The specific calculation formulas (Supplementary File 1) were as follows:

$$\mathrm{LIWTR} = \mathrm{bke}^{-\mathrm{bF}}, \tag{1}$$

$$\mathrm{LIWHC} = \sqrt{(\mathrm{IC})^3}, \tag{2}$$

$$\mathrm{LIWUE} = \frac{\mathrm{d}}{\mathrm{LIWHC}}, \tag{3}$$

where b and k are parameters of the physiological impedance fitting equation, IC (pF) is the leaf physiological capacitance, and d is the specific effective thickness of the leaf.

We detached the leaves from the branches of the naturally growing plants, which were immediately drilled with a hole punch. We then placed the small, drilled disc into a C-52-SF sample chamber; after 6 min balancing, we recorded the Ψ_L data with a dew point microvoltmeter (*Psypro, Wescor*, United States) [34,35]. We measured the Ψ_L at the same position as the abovementioned electrophysiological parameters. We conducted the measurement on five leaves from five different randomly selected plants for each plant species.

2.3. Determination of Photosynthetic Parameters

We measured the diurnal variations in photosynthetic parameters with a portable LI-6400XT photosynthesis measurement system (*LI-COR Inc., Lincoln*, NE, United States). We clamped the leaf with a transparent leaf chamber, which was vertically irradiated with natural light. The photosynthetic photon flux density (PPFD) varied from 36.66 to 1989.99 $\mu\mathrm{mol{\cdot}m^{-2}{\cdot}s^{-1}}$ throughout the day, the air temperature was 33.42 ± 6.46 °C, and the CO_2 concentration was 415.49 ± 15.26 μmol $CO_2{\cdot}\mathrm{mol}^{-1}$. We repeated the in situ and nondestructive measurements of photosynthetic parameters five times by randomly selecting five different plants for each plant species. We recorded the net photosynthetic rate (P_N, $\mu\mathrm{mol{\cdot}m^{-2}{\cdot}s^{-1}}$), stomatal conductance (gs, $\mathrm{mmol{\cdot}m^{-2}{\cdot}s^{-1}}$), and transpiration rate (E, $\mathrm{mmol{\cdot}m^{-2}{\cdot}s^{-1}}$). We calculated the instantaneous water-use efficiency (WUE_i, $\mu\mathrm{mol{\cdot}mmol^{-1}}$) as follows [36]:

$$\mathrm{WUE_i} = P_N/\mathrm{E}, \tag{4}$$

2.4. Determination of Chlorophyll Fluorescence Parameters

We measured the chlorophyll fluorescence parameters using an IMAGING-PAM modulated chlorophyll fluorescence analyzer (*Heinz Walz GmbH, Effeltrich*, Germany). The PSII reaction center was completely opened after 20 min of dark adaptation before measurement. We irradiated the fully dark-adapted leaves with modulation measurement light (about 0.10 $\mu mol \cdot m^{-2} \cdot s^{-1}$), and we recorded the F_o value. Then, we irradiated the leaves with a saturated light flash (usually about 3000 $\mu mol \cdot m^{-2} \cdot s^{-1}$ or higher for less than 1 s). The primary electron acceptor QA was completely restored in a short time, the PSII completely "switched off", and we measured the Fm. We measured the qP and NPQ of the front of plant leaves with an activation light of 800 $\mu mol \cdot m^{-2} \cdot s^{-1}$. We calculated F_v/F_m as $(F_m - F_o)/F_m$. We conducted the measurement on five leaves from five different randomly selected plants for each plant species.

2.5. Statistical Analysis

We analyzed all collected data using SPSS software (*version 22.0, SPSS Inc.*, New York, NY, USA) and SigmaPlot software (*version 10.0, Systat Software Inc.*, California, CA, USA). We fit the coupling relationships between gripping force and the electrophysiological parameters using SigmaPlot software. We compared the parameters at different times with Duncan's multiple comparison at the 5% significance level ($p \leq 0.05$) using SPSS software. The data are reported as the means ± standard errors (n = 5). We prepared the graphs using Origin 2019 (*Northampton*, MA, USA).

3. Results

3.1. Diurnal Variations in Leaf Water Potential and Electrophysiological Parameters

The Ψ_L values of *C. nepalensis* between 8:00 and 10:00 a.m. were not significantly different, but were clearly lower than at 2:00 and 6:00 p.m. (Figure 1). The Ψ_L values of *B. papyrifera* at 8:00 a.m. and 12:00 p.m. were significantly higher than those at 10:00 a.m. and 2:00 p.m., and those at 6:00 p.m. were not notably different from the values at 8:00 a.m. and 12:00 p.m. The Ψ_L value of *E. decipiens* at 10:00 a.m. was clearly higher than that between 12:00 and 6:00 p.m., and the values exhibited no remarkable change after 2:00 p.m. The order of the daily mean Ψ_L value was *C. nepalensis* > *E. decipiens* > *B. papyrifera*.

According to previous experimental results, when the gripping force of the capacitor sensor was 4.1 N, the electrodes were in close contact with the leaf surface without damaging the leaf, and the recorded values remained stable [37]. Therefore, we calculated the corresponding electrophysiological parameters by assuming the gripping force was 4.1 N in this study. The LIWTR values of *C. nepalensis* between 10:00 a.m. and 12:00 p.m. were significantly higher than those at 4:00 and 6:00 p.m., and we found no significant change before 2:00 p.m. (Table 2). We observed no obvious change among the values for *B. papyrifera*. The values for *E. decipiens* at 10:00 a.m. and 2:00 p.m. were significantly higher than at 12:00, 4:00, and 6:00 p.m. The order of the daily mean LIWTR value was *E. decipiens* > *B. papyrifera* > *C. nepalensis*.

The LIWHC values for *C. nepalensis* and *E. decipiens* between 8:00 a.m. and 2:00 p.m. were not significantly different, and the highest values for these two plant species were both at 4:00 p.m. (Table 2). The values for *B. papyrifera* significantly decreased at 10:00 a.m., and then showed no clear change between 10:00 a.m. and 4:00 p.m. The order of the daily mean LIWHC value was *C. nepalensis* > *E. decipiens* > *B. papyrifera*.

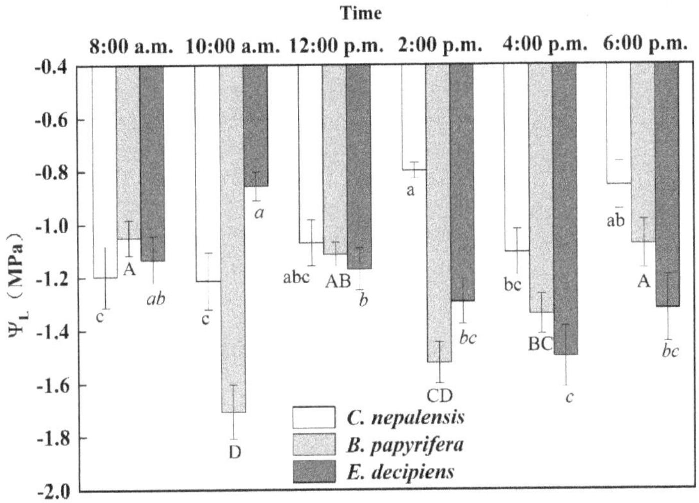

Figure 1. Leaf water potential (Ψ_L, MPa) of three plant species. Different lowercase letters above the error bars of the same plant species indicate that subsequent values significantly differed at $p \leq 0.05$, according to one-way ANOVA. a, b, and c indicate the differences between the values of *C. nepalensis*; A, B, C, and D indicate the differences between the values of *B. papyrifera*; *a*, *b*, and *c* indicate the differences between the values of *E. decipiens*.

Table 2. Leaf intracellular water transport rate (LIWTR, $M\Omega \cdot N^{-1}$), leaf intracellular water-holding capacity (LIWHC), and leaf intracellular water-use efficiency (LIWUE) of *C. nepalensis*, *B. papyrifera*, and *E. decipiens*. Data are shown as the means ± SE (n = 5). Different lowercase letters in the same column indicate significant differences between different timepoints.

Plant Species	Time	LIWTR	LIWHC	LIWUE
C. nepalensis	8:00 a.m.	0.24 ± 0.04^{abc}	288.38 ± 31.05^{b}	0.05 ± 0.00^{bc}
	10:00 a.m.	0.41 ± 0.05^{a}	235.40 ± 31.40^{b}	0.06 ± 0.01^{ab}
	12:00 p.m.	0.43 ± 0.17^{a}	287.49 ± 101.52^{b}	0.09 ± 0.02^{a}
	2:00 p.m.	0.35 ± 0.06^{ab}	226.67 ± 41.76^{b}	0.07 ± 0.00^{ab}
	4:00 p.m.	0.04 ± 0.01^{c}	3039.45 ± 673.30^{a}	0.02 ± 0.00^{c}
	6:00 p.m.	0.09 ± 0.03^{bc}	739.56 ± 175.68^{b}	0.04 ± 0.01^{bc}
B. papyrifera	8:00 a.m.	0.25 ± 0.03^{a}	371.46 ± 26.46^{a}	0.11 ± 0.02^{c}
	10:00 a.m.	0.29 ± 0.02^{a}	58.01 ± 9.22^{c}	0.56 ± 0.12^{a}
	12:00 p.m.	0.46 ± 0.04^{a}	126.41 ± 18.04^{bc}	0.20 ± 0.03^{bc}
	2:00 p.m.	0.27 ± 0.08^{a}	65.69 ± 9.44^{c}	0.34 ± 0.06^{b}
	4:00 p.m.	0.46 ± 0.15^{a}	49.91 ± 0.00^{c}	0.75 ± 0.00^{a}
	6:00 p.m.	0.31 ± 0.01^{a}	183.39 ± 50.75^{b}	0.23 ± 0.09^{bc}
E. decipiens	8:00 a.m.	0.89 ± 0.25^{abc}	198.25 ± 25.86^{bc}	0.06 ± 0.01^{ab}
	10:00 a.m.	1.03 ± 0.20^{ab}	123.98 ± 39.73^{c}	0.13 ± 0.06^{a}
	12:00 p.m.	0.52 ± 0.10^{bcd}	220.49 ± 24.54^{bc}	0.06 ± 0.01^{ab}
	2:00 p.m.	1.18 ± 0.19^{a}	90.41 ± 15.75^{c}	0.12 ± 0.02^{ab}
	4:00 p.m.	0.20 ± 0.06^{d}	534.29 ± 82.44^{a}	0.04 ± 0.00^{b}
	6:00 p.m.	0.47 ± 0.11^{cd}	341.94 ± 54.26^{b}	0.05 ± 0.00^{ab}

The LIWUE values for *C. nepalensis* gradually increased between 8:00 a.m. and 12:00 p.m. and then decreased (Table 2). The values for *B. papyrifera* notably increased at 10:00 a.m., and then decreased at 12:00 and 2:00 p.m.; that at 4:00 p.m. showed no clear difference from that at 10:00 a.m. The LIWUE of *E. decipiens* exhibited no clear change before 2:00 p.m., but the values at 4:00 and 6:00 p.m. were not remarkably different from that at 2:00 p.m. The order of the daily mean LIWUE value was *B. papyrifera* > *E. decipiens* > *C. nepalensis*.

3.2. Diurnal Variations in Photosynthetic Parameters and Instantaneous Water-Use Efficiency

We observed no remarkable change among the P_N values of *C. nepalensis* before 12:00 p.m., but it considerably decreased over time after 2:00 p.m., and the daily mean value was 12.60 μmol·m^{-2}·s^{-1} (Figure 2A). The P_N of *B. papyrifera* was much higher between 10:00 a.m. and 2:00 p.m., and then decreased at 6:00 p.m., the daily mean value was 14.99 μmol·m^{-2}·s^{-1}. The lower P_N of *E. decipiens* was associated with increasing time; the values between 12:00 and 4:00 p.m. were not noticeably different. The daily mean value was 8.34 μmol·m^{-2}·s^{-1}.

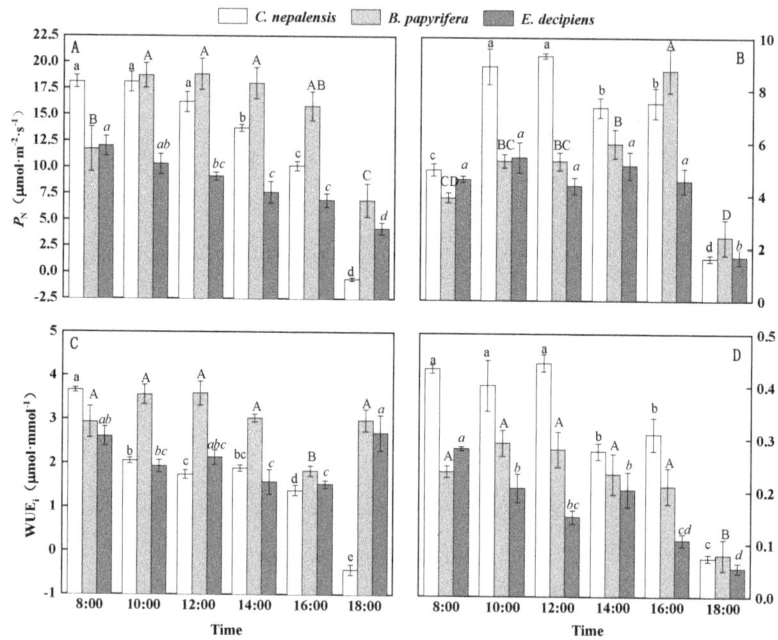

Figure 2. Diurnal variations in (**A**) net photosynthetic rate (P_N, μmol·m^{-2}·s^{-1}); (**B**) transpiration rate (E, mmol·m^{-2}·s^{-1}); (**C**) instantaneous water-use efficiency (WUE$_i$, μmol·mmol^{-1}); (**D**) stomatal conductance (gs, mmol·m^{-2}·s^{-1}). Different lowercase letters appear above error bars of same plant species when subsequent values significantly differed at $p \leq 0.05$ according to one-way ANOVA. a, b, c, etc., indicate differences between values of *C. nepalensis*; A, B, C, etc., indicate the differences between values of *B. papyrifera*; *a, b, c*, etc., indicate the differences between values of *E. decipiens*.

The E values of *C. nepalensis* at 10:00 a.m. and 12:00 p.m. were significantly higher than those at other times; we observed the lowest value at 6:00 p.m. (Figure 2B). The E of *B. papyrifera* did not remarkably change between 10:00 a.m. and 2:00 p.m., and it reached the highest value at 4:00 p.m. The E of *E. decipiens* did not significantly differ between 8:00 a.m. and 4:00 p.m., but decreased to the lowest value at 6:00 p.m.

The WUE$_i$ of *C. nepalensis* decreased over time; the values at 12:00 and 2:00 p.m. were not remarkably different (Figure 2C). The WUE$_i$ of *B. papyrifera* was not notably different between 8:00 a.m. and 2:00 p.m., but was much higher than at 4:00 p.m.; that at 6:00 p.m. was not considerably different from that at 2:00 p.m. The WUE$_i$ of *E. decipiens* gradually decreased from 8:00 a.m. to 4:00 p.m. but did not remarkably change; it noticeably increased at 6:00 p.m.

The gs of *C. nepalensis* between 8:00 a.m. and 12:00 p.m. did not notably differ, and the lowest value appeared at 6:00 p.m. (Figure 2D). We found no significant difference among the gs values of *B. papyrifera* between 8:00 a.m. and 4:00 p.m., and the value at 6:00 p.m. was the lowest. We observed no significant change among the gs values of *E. decipiens* from 10:00 a.m. to 2:00 p.m., and we observed the lowest value at 6:00 p.m.

3.3. Diurnal Variations in Chlorophyll Fluorescence Parameters

The F_o value of *C. nepalensis* was the highest at 12:00 p.m., and then exhibited no significant differences between 2:00 and 6:00 p.m. (Figure 3A). We observed higher F_o values in *B. papyrifera* at 8:00 a.m., 2:00, and 6:00 p.m., but these values did not significantly differ. The F_o of *E. decipiens* was the highest at 2:00 p.m. and then notably decreased over time.

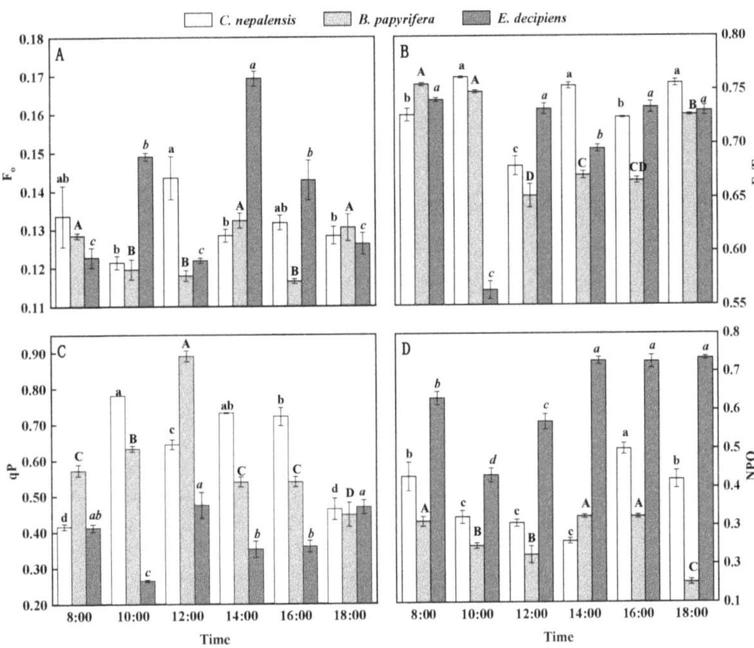

Figure 3. (**A**) Minimal fluorescence (F_o); (**B**) primary light energy conversion efficiency (F_v/F_m); (**C**) photochemical quenching(qP); (**D**) non-photochemical quenching (NPQ). Different lowercase letters appear above error bars of the same plant species when subsequent values significantly differ at $p \leq 0.05$ according to one-way ANOVA. A, b, c, etc., indicate differences between values of *C. nepalensis*; A, B, C, etc., indicate differences between values of *B. papyrifera*; *a*, *b*, *c*, etc., indicate differences between values of *E. decipiens*.

We observed higher F_v/F_m values in *C. nepalensis* at 10:00 a.m., 2:00, and 6:00 p.m., and the value at 12:00 p.m. was the lowest (Figure 3B). The F_v/F_m values of *B. papyrifera* between 8:00 and 10:00 a.m. were significantly higher than at other times, and those between 12:00 and 4:00 p.m. were lower than at 6:00 p.m. We observed the lowest F_v/F_m value for *E. decipiens* at 10:00 a.m., and we found no significant difference between the values at 12:00, 4:00 and 6:00 p.m.

The qP values of *C. nepalensis* between 10:00 a.m. and 4:00 p.m. were significantly higher than those at 8:00 a.m. and 6:00 p.m. (Figure 3C). The qP values of *B. papyrifera* increased over time before 12:00 p.m., and then remarkably decreased at 2:00 p.m. The lowest qP value of *E. decipiens* occurred at 10:00 a.m., and the value at 12:00 p.m. was much higher than at 2:00 and 4:00 p.m.

The NPQ values of *C. nepalensis* considerably decreased at 10:00 a.m., and then showed no remarkable change between 10:00 a.m. and 2:00 p.m. The value at 4:00 p.m. was the highest (Figure 3D). The NPQ values of *B. papyrifera* at 10:00 a.m. and 12:00 p.m. were much lower than at 2:00 and 4:00 p.m. We noted the lowest NPQ value of *E. decipiens* at 10:00 a.m. The values between 2:00 and 6:00 p.m. exhibited no clear difference but were higher than those at other times.

3.4. *Difference among Intracellular Water and Photosynthesis in C. nepalensis, B. papyrifera, and E. decipiens*

C. japonica exhibited high LIWHC, P_N, and E, but low LIWTR, LIWUE, and WUE_i (Table 3). *B. papyrifera* showed a high LIWUE, P_N, and WUE_i, but low LIWTR, LIWHC, and E. *E. sinensis* showed a relatively high LIWTR, medium LIWHC, and low LIWUE, P_N, E, and WUE_i.

Table 3. Differences among intracellular water use (leaf intracellular water transport rate (LIWTR), water-holding capacity (LIWHC), and water-use efficiency (LIWUE)) and photosynthesis (net photosynthetic rate (P_N), transpiration rate (E), and instantaneous water-use efficiency (WUE_i)) in *C. nepalensis*, *B. papyrifera*, and *E. decipiens*.

Plant Species	Leaf Intracellular Water Traits			Photosynthesis		
	LIWTR	LIWHC	LIWUE	P_N	E	WUE_i
C. nepalensis	low	high	low	high	high	low
B. papyrifera	low	Low	high	high	low	high
E. decipiens	high	middle	low	low	low	low

4. Discussion

4.1. *Intracellular Water Use vs. Photosynthesis*

Plant electrophysiology has been successfully used to study intracellular leaf water [14,23]. Further exploring the photosynthetic response mechanism under adversity is required by jointly analyzing the photosynthetic and electrophysiological traits [24]. In this study, we aimed to investigate the intracellular water transport and use strategies of the three plant species in a karst environment, and to analyze the mechanism through which those strategies influence photosynthetic characteristics. Plants can adapt to karst environments by applying different photosynthetic patterns [38]. *C. nepalensis* is a deciduous plant that have higher root water uptake ability than *E. decipiens*, which is an evergreen plant [39]. The abundant roots of *C. nepalensis* plants also supported their strong water uptake. The high LIWHC in *C. nepalensis* was conducive to maintaining high levels of photosynthesis when the stomata were opened in the morning. Consequently, *C. nepalensis* plants exhibited high photosynthesis, high transpiration, and low WUE_i. Unlike *C. nepalensis*, *B. papyrifera* plants efficiently used the inter- and intracellular leaf water, exhibiting high photosynthesis, low transpiration, and high WUE_i. This result is consistent with that reported by Li and Wu [40]. Tree species with leathery leaves naturally have a lower water adsorption ability than those with papery leaves [39]. Although the water uptake ability of *E. decipiens* was lower than that of *B. papyrifera*, the leathery leaves helped them store water within leaves and maintain a high LIWTR. *E. decipiens* plants showed a consistently low inter- and intracellular leaf water-use efficiency, and exhibited low photosynthesis, low transpiration, and low WUE_i. Therefore, *B. papyrifera* exhibited better adaptability to the karst environment, but *E. decipiens* showed lower water use and photosynthetic efficiency compared with *C. nepalensis* and *B. papyrifera*.

4.2. *Dynamic Traits of Leaf Intracellular Water*

The dynamic traits of leaf intracellular water differed among plant species. From 8:00 to 10:00 a.m., the leaf water loss in *C. nepalensis* enhanced the transport of intracellular water, which was conducive to maintaining the LIWUE and photosynthesis (Figure 4). Therefore, the water was retained within the cells, which was also attributed to the developed leaf cuticle and strong water uptake ability of the root system, which has many horizontal and oblique roots [39]. *C. nepalensis* plants had a low Ψ_L value, favoring the maintenance of water movement from soil to plant [41]. The higher transpiration and photosynthesis of *B. papyrifera* improved inter- and intracellular leaf water use and consumption. However, the LIWTR of *B. papyrifera* remained stable, which was attributed to the strong water uptake ability of the shallow root system, which had a wide lateral distribution. Consequently, the PSII reaction center maintained stable. The LIWTR, LIWHC, and LIWUE of *E. decipiens* remained stable due to the unchanged transpiration and photosynthesis; the deep root system and leathery leaves also prevented the dissipation of water from the leaf surface [31]. However, the activity of the PSII reaction center of *E. decipiens* was influenced. In this period, *C. nepalensis* showed higher water transport within cells than *B. papyrifera* and *E. decipiens*, whereas *B. papyrifera* efficiently used the intracellular water.

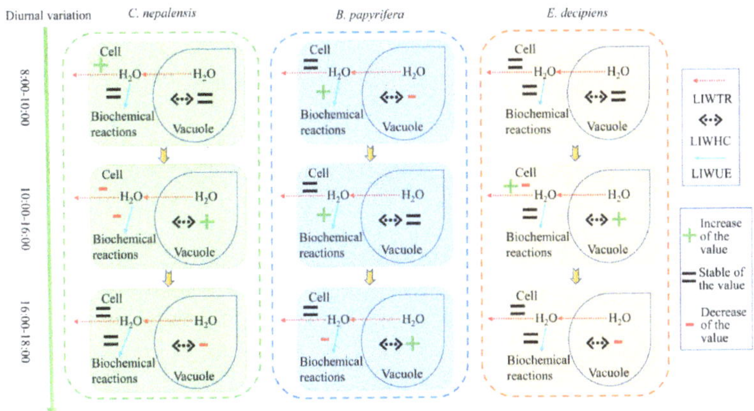

Figure 4. Dynamic traits of leaf intracellular water in *C. nepalensis*, *B. papyrifera*, and *E. decipiens*. LIWTR, leaf intracellular water transport rate; LIWHC, leaf intracellular water-holding capacity; LIWUE, leaf intracellular water-use efficiency.

From 10:00 a.m. to 4:00 p.m., water was mainly retained in cells; the transport and use of inter- and intracellular leaf water was limited, which depressed the photosynthesis of *C. nepalensis*. The strong water-holding capacity of cells helped to maintain high light energy use efficiency in *C. nepalensis*, which was indicated by the variations in qP and NPQ [42]. The strong water uptake ability of the roots and notably decreased water consumption of *C. nepalensis* increased the intracellular water storage at 4:00 p.m., which also helped *B. papyrifera* to maintain a stable inter- and intracellular water status. The increased use of intracellular water supported its high photosynthetic efficiency. The remarkably decreased F_v/F_m in *B. papyrifera* indicated that the photosynthetic structure was damaged, and the plants suffered from water deficit [22]. However, the conversion and use of light energy were increased to maintain the stability of photosynthesis [20]. We mainly attributed the fluctuation in the LIWTR of *E. decipiens* in this period to the slightly changed transpiration, whereas the slowly declined photosynthesis reduced the water consumption within the leaves and increased the LIWHC. The stability of the photosynthetic structure and the activity of PSII reaction center recovered to the level at 8:00 a.m. due to the conservation of intracellular water [43]. In this period, we mainly observed increased water-holding capacities for the leaves of *C. nepalensis* and *E. decipiens*, but *B. papyrifera* exhibited higher intracellular water-use efficiency and photosynthetic capacity than *C. nepalensis* and *E. decipiens*.

From 4:00 to 6:00 p.m., the closed stomatal aperture decreased the transpiration pull and kept the water within cells [15]. The reduction in the P_N of *C. nepalensis* mainly occurred owing to stomatal closure in the afternoon, which inhibited the transpiration and retained the water within the leaves [44]. As a result, the activity of PSII reaction center recovered. Although lower at 6:00 p.m., the P_N of *B. papyrifera* was still not lower than those of *C. nepalensis* and *E. decipiens* at 4:00 p.m. The water consumption and WUE_i of *B. papyrifera* remained high, whereas the decreased use of intracellular water increased the LIWHC of *B. papyrifera*. Consequently, the stability of the PSII reaction center in *B. papyrifera* also recovered. The considerably increased WUE_i and stable LIWUE slowed the decrease in the photosynthesis in *E. decipiens*, whereas the lower water uptake ability by roots compared with that of the other two plant species decreased the LIWHC. The water status in the three plant species all entered to the recovery phase after 6:00 p.m. Water transport within the leaf cells of *C. nepalensis*, *B. papyrifera*, and *E. decipiens* stabilized in this period.

Consequently, *B. papyrifera* will be more adaptable to the karst drought caused by future climate change than the other two plant species, because it can alternatively and efficiently use the inter- and intracellular leaf water with changing surroundings. *C. nepalensis* can efficiently conserve the intracellular leaf water and exhibits an efficient photosynthetic capacity, which helps this species to better adapt to the karst droughts than *E. decipiens*.

5. Conclusions

Plants adjust their photosynthesis by changing the water transport and use within leaf cells in karst environments. With strong water uptake ability by the roots and high leaf intracellular water-holding capacity, *C. nepalensis* plants maintained high photosynthesis, and the stomata apertures quickly closed to conserve intracellular water. *C. nepalensis* plants exhibited high photosynthesis, high transpiration, and low WUE_i. *B. papyrifera* maintained the stable transport of intracellular water, and their increased intracellular water-use efficiency improved their photosynthetic efficiency. The photosynthesis of *B. papyrifera* was also maintained by efficiently using the transpired water when the intracellular water-holding capacity increased. *B. papyrifera* showed high photosynthesis, low transpiration, and high WUE_i. The inter- and intracellular leaf water of *E. decipiens* remained stable due to its strong water-holding capacity, which could be attributed to the leathery leaves. The photosynthesis of *E. decipiens* remained stable, but the photosynthetic structure and PSII reaction center were notably influenced. *E. decipiens* exhibited low photosynthesis, low transpiration, and low WUE_i. *B. papyrifera* exhibited better adaptability to the karst drought than *C. nepalensis* and *E. decipiens*. Our results provide a reference for accurately analyzing plant photosynthetic adaptability in heterogeneous karst environments, and they can be used for improving the revegetation efficiency in these fragile habitats.

Supplementary Materials: The following supporting information can be downloaded at https://www.mdpi.com/article/10.3390/agronomy12112758/s1. Supplementary File 1: The calculation of leaf intracellular water transport rate, water-holding capacity and water-use efficiency based on electrophysiological parameters.

Author Contributions: Conceptualization, D.X. and Y.W.; methodology, Y.W., D.X., and X.Q.; data curation, X.Q. and W.W.; writing—original draft preparation, X.Q. and D.X.; writing—review and editing, M.L. and K.S.; funding acquisition Y.W. All authors read and agreed to the published version of the manuscript.

Funding: This research was funded by the Support Plan Projects of Science and Technology of Guizhou Province [No. (2021) YB453], the National Key Research and Development Program of China [No. 2021YFD1100300], and the Priority Academic Program Development (PAPD) of Jiangsu Higher Education Institutions.

Data Availability Statement: The datasets analyzed during the current study are available from the corresponding author upon reasonable request.

Conflicts of Interest: The authors declare no conflict of interest.

References

1. Zhu, X.R.; Liu, H.Y.; Wu, L.; Liang, B.Y.; Liu, F.; He, W.Q. Impact of Bedrock Geochemistry on Vegetation Productivity Depends on Climate Dryness in the Guizhou Karst of China. *Pro. Phys. Geog.* **2021**, *45*, 20–32. [CrossRef]
2. Peng, T.; Wang, S.J. Effects of Land Use, Land Cover and Rainfall Regimes on the Surface Runoff and Soil Loss on Karst Slopes in Southwest China. *Catena* **2012**, *90*, 53–62. [CrossRef]
3. Wang, Z.H.; Luo, D.; Xiong, K.N.; Gu, X.; Zhu, Z.Z. Studies on Hydrological Processes on Karst Slopes for Control of Soil and Water Loss. *Sustainability* **2022**, *14*, 5789. [CrossRef]
4. Nie, Y.P.; Chen, H.S.; Wang, K.L.; Tan, W.; Deng, P.Y.; Yang, J. Seasonal Water Use Patterns of Woody Species Growing on the Continuous Dolostone Outcrops and Nearby Thin Soils in Subtropical China. *Plant Soil* **2011**, *341*, 399–412. [CrossRef]
5. Liu, Z.Q.; She, R.; Xiong, K.N.; Li, Y.; Cai, L.L. Effect of Vegetation Restoration on Soil Hydrology in Karst Area of Southwest China: Inspiration from Barrel Planting Experiments. *Water* **2021**, *13*, 1719. [CrossRef]
6. Bonacci, O.; Pipan, T.; Culver, D.C. A Framework for Karst Ecohydrology. *Environ. Geol.* **2009**, *56*, 891–900. [CrossRef]
7. Yue, Y.M.; Wang, K.L.; Xiong, Y. Feasibility of Monitoring Karst Standing Conditions with Vegetation Spectra. *Spectrosc. Spectr. Anal.* **2012**, *32*, 1891–1894.
8. Chen, H.S.; Zhang, W.; Wang, K.L.; Fu, W. Soil Moisture Dynamics under Different Land Uses on Karst Hillslope in Northwest Guangxi, China. *Environ. Earth Sci.* **2010**, *61*, 1105–1111. [CrossRef]
9. Dawson, T.E. Fog in the California Redwood Forest: Ecosystem Inputs and Use by Plants. *Oecologia* **1998**, *117*, 476–485. [CrossRef]
10. Ehleringer, J.R.; Phillips, S.L.; Schuster, W.S.F.; Sandquist, D.R. Differential Utilization of Summer Rains by Desert Plants. *Oecologia* **1991**, *88*, 430–434. [CrossRef]
11. Panda, D.; Mishra, S.S.; Behera, P.K. Drought Tolerance in Rice: Focus on Recent Mechanisms and Approaches. *Rice Sci.* **2021**, *28*, 119–132. [CrossRef]
12. Lawson, T.; Blatt, M.R. Stomatal Size, Speed, and Responsiveness Impact on Photosynthesis and Water Use Efficiency. *Plant Physiol.* **2014**, *164*, 1556–1570. [CrossRef] [PubMed]
13. Henry, C.; John, G.P.; Pan, R.; Bartlett, M.K.; Fletcher, L.R.; Scoffoni, C.; Sack, L. A Stomatal Safety-Efficiency Trade-off Constrains Responses to Leaf Dehydration. *Nat. Commun.* **2019**, *10*, 3398. [CrossRef] [PubMed]
14. Yu, R.; Wu, Y.Y.; Xing, D.K. The Differential Response of Intracellular Water Metabolism Derived from Intrinsic Electrophysiological Information in *Morus Alba* L. and *Broussonetia Papyrifera* (L.) Vent. Subjected to Water Shortage. *Horticulturae* **2022**, *8*, 182. [CrossRef]
15. Ozeki, K.; Miyazawa, Y.; Sugiura, D. Rapid Stomatal Closure Contributes to Higher Water Use Efficiency in Major C4 Compared to C3 Poaceae Crops. *Plant Physiol.* **2022**, *189*, 188–203. [CrossRef] [PubMed]
16. Yu, W.Y.; Ji, R.P.; Feng, R.; Zhao, X.L.; Zhang, Y.S. Response of Water Stress on Photosynthetic Characteristics and Water Use Efficiency of Maize Leaves in Different Growth Stage. *Acta Ecol. Sin.* **2015**, *35*, 2902–2909.
17. Jin, T.T.; Fu, B.J.; Liu, G.H.; Hu, C.J.; Su, C.H.; Liu, Y. Diurnal Changes of Photosynthetic Characteristics of *Hippophae Rhamnoides* and the Relevant Environment Factors at Different Slope Locations. *Acta Ecol. Sin.* **2011**, *31*, 1783–1793.
18. Hájková, M.; Kummerová, M.; Zezulka, Š.; Babula, P.; Váczi, P. Diclofenac as an Environmental Threat: Impact on the Photosynthetic Processes of Lemna Minor Chloroplasts. *Chemosphere* **2019**, *224*, 892–899. [CrossRef]
19. Nedbal, L.; Soukupová, J.; Kaftan, D.; Whitmarsh, J.; Trtílek, M. Kinetic Imaging of Chlorophyll Fluorescence Using Modulated Light. *Photosynth. Res.* **2000**, *66*, 3–12. [CrossRef]
20. Kooten, O.V.; Snel, J.F.H. The Use of Chlorophyll Fluorescence Nomenclature in Plant Stress Physiology. *Photosynth. Res.* **1990**, *25*, 147–150. [CrossRef]
21. Genty, B.; Briantais, J.M.; Baker, N.R. The Relationship between the Quantum Yield of Photosynthetic Electron Transport and Quenching of Chlorophyll Fluorescence. *BBA—Gen. Subj.* **1989**, *990*, 87–92. [CrossRef]
22. Maxwell, K.; Johnson, G.N. Chlorophyll Fluorescence—A Practical Guide. *J. Exp. Bot.* **2000**, *51*, 659–668. [CrossRef] [PubMed]
23. Zhang, C.; Wu, Y.Y.; Su, Y.; Xing, D.K.; Dai, Y.; Wu, Y.S.; Fang, L. A Plant's Electrical Parameters Indicate Its Physiological State: A Study of Intracellular Water Metabolism. *Plants* **2020**, *9*, 1256. [CrossRef] [PubMed]
24. Xing, D.K.; Chen, L.; Wu, Y.Y.; Zwiazek, J.J. Leaf Physiological Impedance and Elasticity Modulus in *Orychophragmus Violaceus* Seedlings Subjected to Repeated Osmotic Stress. *Sci. Hortic.* **2021**, *276*, 109763. [CrossRef]
25. Buckley, D.J.; Lefebvre, M.; Meijer, E.G.M.; Brown, D.C.W. A Signal Generator for Electrofusion of Plant Protoplasts. *Comput. Electron. Agr.* **1990**, *5*, 179–185. [CrossRef]
26. Jócsák, I.; Végvári, G.; Vozáry, E. Electrical Impedance Measurement on Plants: A Review with Some Insights to Other Fields. *Theor. Exp. Plant Physiol.* **2019**, *31*, 359–375. [CrossRef]
27. Zhang, M.M.; Wu, Y.Y.; Xing, D.K.; Zhao, K.; Yu, R. Rapid Measurement of Drought Resistance in Plants Based on Electrophysiological Properties. *T. ASABE* **2015**, *58*, 1441–1446.
28. Mourya, N.R.; Bargali, K.; Bargali, S.S. Impacts of *Coriaria Nepalensis* Colonization on Vegetation Structure and Regeneration Dynamics in a Mixed Conifer Forest of Indian Central Himalaya. *J. For. Res.* **2019**, *30*, 305–317. [CrossRef]
29. Guo, L.; Qiang, T.; Ma, Y.; Wang, K.; Du, K. Optimisation of Tannin Extraction from *Coriaria Nepalensis* Bark as a Renewable Resource for Use in Tanning. *Ind. Crop. Prod.* **2020**, *149*, 112360. [CrossRef]

30. Guo, L.X.; Qiang, T.T.; Ma, Y.M.; Ren, L.F.; Dai, T.T. Purification and Characterization of Hydrolysable Tannins Extracted from *Coriaria Nepalensis* Bark Using Macroporous Resin and Their Application in Gallic Acid Production. *Ind. Crop. Prod.* **2021**, *162*, 113302. [CrossRef]
31. Lv, M.; Tan, M.H.; Lu, L.W.; Bao, S.S.; Guo, Z.Y.; Deng, Z.S.; Zou, K. Secondary Metabolites from Endophytes of *Elaeocarpus Decipiens* Hemsl. with Co-Cultivation Method. *J. China Three Gorges Univ. (Nat. Sci.)* **2018**, *40*, 108–112.
32. Liu, Z.B.; Xiao, D.M.; Zhang, G.L.; Li, X. Cutting Propagation Technology for *Elaeocarpus Decipiens* Hems1. *North. Hortic.* **2010**, *3*, 83–85.
33. Hang, H.T.; Wang, R.; Xing, D.K.; Wu, Y.Y.; Zhang, K.Y.; Rao, S.; Zhao, L.H. Photosynthetic Capacity and Adaptability of Three Herbaceous Energy Plants in Karst Habitats. *Jiangsu Agr. Sci.* **2018**, *46*, 248–254.
34. Laiming, H.; Wen, Z.; Ming'an, S. Response of Plant Physiological Parameters to Soil Water Availability During Prolonged Drought is Affected by Soil Texture. *J. Arid Land* **2021**, *13*, 688–698.
35. Yang, Z.; Tian, J.; Wang, Z.; Feng, K. Monitoring the Photosynthetic Performance of Grape Leaves Using a Hyperspectral-Based Machine Learning Model. *Eur. J. Agron.* **2022**, *140*, 126589. [CrossRef]
36. Xing, D.K.; Mao, R.L.; Li, Z.Y.; Wu, Y.Y.; Qin, X.J.; Fu, W.G. Leaf Intracellular Water Transport Rate Based on Physiological Impedance: A Possible Role of Leaf Internal Retained Water in Photosynthesis and Growth of Tomatoes. *Front. Plant Sci.* **2022**, *13*, 845628. [CrossRef]
37. Li, M.H. Response of electrophysiological properties of *Broussonetia papyrifera* (Linn L'Hér. ex Vent.and *Morus alba* L. to drought. Master's Thesis, Jiangsu University, Zhenjiang, China, 2018.
38. Chi, Y.K.; Xiong, K.N.; Zhang, J.H.; Wang, Y.S.; Zhang, Y. Study on the Photosynthetic Rate and Water Use Efficiency of Three Leguminous Grass Species in Karst Rocky Desertification Area. *Chin. J. Grassl.* **2014**, *36*, 116–120.
39. Yan, T.; Wang, Z.; Liao, C.; Xu, W.; Li, W. Effects of the Morphological Characteristics of Plants on Rainfall Interception and Kinetic Energy. *J. Hydrol.* **2021**, *592*, 125807. [CrossRef]
40. Li, H.; Wu, Y. The Responses of Leaf Photosynthesis and Photorespiration to the Simulated Drought by Two Moraceae Plants. *Earth Environ.* **2019**, *47*, 141–150.
41. Bargali, K.; Tewari, A. Growth and Water Relation Parameters in Drought-Stressed *Coriaria nepalensis* Seedlings. *J. Arid Environ.* **2004**, *58*, 505–512. [CrossRef]
42. Bussotti, F.; Gerosa, G.; Digrado, A.; Pollastrini, M. Selection of Chlorophyll Fluorescence Parameters as Indicators of Photosynthetic Efficiency in Large Scale Plant Ecological Studies. *Ecol. Indic.* **2020**, *108*, 105686. [CrossRef]
43. Huang, W.; Yang, Y.J.; Zhang, S.B. Specific Roles of Cyclic Electron Flow around Photosystem I in Photosynthetic Regulation in Immature and Mature Leaves. *J. Plant Physiol.* **2017**, *209*, 76–83. [CrossRef] [PubMed]
44. Mittelheuser, C.J.; Steveninck, R.F.M.V. Stomatal Closure and Inhibition of Transpiration Induced by (RS)-Abscisic Acid. *Nature* **1969**, *221*, 281–282. [CrossRef]

agronomy

MDPI

Article

Comparison of Aboveground Vegetation and Soil Seed Bank Composition among Three Typical Vegetation Types in the Karst Regions of Southwest China

Yili Guo [1], Yufei Li [1,2], Jianxing Li [1], Jiaqi Li [1,2], Shujun Wen [1], Fuzhao Huang [1], Wen He [1], Bin Wang [1], Shuhua Lu [1], Dongxing Li [1], Wusheng Xiang [1] and Xiankun Li [1,*]

1 Guangxi Key Laboratory of Plant Conservation and Restoration Ecology in Karst Terrain, Guangxi Institute of Botany, Guangxi Zhuang Autonomous Region and Chinese Academy of Sciences, Guilin 541006, China

2 College of Tourism and Landscape Architecture, Guilin University of Technology, Guilin 541006, China

* Correspondence: xiankunli@163.com

Abstract: Rural agricultural activity generates cropland, secondary vegetation and straggling primary forest and can modify the soil seed bank (SSB), potentially impacting the restoration of preferred species. The interaction between vegetation and seed banks during the recovery process is dependent on management practices and recovery pathways. This study was carried out in Guilin of southwest China to assess the variation in plant diversity and species composition of both aboveground and soil seed banks across three typical vegetation types with different human interventions: orchard, bamboo shrub and primary forest. The results show that there were significant differences in the species composition and diversity of aboveground vegetation and SSB, as well as in soil properties among three typical vegetation types. The primary forest had the highest aboveground species diversity, while the orchard had the highest species diversity and seed density of SSB. In addition, principal component analysis (PCA) and canonical correspondence analyses (CCAs) showed that the species composition and plant life forms of the three typical vegetation types were significantly influenced by soil properties. Based on these findings, the characteristics of aboveground vegetation and the soil seed bank and their correlations with soil properties are expected to drastically change with human intervention. These results imply that unsustainable land use has greatly impacted soil properties, and consequently, the aboveground vegetation and SSB. Nevertheless, vegetation will recover quickly after farming is abandoned. The successful restoration of fragmented ecosystems requires the addition of seeds and seedlings of target species, especially perennial woody plants from the relevant natural ecosystems, to accelerate succession from bamboo shrub to forest.

Keywords: aboveground vegetation; soil seed bank; soil property; human intervention; karst

Citation: Guo, Y.; Li, Y.; Li, J.; Li, J.; Wen, S.; Huang, F.; He, W.; Wang, B.; Lu, S.; Li, D.; et al. Comparison of Aboveground Vegetation and Soil Seed Bank Composition among Three Typical Vegetation Types in the Karst Regions of Southwest China. *Agronomy* **2022**, *12*, 1871. https://doi.org/10.3390/agronomy12081871

Academic Editor: Yanyou Wu

Received: 31 March 2022
Accepted: 30 July 2022
Published: 8 August 2022

1. Introduction

Southwest China is one of the largest continuous karsts in the world, and it is known for its unique landscapes and rich biodiversity. However, it has experienced rapid turnover within the past few decades due largely to intense human activity on the fragile karst ecosystems [1–7]. Farming leads to diverse land use patterns, with a range of specific types, such as farmlands, orchards, natural vegetation, and other, more complex landscapes. The 20th century intensification in grain and fruit production through agriculture resulted in severe vegetation cover loss and soil erosion. This unsustainable land use in the karst ecosystem could cause a catastrophic impact on ecosystem function and, therefore, the ecosystem may not recover on timescales relevant to land management decisions.

Since the end of the 1990s, the Chinese government at various levels has initiated ecological engineering projects to mitigate desertification and restore ecosystem services [1,8]. The karst environment has remarkably improved in the past 20 years [3]. Vegetation is one of the most important components in the karst ecosystem, and it has changed greatly over the years. Nowadays, orchards, secondary vegetation and a low percentage of primary forest are the three typical vegetation types in most karst areas. After 20 years of recovery, abandoned orchards were replaced by secondary vegetation, which was dominated by bamboo shrubs. However, it may take decades or centuries for a site to return to a state similar to primary forest, even if it is abandoned without any human disturbance. Ecologists have paid increasing attention to evaluating the improvement in the ecosystem services of karst degradation areas, mainly by remote sensing technology [3,9–12]. However, the changes in soil properties and vegetation variation in the process of vegetation restoration remain unclear.

As two major components of terrestrial ecosystems, vegetation and soil are strongly interlinked. Successional changes in vegetation are usually associated with the physical, chemical and biological alteration of soil properties. The effects of vegetation recovery on soil nutrients are still poorly understood and contradictory results have been reported in different studies [13]. This may be due to different management practices or recovery pathways. The usage of chemical fertilizers, pesticides, herbicides, and hormones in croplands has changed the physical, chemical and biological properties of soil [14]. The recovery of vegetation community structure may encourage the restoration of soil function after eliminating external interference [9]. Understanding how vegetation and soil properties are correlated and how their relationship develops in the process of vegetation restoration is key for effective forest restoration and management.

Soil seed banks (SSBs) play a key role in the post-disturbance recruitment of plant species. It provides a memory of past vegetation and an indication of the likely composition of future vegetation [15,16]. In addition, it has been suggested that SSBs could be one of the major sources that facilitate the recovery of vegetation after disturbance in degraded communities [17]. When assessing the restoration potential of degraded areas, viable SSBs may be useful for restoration after removing disturbances or stressors [18,19]. As a result, the evaluation of SSBs can therefore provide an index of the recovery potential of degraded plant communities. Knowing the relationships between SSB and aboveground vegetation may help to develop an effective conservation plan for restoring diversity, and for better understanding the resilience of plant communities after disturbances [16]. Vegetation dynamics, including the role of SSBs, in regeneration phases in karst areas are still poorly understood.

The influence of soil properties on the variation in the plant diversity and species composition of aboveground vegetation and the SSB of the three typical vegetation types are still unknown, despite their importance for understanding ecological patterns and processes, as well as for guiding management decisions for conservation or restoration measures. We therefore conducted a series of field surveys and specifically asked the following questions: (1) How do the soil properties differ among three typical vegetation types? (2) Do the characteristics of SSBs differ among three typical vegetation types? (3) How do the relationships between aboveground vegetation, SSB and soil properties change within different vegetation types? Finally, we discuss our results within the framework of future conservation management plans for karst ecosystems.

2. Materials and Methods

2.1. Study Area

The study was conducted in the Gongcheng Yao Autonomous County (hereinafter referred to as Gongcheng) (24°37′ N to 25°17′ N, 110°36′ E to 111°10′ E), Guangxi Zhuang Autonomous Region, in Southern China (Figure 1). Topographically, the area is a hilly area of karst terrain, with a valley, flat land, and hilly crisscrossing zone in the middle, among which mountains and hills account for more than 70% of the total land area. Typically, this area keeps a temperate climate throughout the year, belonging to the subtropical monsoon climate zone. The average annual temperatures and the average annual precipitation are 20.0 °C and 1517.5 mm, respectively, for the years 1971−2020. The annual accumulated temperature of measurements higher than 10 °C ranges from 5718.6 to 6382.6 °C.

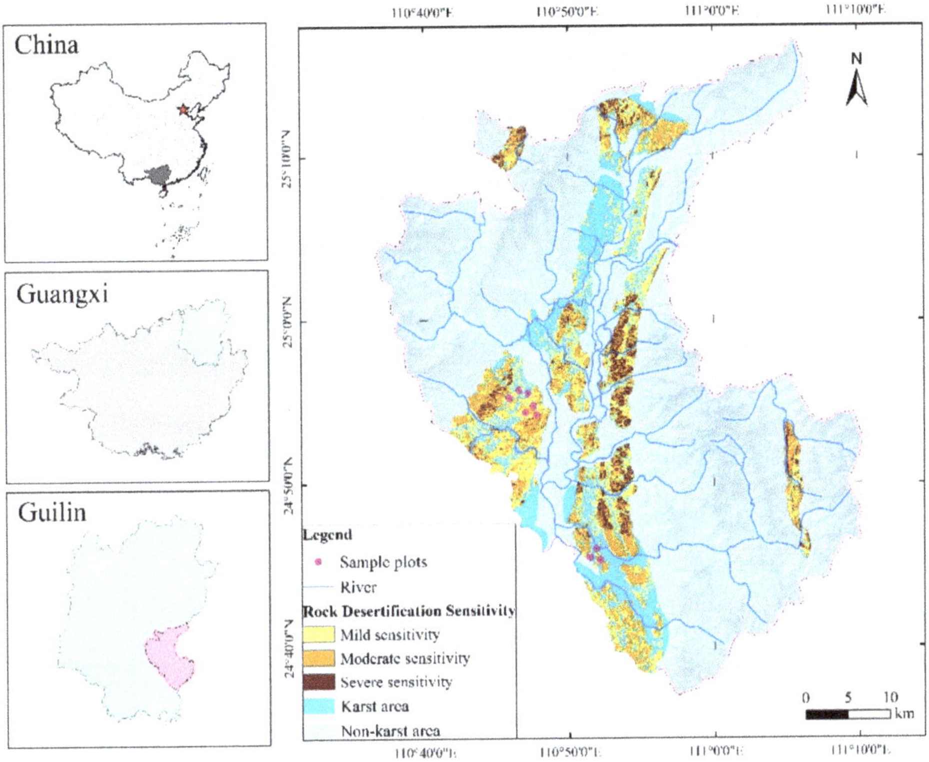

Figure 1. Sampling points distribution of plant communities.

The karst land area in Gongcheng is large. The rocky desertification area is mainly distributed across six townships (towns), Lianhua, Pingan, Gongcheng town, Xiling, Limu and Jiahui, with a total area of 25,300 hm^2, accounting for 12% of the total land area of the county. Preliminary investigation showed that the vegetation has been completely removed in parts of the area by the intensive planting of fruit trees (e.g., peach, plum and persimmon), which caused the emergence of bare soil patches. The fruit industry has become the leading industry fueling local economic development. However, the area of rocky desertification caused by orchard planting accounted for 42.5% of the total area of rocky desertification, according to the 2019 technical reports from local authorities.

Bamboo shrubs (e.g., *Phyllostachys sulphurea*) are distributed either individually or in clumps in most of the karst mountainside. It became the main type of secondary vegetation after orchards were abandoned in this karst area. Vegetation species have drought-resistant characteristics because of their extensive exposure to the basement rocks and thin soil. The bamboo shrub is generally dominated by vines; small, almost leather-like tree species; and single and mini-type leaf plants. Long-term observations show that abandoned orchards turn into bamboo shrubs after 10 years of natural restoration, but it is difficult for them to turn into a shrub or forest that is dominated by tree species. The primary forest ecosystem characteristic of the mid-subtropics is dotted around a housing colony, which is preserved in the form of a geomantic forest. The local people claim that almost all primary forest was stripped in the "Great Leap Forward" from 1958 to 1961, after which the area was reclaimed repeatedly for farming or restored naturally to bamboo shrub (Figure 2).

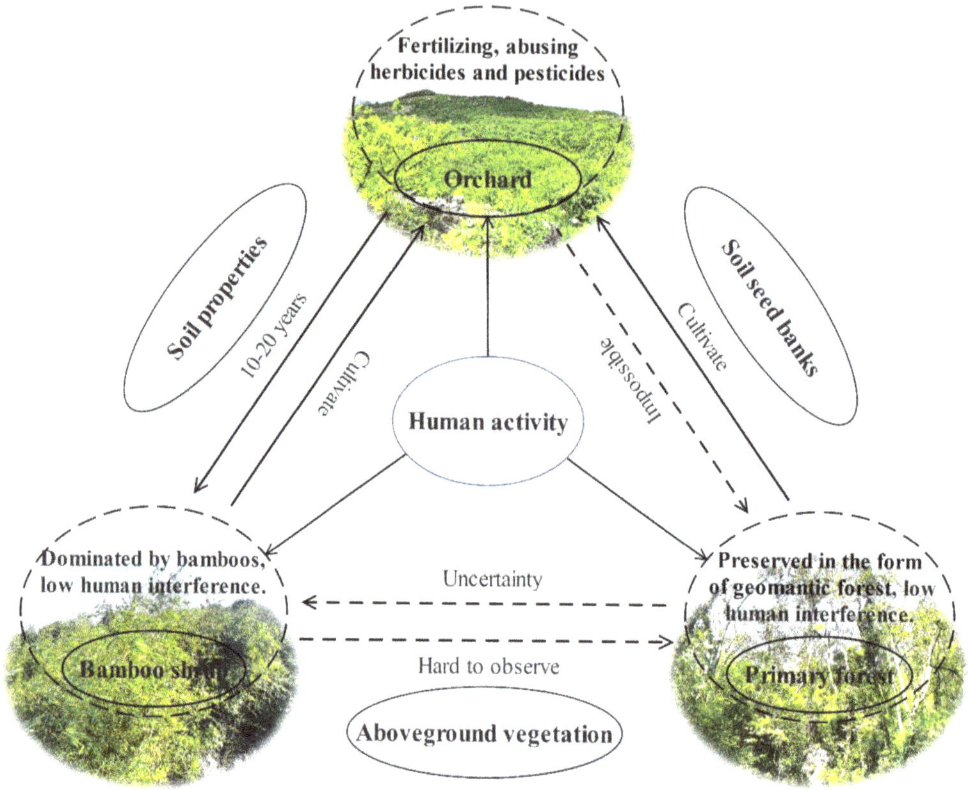

Figure 2. The characteristics of the three typical vegetation types and the interrelations between them.

2.2. Sampling of Vegetation and Site Conditions

We identified three typical vegetation types in our study area: orchard, bamboo shrub and primary forest. For each type, three 20 × 20 m plots were randomly selected (Table 1). At each plot, five 1 × 1 m quadrats were chosen in the corners and center. All woody stems with a diameter at breast height (DBH) of $\geq$ 1 cm in each plot were counted, measured and identified to the species level. The composition and abundance of vascular plant species in the herb layer (including seedlings of woody plants) were recorded in each quadrat. In addition, vegetation and litter height, percentage cover of bare rock, cover of bryophytes and litter, species composition and percentage cover of the woody overstorey were recorded. Plant specimens were identified according to Flora of China.

Parallel to the vegetation census, in August 2020, soil samples of 200 g each were taken at each quadrat at an average depth of 10 cm. Each soil sampling quadrat was marked with a PVC tube after sampling for later soil seed bank sampling. After air-drying and sieving to 2 mm to remove stones, roots and rhizomes, soil chemical properties (organic carbon (OCC), total nitrogen (TNC), total phosphorus (TPHC), total potassium (TPOC), available nitrogen (Alkeline-N), available phosphorus (Olsen-P), rapidly available potassium (Olsen-K), calcium (CAC) and magnesium (MAC)) were analyzed in October 2020, by the Shaanxi Biorace Biological Technologies Company using the procedures described by Bao [20].

Table 1. Sample plot survey.

Vegetation Types	Sample Number	Longitude	Latitude	Altitude (m)	Rock Exposed Rate	Slope (°)	Edificator
Orchard	O-1	110°47′16″ E	24°55′00″ N	208	75	5	*Prunus salicina*
	O-2	110°47′13″ E	24°55′05″ N	235	80	2	
	O-3	110°47′15″ E	24°55′06″ N	234	70	3	
Bamboo shrub	B-1	110°47′20″ E	24°55′16″ N	198	85	6	*Phyllostachys sulphurea*
	B-2	110°47′11″ E	24°55′07″ N	229	85	9	
	B-3	110°47′23″ E	24°55′17″ N	192	80	8	
Primary forest	S-1	110°51′50″ E	24°45′49″ N	173	65	10	*Cyclobalanopsis glauca*, *Pterospermum heterophyllum*
	S-2	110°51′49″ E	24°45′46″ N	155	75	17	
	S-3	110°52′07″ E	24°45′18″ N	150	70	5	

2.3. Seed Bank Sampling

The SSB experiment was conducted in November 2020 prior to the germination of seeds in the field. Soil samples were manually collected using a 10 × 10 cm frame at each quadrat. Each sample was divided into three soil depths (0–5, 5–10 and 10–15 cm) (Figure S1). A total of 135 soil samples (3 vegetation types × 3 plots × 5 quadrats × 3 soil depths) were collected. Soil samples were stored at 4 °C for approximately one week. Soil samples were then sieved with a 2 mm screen to remove stones, roots and rhizomes. After sieving, each soil sample was manually homogenized and spread on 10 cm of sterilized sand in a 50 cm length × 20 cm width × 10 cm depth plastic tray with drainage holes in early January 2021. Trays were randomly arranged on benches lined with plastic in the greenhouse. Samples were generally kept at field moisture conditions by surface watering every 2 days. Newly emerged seedlings were removed from pots after identification to prevent crowding. Seed germination assays continued for six months after germination had ceased. The mean density of the total number of emerging seedlings per sample was expressed on a per m^2 basis.

2.4. Statistical Analyses

This study employed the Shannon–Wiener diversity, Simpson diversity, species richness, Pielou evenness and Sørensen's quotient of similarity indices to demonstrate the different cover of the distinguished vegetation types and the characteristics of changes in the species composition of the seed bank:

(1) Diversity index (H), using the Shannon–Wiener index [21]:

$$H = -\sum_{i=1}^{S}(P_i \ln P_i)$$

(2) Dominance index (D), using the Simpson dominance index [22]:

$$D = 1 - \sum_{i=1}^{S} P_i^2 = 1 - \sum_{i=1}^{S} \left({}^{N_i}/_{n} \right)^2$$

(3) Richness index (R):

$$R = S$$

(4) Pielou index (E) [23]:

$$E = {}^{H}/_{\ln S}$$

(5) Similarity index, using Sørensen's quotient index (SQ) [24]:

$$SQ = \left[{}^{2C}/_{A+B}\right] \times 100\%$$

where N is the total number of plants, N_i is the number of individuals of species "i", P_i is the relative abundance of species "i", S is the total number of species, A is the number of species in the soil seed bank, B is the number of species in aboveground vegetation and C is the number of species common to the seed bank and aboveground vegetation.

One-way analysis of variance (ANOVA) was used to identify significant differences among three vegetation types for the different soil physiochemical properties. Principal component analysis (PCA) and canonical correlation analysis (CCA) were used to evaluate similarity in soil chemical properties, species composition of seed banks and aboveground community among three vegetation types. The species diversity, PCA and CCA were implemented using the *vegen* package [25] in R 4.1.2 [26].

3. Results

3.1. Soil Chemical Properties

There were significant differences in the soil chemical properties among different vegetation types (Figure 3). The TPHC, TPOC, Olsen-P and Olsen-K were higher in the orchard than bamboo shrub and primary forests. There were significantly higher contents in OCC, TNC and Alkeline-N in the bamboo shrub than the other two vegetation types, but CAC and MAC did not differ among the three vegetation types. According to the classification standards of the second national soil census, the TNC and Alkeline-N were at an extremely high level in the three vegetation types, while the TPHC and Olsen-P were at a depressed level in bamboo shrub and primary forest.

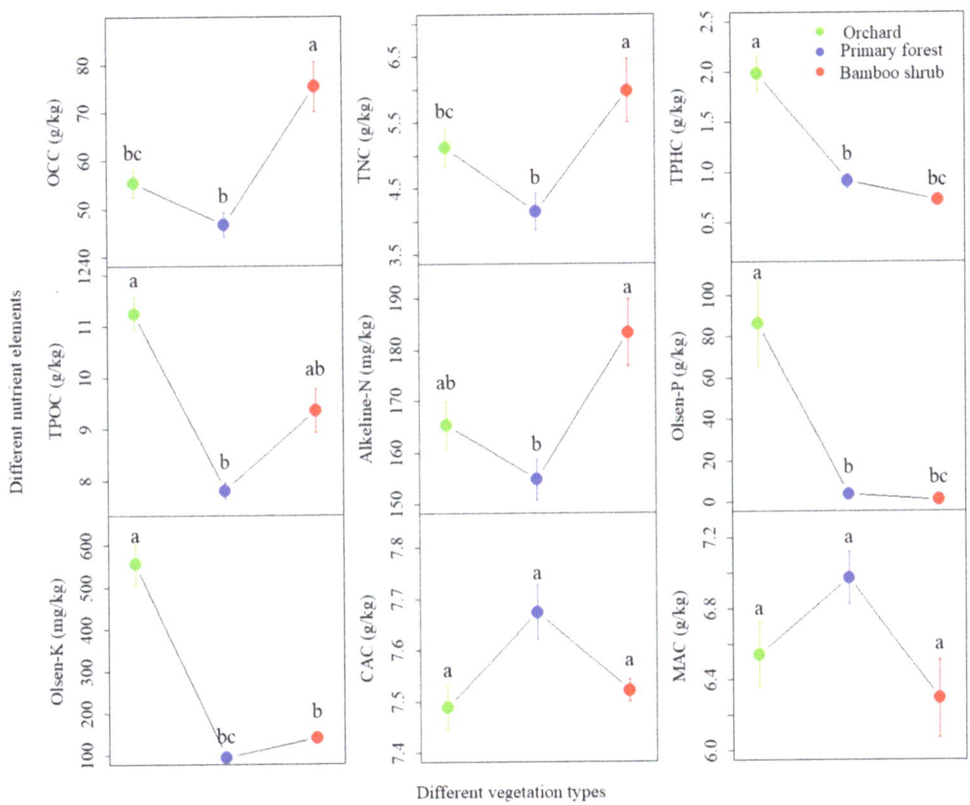

Figure 3. Difference analysis of soil physiochemical properties in each vegetation type (mean ± standard error, $n = 15$). Different capital letters indicate significant differences (ANOVA, LSD test, $p < 0.05$).

The first two principal components in PCA of the nine variables accounted for 68.63% of the total variation among sites (Table 2). Comp 1 (36.05% of variation) was positively correlated with eight of the nine variables, and negatively with MAC; Comp 2 (32.58% of variation) was highly correlated with eight of the nine variables but had no significant correlation with MAC. In addition, the PCA revealed a separation among different vegetation types, but with an obvious overlap between the bamboo shrub and primary forest (Figure 4a).

Table 2. Coefficients of the linear combinations of variables making up the principal components of nine soil properties.

Variable	Comp 1 (36.05%)	Comp 2 (32.58%)
Organic carbon (OCC)	0.369	0.387
Total nitrogen (TNC)	0.442	0.341
Total phosphorus (TPHC)	0.392	−0.379
Total potassium (TPOC)	0.14	−0.399
Available nitrogen (Alkeline-N)	0.417	0.361
Available phosphorus (Olsen-P)	0.353	−0.356
Rapidly available potassium (Olsen-K)	0.379	−0.388
Calcium (CAC)	0.191	0.145
Magnesium (MAC)	−0.132	

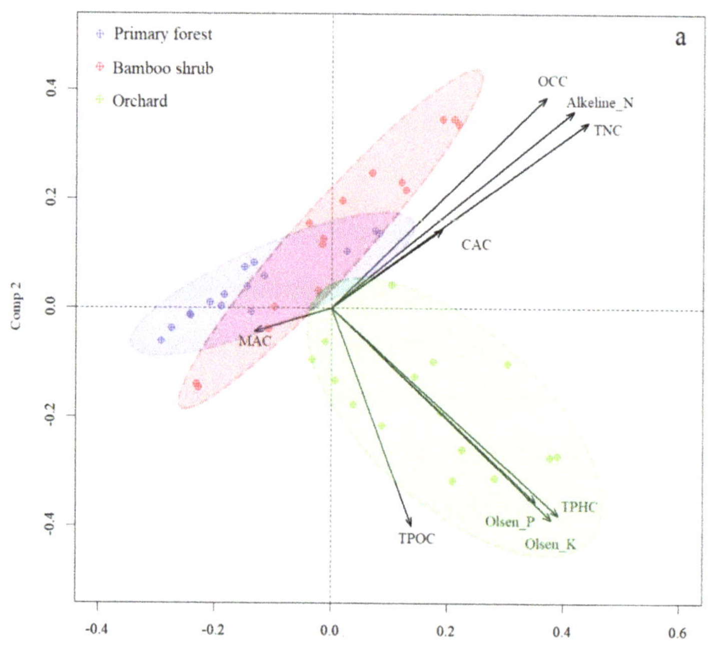

Figure 4. *Cont.*

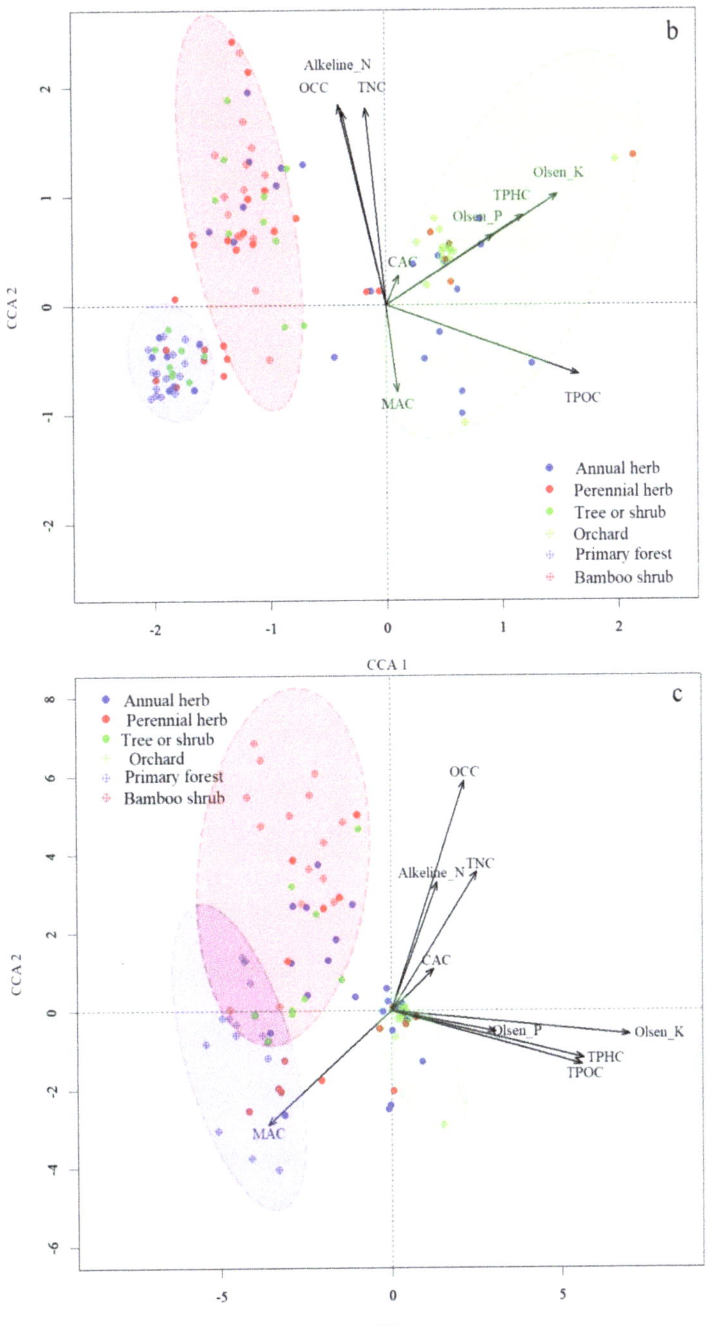

Figure 4. Principal component analysis (PCA, (**a**)) and canonical correlation analysis (CCA, (**b**,**c**)) of the correlation between aboveground vegetation, soil seed bank and soil chemical properties.

3.2. Species Composition of Aboveground Vegetation

A total of 80 species occurred in the 45 quadrats of the three vegetation types (Table S1). The species composition was quite different among the three vegetation types. It was dominated by annual herbs in the orchard. Perennial herbs and perennial woody vines (Scandent shrub) dominated the plant communities of the bamboo shrub, but perennial herbs and tree or shrub species dominated in the primary forest. The four indices of diversity, *H*, *D*, *R* and *E*, all indicated the highest species diversity in the primary forest and the lowest species diversity in the orchard (Figure 5). However, there was no significant difference between primary forest and bamboo shrub for the four diversity indices, while the orchard had a significantly lower diversity among the three vegetation types for *H*, *D* and *E* indices.

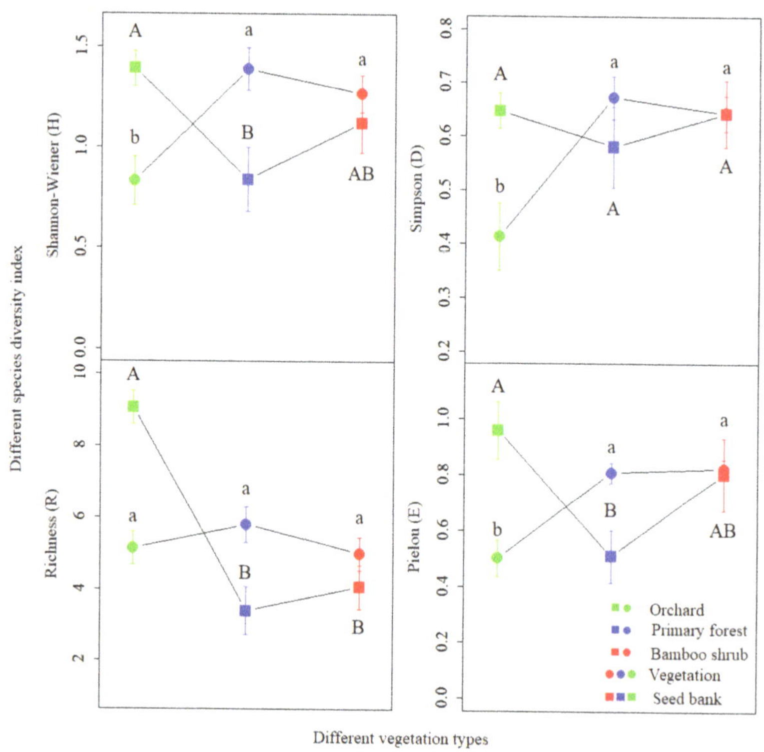

Figure 5. Species diversity index of vegetation and soil seed bank in each vegetation type (mean ± standard error, $n = 15$). Different capital letters indicate significant differences (ANOVA, LSD test, $p < 0.05$): lowercase letters for aboveground vegetation and uppercase letters for soil seed bank.

3.3. Species Composition of Soil Seed Bank

A total of 44 species occurred in the 135 soil samples of the three vegetation types (Table 3). The four indices of diversity, *H*, *D*, *R* and *E*, all indicated a higher species diversity in the orchard but a lower diversity in the primary forest. Although there was no significant difference between primary forest and bamboo shrub for the *H*, *R* and *E* indices of SSB, the SSB of the orchard had a significantly higher diversity among the three vegetation types, which is in contrast with aboveground vegetation (Figure 5). In addition, the mean Sørensen's quotient of similarity between the SSB and aboveground vegetation was less than 10% in the three typical vegetation types, and almost all the similarity indices of communities were zero in bamboo shrub and primary forest (Figure 6).

Table 3. Average soil seed bank density (m^2) of each plant species in different vegetation types.

	Latin Name	Family	Life Form	Vegetation Type		
				Orchard	Bamboo Shrub	Primary Forest
1	*Galium aparine*	Rubiaceae	Annual herb	10,733.34	-	-
2	*Stellaria media*	Caryophyllaceae	Annual herb	6099.99	9.99	-
3	*Oxalis corniculata*	Oxalidaceae	Annual herb	1253.34	20.01	-
4	*Clinopodium chinense*	Labiatae	Perennial herb	860.01	-	-
5	*Gnaphalium affine*	Compositae	Annual herb	826.68	-	-
6	*Youngia japonica*	Compositae	Annual herb	346.68	69.99	-
7	*Paspalum thunbergii*	Gramineae	Perennial herb	246.66	-	-
8	*Cardamine hirsuta*	Brassicaceae	Herbs annual	146.67	-	-
9	*Capsella bursa-pastoris*	Brassicaceae	Annual or biennial herb	133.32	-	-
10	*Talinum paniculatum*	Portulacaceae	Perennial herb	60	-	-
11	*Thyrocarpus sampsonii*	Boraginaceae	Annual herb	33.33	-	-
12	*Lygodium japonicum*	Lygodiaceae	Perennial herbaceous vines	26.67	6.6	-
13	*Agastache rugosa*	Labiatae	Perennial herb	13.32	-	-
14	*Centella asiatica*	Umbelliferae	Perennial herb	13.32	-	-
15	*Eclipta prostrata*	Compositae	Annual herb	13.32	-	-
16	*Dendranthema indicum*	Compositae	Perennial herb	13.32	110.01	-
17	*Solanum lyratum*	Solanaceae	Perennial herbaceous vines	6.66	39.99	-
18	*Parathelypteris glanduligera*	Thelypteridaceae	Perennial herb	6.66	-	-
19	*Cichorium endivia*	Compositae	Annual herb	6.66	-	-
20	*Clematis florida*	Ranunculaceae	Perennial herbaceous vines	6.66	-	-
21	*Selaginella uncinata*	Selaginellaceae	Perennial herb	-	20.01	-
22	*Conandron ramondioides*	Gesneriaceae	Perennial herb	-	200.01	-
23	*Botrychium ternatum*	Botrychiaceae	Perennial herb	279.99	-	530.01
24	*Woodsia ilvensis*	Woodsiaceae	Perennial herb	33.33	30	279.99
25	*Broussonetia papyifera*	Moraceae	Trees or shrubs	33.33	50.01	270
26	*Solanum nigrum*	Solanaceae	Annual herb	53.34	80.01	60
27	*Carex doniana*	Cyperaceae	Annual herb	6.66	129.99	50.01
28	*Oplismenus compositus*	Gramineae	Annual herb	-	-	50.01
29	*Stephania tetrandra*	Menispermaceae	Perennial herbaceous vines	-	39.99	39.99
30	*Arthraxon hispidus*	Gramineae	Annual herb	-	-	39.99
31	*Achyranthes bidentata*	Amaranthaceae	Perennial herb	-	-	39.99
32	*Setaria viridis*	Gramineae	Annual herb	20.01	170.01	30
33	*Erigeron acer*	Compositae	Biennial herb	13.32	39.99	20.01
34	*Commelina bengalensis*	Commelinaceae	Perennial herb	1113.33	9.99	9.99
35	*Pilea notate*	Urticaceae	Perennial herb	26.67	9.99	9.99

Table 3. *Cont.*

	Latin Name	Family	Life Form	Vegetation Type		
				Orchard	Bamboo Shrub	Primary Forest
36	*Artemisia carvifolia*	Compositae	Annual or biennial herb	13.32	9.99	9.99
37	*Fatoua villosa*	Moraceae	Annual herb	6.66	-	9.99
38	*Mallotus paniculatus*	Euphorbiaceae	Trees or shrubs	-	-	9.99
39	*Vernonia esculenta*	Compositae	Trees or shrubs	-	20.01	9.99
40	*Pistacia chinensis*	Anacardiaceae	Trees or shrubs	-	-	9.99
41	*Praxelis clematidea*	Compositae	Annual	-	20.01	9.99
42	*Celtis sinensis*	Ulmaceae	Trees or shrubs	-	-	9.99
43	*Cyclobalanopsis glauca*	Fagaceae	Trees or shrubs	-	-	9.99
44	*Cayratia japonica*	Vitaceae	Perennial herbaceous vines	-	-	9.99

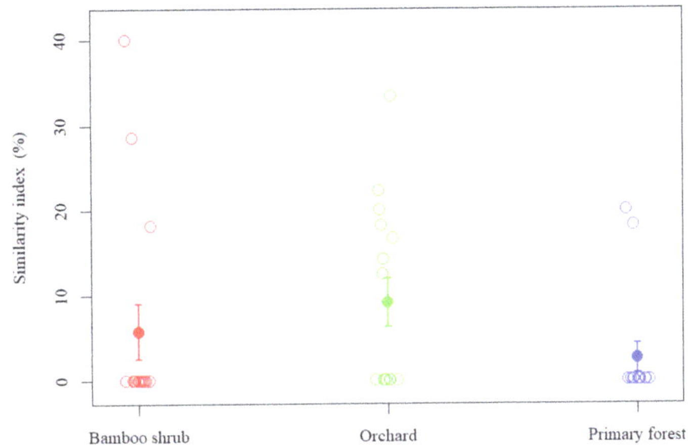

Figure 6. The Sørensen's quotient of similarity between the SSB and aboveground vegetation in each vegetation type (mean ± standard error, n = 15).

The soil seed bank density was significantly different among the three vegetation types (Table 3). The average SSB density was 22,446.67 ± 14,315.18 seed/m^2 in the orchard, 1086.6 ± 213.22 seed/m^2 in the bamboo shrub and 1519.89 ± 383.65 seed/m^2 in the primary forest. Besides the SSB density being higher in the orchard than in the bamboo shrub and primary forest, the standard deviation was also higher in the orchard than in the other vegetation types, indicating the high volatility of the SSB density in the orchard. The SSB density was gradually decreased with soil depth in the three typical vegetation types (Table 4).

Table 4. Average soil seed bank density (m^2; mean ± standard deviation) of each vegetation type in different soil depths.

Soil Depth	Orchard	Bamboo Shrub	Primary Forest
0–5 cm	12,906.67 ± 8995.6	406.67 ± 55.87	826.67 ± 197.66
5–10 cm	6033.33 ± 5357.29	353.33 ± 68.24	493.33 ± 106.77
10–15 cm	3526.67 ± 1398.33	326.6 ± 67.21	200 ± 43.38
Total	22,446.67 ± 14,315.18	1086.6 ± 213.22	1519.89 ± 383.65

The species composition and the dominant species of the SSB were significantly different among the three vegetation types (Table 3). In the orchard, a total of 31 plant species belonging to 20 families were observed in the SSB. The dominant species were *Galium aparine* (47.82%), *Stellaria media* (27.18%) and *Oxalis corniculata* (5.58%). In the bamboo shrub, only 19 species belonging to 12 families were observed in the soil bank. The dominant species were *Conandron ramondioides* (18.52%), *Setaria viridis* (15.74%) and *Carex doniana* (12.04%). In the primary forest, only 22 species belonging to 16 families were observed in the soil bank. The dominant species were *Botrychium ternatum* (34.87%), *Woodsia ilvensis* (18.42%) and *Broussonetia papyifera* (17.76%).

Annual herbs dominated the SSB in the orchard and included 18 species, accounting for more than 75% of life forms. Only one tree species, a native pioneer species, *B. papyifera*, was found in the orchard in two soil samples. Perennial plants dominated the species composition of both bamboo shrub and primary forest, and included 9 and 14 species, accounting for approximately 60% of life forms. However, the primary forest had a higher percentage of trees or shrubs than bamboo shrubs (Figure 7, Table 4).

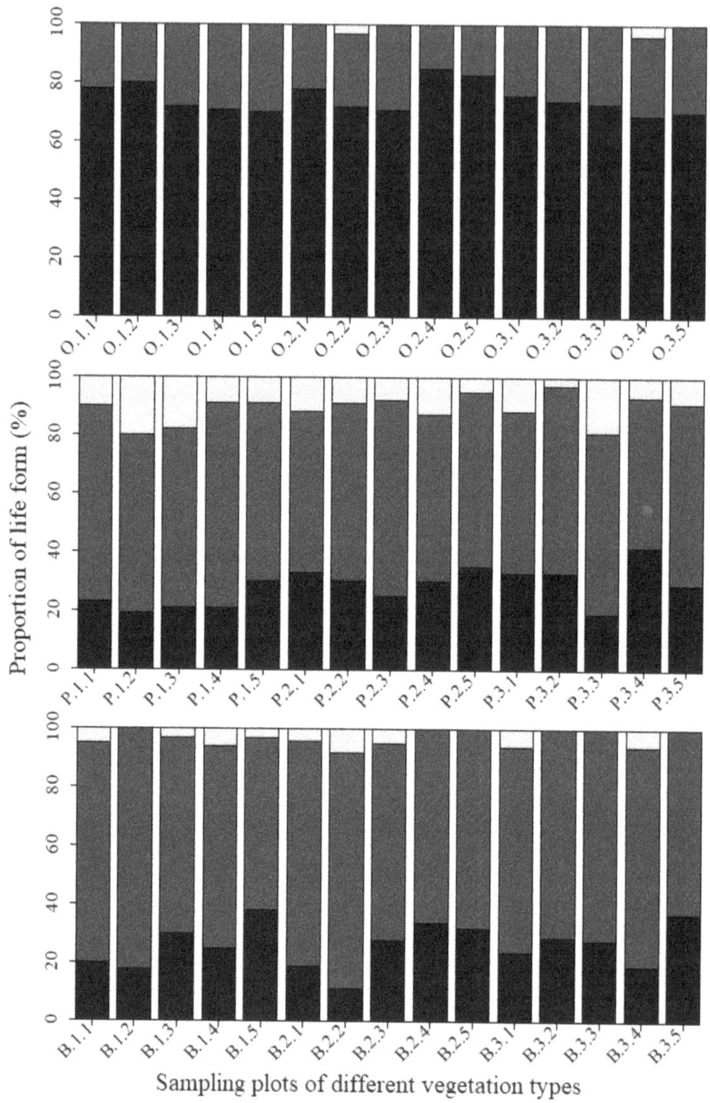

Figure 7. Soil seed bank life form spectrum of sampling points among different vegetation types. O, P and B represent the orchard, primary forest and bamboo shrub, respectively. The black column indicates annual herbs, the grey column indicates perennial plants (herbs and woody vines), and the white column indicates trees or shrubs.

3.4. Correlation between Vegetation, Soil Seed Banks and Soil Properties

The CCA identified the species composition groupings of the aboveground vegetation and SSBs among the three vegetation types (Figure 4b,c). The established vegetation and SSBs in the orchard were positively correlated with TPHC, TPOC, Olsen-P and Olsen-K, while the established vegetation and soil seed banks in the bamboo shrub had no significant correlations with any soil properties. Meanwhile, the CCA results indicate that there was a converse trend between the species composition of vegetation and SSB in the orchard and primary forest. The species composition of vegetation had a positive correlation with

TPHC, TPOC, Olsen-P and Olsen-K, but the soil seed bank had a negative correlation with these variables. The SSBs in the primary forest had a positive correlation with MAC (Figure 4c).

4. Discussion

Our results show that the amounts of basic elements and available ingredients in the soil, such as TPHC, TPOC, Olsen-P and Olsen-K, were significantly higher in the orchard than in the other vegetation types. After 20 years of recovery, OCC, TNC and Alkeline-N showed a significant increase. This was mainly due to the well-developed fibrous roots of bamboo, which hold the soil in place, increasing the stability of soil matrixes as succession proceeds. The accumulated litter covers the soil and increases soil organic matter content and permeability, reduces runoff and extends runoff time [27]. The interlocked roots of bamboo and the dry micro-climate played a limiting role in litter decomposition. The PCA revealed a separation among different communities and a greater separation between the orchard and the other two vegetation types, and the typical vegetation types were influenced by different soil properties. These results indicate that aspects of intensive planting such as fertilizing, abusing herbicides and pesticides dramatically changed the soil chemical properties. In addition, the CCA also revealed a separation among different vegetation types but a modicum of overlap between bamboo shrub and primary forest in the SSBs. Almost all soil samples were gathered together, which indicated that the species composition of SSBs had incredible similarity in the orchard. In contrast, in bamboo shrub and primary forest, all soil samples were spread out, which indicated that the species composition of SSBs had a striking contrast.

Soil physiochemical properties are the most important factors influencing the vegetation restoration in karst regions [28,29]. As nutrients and organic matter accumulate with time, they increase species complexity along the following successional stages [30]. Available P, rather than N, in calcareous soils is an effective indicator of nutrient limitation in karst regions [31]. The secondary and primary forests are P-limited, while the shrub is constrained by N and P or by other nutrients [5]. The low availability of P is expected to be due to the very low solubility of Ca-phosphate minerals at a neutral to alkaline pH in karst soils [9]. There are extremely high levels of TNC and Alkaline-N but depressed levels of TPHC and Olsen-P in the bamboo shrub and primary forest in our study. Thus, the bamboo shrub and primary forest within our region may be limited by P rather than N. This could be an intrinsic factor making it difficult for an area to succeed from bamboo shrub to primary forest.

It was found that the total seed density in the seed banks of the orchard was much greater than that in the bamboo shrub and primary forest. The total seed density in the orchard was much greater than that in plant communities in which there was no bare rock. For example, a density of approximately 3500 seeds m^{-2} in SSBs at a depth of 0–10 cm was observed in wetlands of the Yellow River Delta [32]. However, the number of seeds produced by malignant weeds that adapted to herbicides was great, which resulted in very high seed bank densities. For example, the density of *G. aparine* and *S. media* reached 10,733.34 and 6099.99 seeds m^{-2}, which accounted for approximately 80% of the total seed density in the orchard. A density of 1519.89 seeds m^{-2} was observed in SSBs in the primary forest of this study, which was much lower than that found in another tropical karst region of China [33]. The primary forest is located next to the village, and the understory herbaceous layer and litter layer might have therefore been disturbed by free-range poultry in this study. In addition, seed density declines with increasing soil depth, with the majority of seeds being distributed in the surface soil.

The biological characteristics of the species that present in the SSB influence its similarity to aboveground vegetation. There were significantly different proportions of plant life forms among the three typical vegetation types. Seeds of annual herbs constituted almost all SSBs in the orchard, while perennial plants constituted most of the SSBs in the bamboo shrub and primary forest in this study. These results are similar to those previously found regarding the level of life forms reported in secondary forests in karst areas of southwest China [34]. Of particular interest here is that the primary forest had a higher percentage of trees or shrubs, but the bamboo shrub had a higher proportion of perennial woody vines. In the intermediate succession stage (herein referred to as bamboo shrubs), there were few shrubs and trees, but perennial plants, especially woody vines, were still abundant, which caused some shade, although the habitats received abundant sunlight and had drought-resistant characteristics because of their extensive exposure to the basement rocks and thin soil. These conditions made them unfavorable for seed viability for some plant species. In addition, creeping rootstocks of bamboo were unfavorable for the maintenance of the seed viability of trees or shrubs, and climbing perennial woody vines are the main restricting factor affecting the vertical growth and development of trees.

The species diversity of the SSBs is much higher than that of the aboveground vegetation in the orchard, but the opposite patterns were found in the primary forest. The seed bank and aboveground vegetation assemblages differed substantially, with less than 10% similarity in species composition in the three vegetation types. In the current study, approximately 76.33% of the quadrats have no species common to the seed bank and aboveground vegetation. Some studies have demonstrated that SSBs consisting mainly of annual plants have a high degree of similarity in terms of species composition to the aboveground vegetation [32]. Seed banks that consist mainly of perennial plants often have a lower degree of similarity in terms of species composition to the aboveground vegetation [35]. Differences in the species composition of the seed bank and aboveground vegetation can be caused by many factors. A possible reason is that the aboveground vegetation utilizes a seasonal strategy in the orchard, but our vegetation census was conducted in August 2020, when some spring-appearing herbs had been removed by herbicides, while the seeds of some species delay germination and remain viable in the soil to await suitable conditions for germination in the bamboo shrub and primary forest. The perennial plant population usually produces a large number of fruits or seeds intermittently and synchronously in different years.

5. Implications for Ecological Restoration in Karst Rocky Desertification Regions

In the karst region of southwestern China, the indices of population density, the degree of gathering of low-income individuals and the size of the workforce are relatively lower. The growing population and low land productivity have triggered agricultural expansion to marginal cropland on slopes and ridges and have resulted in severe rocky desertification. If management practices are adjusted towards sustainable land use and the promotion of active restoration efforts, the karst in southwest China stands a chance to recover to its state of function in earlier centuries [4,12]. The karst environment has remarkably improved after more than 20 years of effort by the Chinese government at different levels.

With the development of the economy and social transition, many youths migrate to urban areas for better job opportunities, which helps to lessen regional land use pressure. However, at the same time, advanced tools have increased farmers' production efficiency, and so bamboo shrub faces a greater risk of being reclaimed again in this region. In addition, the overuse of fertilizers, herbicides and pesticides has stripped the soil of nutrients, making it too acidic, and has resulted in the invasion of exotic noxious weeds, as well as severe agricultural diffuse pollution. Therefore, stereoscopic agriculture and/or complex ecological agriculture are sorely needed to increase the per-unit labor and consequently raise the per-unit profit. Further study is needed to explore the stereoscopic planting modes of interplanting in orchards, such as Chinese medicinal herbs and cash crops. How to comprehensively apply and evaluate stereoscopic agriculture and/or complex ecological agriculture is the most important aspect to consider when discussing ecosystem services in karst landscapes in the future.

Ecological restoration projects have greatly decreased the aerial cover of karst rocky desertification in southwestern China [1]. However, land use/land cover changes are only the first step toward ecological restoration, and further close-to-nature restoration methods are necessary to promote ecological functions, e.g., biodiversity and carbon stock. Bamboo shrub is a typical vegetation type with low species diversity and carbon stock, and it is difficult to turn this type of vegetation into shrub or forest dominated by tree species by natural succession. Close-to-nature restoration methods could accelerate the succession of the bamboo shrub into forest. Close-to-nature restoration adopts traditional artificial restoration approaches and relies on natural ecological processes to achieve sustainable ecological restoration. It focuses on a return to nature and realizes sustainable restoration through the self-regulating function of the natural ecosystem. Therefore, ecosystems that are restored through close-to-nature restoration may maintain higher biodiversity and productivity and provide more ecosystem functions and services. The addition of seeds and seedlings of target species from natural ecosystems, especially perennial woody plants, is needed to accelerate succession from bamboo shrub to forest.

Supplementary Materials: The following supporting information can be downloaded at: https://www.mdpi.com/article/10.3390/agronomy12081871/s1, Figure S1: A sketch of the sampling method; Table S1: Abundance of each plant species in different vegetation types.

Author Contributions: Data curation: Y.G. and Y.L.; Investigation: J.L. (Jianxing Li), J.L. (Jiaqi Li), S.W., F.H., S.L., D.L. and W.X.; Methodology: Y.G. and S.W.; Project administration: Y.G. and X.L.; Resources: Y.G.; Software: Y.G., B.W. and W.H.; Supervision: Y.G. and X.L.; Visualization: Y.G. and W.H.; Writing—original draft: Y.G.; Writing—review & editing: Y.G. All authors have read and agreed to the published version of the manuscript.

Funding: The research was funded by "National key Research and development program, grant number 2019YFC0507503-05" and "National Natural Science Foundation of China, grant number 32071540, 31760141".

Data Availability Statement: Not applicable.

Conflicts of Interest: The authors declare no conflict of interest.

References

1. Wang, K.; Zhang, C.; Chen, H.; Yue, Y.; Zhang, W.; Zhang, M.; Qi, X.; Fu, Z. Karst landscapes of China: Patterns, ecosystem processes and services. *Landsc. Ecol.* **2019**, *34*, 2743–2763. [CrossRef]
2. Yue, Y.; Wang, K.; Zhang, B.; Jiao, Q.; Liu, B.; Zhang, M. Remote sensing of fractional cover of vegetation and exposed bedrock for karst rocky desertification assessment. *Procedia Environ. Sci.* **2012**, *13*, 847–853. [CrossRef]
3. Hu, Z.; Wang, S.; Bai, X.; Luo, G.; Li, Q.; Wu, L.; Yang, Y.; Tian, S.; Li, C.; Deng, Y. Changes in ecosystem service values in karst areas of China. *Agric. Ecosyst. Environ.* **2020**, *301*, 107026. [CrossRef]
4. Liao, C.; Yue, Y.; Wang, K.; Fensholt, R.; Tong, X.; Brandt, M. Ecological restoration enhances ecosystem health in the karst regions of southwest China. *Ecol. Indic.* **2018**, *90*, 416–425. [CrossRef]
5. Zhang, W.; Zhao, J.; Pan, F.; Li, D.; Chen, H.; Wang, K. Changes in nitrogen and phosphorus limitation during secondary succession in a karst region in southwest China. *Plant Soil* **2015**, *391*, 77–91. [CrossRef]
6. Allen, C.D.; Macalady, A.K.; Chenchouni, H.; Bachelet, D.; McDowell, N.; Vennetier, M.; Kitzberger, T.; Rigling, A.; Breshears, D.D.; Hogg, E.H.; et al. A global overview of drought and heat-induced tree mortality reveals emerging climate change risks for forests. *For. Ecol. Manag.* **2010**, *259*, 660–684. [CrossRef]
7. Winbourne, J.B.; Brewer, S.W.; Houlton, B.Z. Iron controls over di-nitrogen fixation in karst tropical forest. *Ecology* **2017**, *98*, 773–781. [CrossRef] [PubMed]
8. Ouyang, Z.; Zheng, H.; Xiao, Y.; Polasky, S.; Liu, J.; Xu, W.; Wang, Q.; Zhang, L.; Xiao, Y.; Rao, E.; et al. Improvements in ecosystem services from investments in natural capital. *Science* **2016**, *352*, 1455–1459. [CrossRef] [PubMed]
9. Green, S.M.; Dungait, J.A.; Tu, C.; Buss, H.L.; Sanderson, N.; Hawkes, S.J.; Xing, K.; Yue, F.; Hussey, V.L.; Peng, J.; et al. Soil functions and ecosystem services research in the Chinese karst Critical Zone. *Chem. Geol.* **2019**, *527*, 119107. [CrossRef]
10. Quine, T.; Guo, D.; Green, S.M.; Tu, C.; Hartley, I.; Zhang, X.; Dungait, J.; Wen, X.; Song, Z.; Liu, H.; et al. Ecosystem service delivery in Karst landscapes: Anthropogenic perturbation and recovery. *Acta Geochim.* **2017**, *36*, 416–420. [CrossRef]
11. Jiang, C.; Zhang, H.; Zhang, Z. Spatially explicit assessment of ecosystem services in China's Loess Plateau: Patterns, interactions, drivers, and implications. *Glob. Planet. Chang.* **2018**, *161*, 41–52. [CrossRef]
12. Tong, X.; Brandt, M.; Yue, Y.; Horion, S.; Wang, K.; De Keersmaecker, W.; Tian, F.; Schurgers, G.; Xiao, X.; Luo, Y.; et al. Increased vegetation growth and carbon stock in China karst via ecological engineering. *Nat. Sustain.* **2018**, *1*, 44–50. [CrossRef]
13. Huang, F.; Zhang, W.; Gan, X.; Huang, Y.; Guo, Y.; Wen, X. Changes in vegetation and soil properties during recovery of a subtropical forest in South China. *J. Mt. Sci.* **2018**, *15*, 46–58. [CrossRef]

14. Matson, P.A.; Parton, W.J.; Power, A.G.; Swift, M.J. Agricultural Intensification and Ecosystem Properties. *Science* **1997**, *277*, 504–509. [CrossRef]
15. Fisher, J.; Loneragan, W.A.; Dixon, K.; Veneklaas, E.J. Soil seed bank compositional change constrains biodiversity in an invaded species-rich woodland. *Biol. Conserv.* **2009**, *142*, 256–269. [CrossRef]
16. Hopfensperger, K.N. A review of similarity between seed bank and standing vegetation across ecosystems. *Oikos* **2007**, *116*, 1438–1448. [CrossRef]
17. Shang, Z.; Yang, S.; Wang, Y.; Shi, J.; Ding, L.; Long, R. Soil seed bank and its relation with above-ground vegetation along the degraded gradients of alpine meadow. *Ecol. Eng.* **2016**, *90*, 268–277. [CrossRef]
18. Sanou, L.; Savadogo, P.; Zida, D.; Thiombiano, A. Contrasting land use systems influence soil seed bank composition and density in a rural landscape mosaic in West Africa. *Flora* **2018**, *250*, 79–90. [CrossRef]
19. Niknam, P.; Erfanzadeh, R.; Ghelichnia, H.; Cerdà, A.; Erfanzadah, R.; Ghelichhnia, H. Spatial Variation of Soil Seed Bank under Cushion Plants in a Subalpine Degraded Grassland. *Land Degrad. Dev.* **2017**, *29*, 4–14. [CrossRef]
20. Bao, S. *Soil Agrochemical Analysis*, 3rd ed.; China Agriculture Press: Beijing, China, 2000.
21. Peet, R.K. The Measurement of Species Diversity. *Annu. Rev. Ecol. Syst.* **1974**, *5*, 285–307. [CrossRef]
22. Simpson, E.H. Measurement of diversity. *Nature* **1949**, *163*, 668. [CrossRef]
23. Pielou, E.C. *An Introduction to Mathematical Ecology*; John Wiley: New York, NY, USA, 1969.
24. Ikeda, A.; Matsuoka, S.; Masuya, H.; Mori, A.; Hirose, D.; Osono, T. Comparison of the diversity, composition, and host recurrence of xylariaceous endophytes in subtropical, cool temperate, and subboreal regions in Japan. *Popul. Ecol.* **2013**, *56*, 289–300. [CrossRef]
25. Jari, O.F.; Guillaume, B.; Michael, F.; Roeland, K.; Pierre, L.; Dan, M.; Peter, R.M.; O'Hara, R.B.; Gavin, L.S.; Peter, S.; et al. Vegan: Community Ecology Package. R package version 2.5-7. 2020. Available online: https://CRAN.R-project.org/package=vegan (accessed on 15 December 2021).
26. R Development Core Team. R: A Language and Environment for Statistical Computing. R Foundation for Statistical Computing, Vienna, Austria. 2020. Available online: https://www.R-project.org/ (accessed on 11 August 2021).
27. Wang, Y.; Chu, L.; Daryanto, S.; Lü, L.; Ala, M.; Wang, L. Sand dune stabilization changes the vegetation characteristics and soil seed bank and their correlations with environmental factors. *Sci. Total Environ.* **2018**, *648*, 500–507. [CrossRef] [PubMed]
28. Hofmeister, J.; Mihaljevič, M.; Hošek, J.; Sádlo, J. Eutrophication of deciduous forests in the Bohemian Karst (Czech Republic): The role of nitrogen and phosphorus. *For. Ecol. Manag.* **2002**, *169*, 213–230. [CrossRef]
29. Rivera, L.W.; Aide, T. Forest recovery in the karst region of Puerto Rico. *For. Ecol. Manag.* **1998**, *108*, 63–75. [CrossRef]
30. Tackett, N.W.; Craft, C.B. Ecosystem Development on a Coastal Barrier Island Dune Chronosequence. *J. Coast. Res.* **2010**, *264*, 736–742. [CrossRef]
31. Wen, L.; Li, D.; Yang, L.; Luo, P.; Chen, H.; Xiao, K.; Song, T.; Zhang, W.; He, X.; Chen, H.; et al. Rapid recuperation of soil nitrogen following agricultural abandonment in a karst area, southwest China. *Biogeochemistry* **2016**, *129*, 341–354. [CrossRef]
32. Guan, B.; Chen, M.; Elsey-Quirk, T.; Yang, S.; Shang, W.; Li, Y.; Tian, X.; Han, G. Soil seed bank and vegetation differences following channel diversion in the Yellow River Delta. *Sci. Total Environ.* **2019**, *693*, 133600. [CrossRef]
33. Shen, Y.; Liu, W.; Cao, M.; Li, Y. Seasonal variation in density and species richness of soil seed-banks in karst forests and degraded vegetation in central Yunnan, SW China. *Seed Sci. Res.* **2007**, *17*, 99–107. [CrossRef]
34. He, X.; Yuan, L.; Wang, Z.; Zhou, Z.; Wan, L. A study of soil seed banks across one complete chronosequence of secondary succession in a karst landscape. *PeerJ* **2020**, *8*, e10226. [CrossRef]
35. Touzard, B.; Amiaud, B.; Langlois, E.; Lemauviel, S.; Clément, B. The relationships between soil seed bank, aboveground vegetation and disturbances in an eutrophic alluvial wetland of Western France. *Flora* **2002**, *197*, 175–185. [CrossRef]

agronomy

Article

Contrasting Adaptation Mechanisms of Golden *Camellia* Species to Different Soil Habitats Revealed by Nutrient Characteristics

Xianliang Zhu [1], Jianmin Tang [1], Huizhen Qin [1], Kundong Bai [2], Zongyou Chen [1], Rong Zou [1], Shengyuan Liu [3], Quanguang Yang [4], Xiao Wei [1] and Shengfeng Chai [1,*]

1 Guangxi Key Laboratory of Functional Phytochemicals Research and Utilization, Guangxi Institute of Botany, Guangxi Zhuang Autonomous Region and Chinese Academy of Sciences, Guilin 541006, China; xianliangzhu2021@126.com (X.Z.); tjm@gxib.cn (J.T.); qhz0122@sina.com (H.Q.); chenzongyou@gxib.cn (Z.C.); zourong@gxib.cn (R.Z.); weixiao@gxib.cn (X.W.)

2 College of Life Science, Guangxi Normal University, Guilin 541006, China; bkd008@126.com

3 Administration of Nonggang National Nature Reserve of Guangxi, Chongzuo 532400, China; nonggang621@163.net

4 Golden Camellia National Nature Reserve Management Center, Fangchenggang 538001, China; yangquanguang@sina.com

* Correspondence: sfchai@163.com

Abstract: Golden *Camellia* species are highly specific to certain soil environments. Most species are only native to calcareous soils in karst regions, except for a few that grow only in acidic soils. Our aim is to elucidate the adaptation mechanisms of the species of calcareous-soil golden *Camellia* (CSC) and acidic-soil golden *Camellia* (ASC) to habitat soils through plant–soil nutrient characteristics and their relationships. We investigated 30 indices for soils and plants. Compared with ASC, CSC had more fertile soil, while their plant tissues exhibited stronger Ca, P, and Mn and weaker K storage, which may be important mechanisms for adapting to habitat soils. However, ASC showed a higher biological absorption coefficient (BAC) for nutrients, which may contribute to the adaptation of ASC to relatively barren acidic soils. Both CSC and ASC showed much higher BAC and accumulation of Ca than other nutrients. We also found that the concentrations of nutrients in the different tissues varied considerably between species. Correlation analysis revealed 135 significant relationships between the 30 indices, with the soil pH and soil Ca levels being the most important factors influencing the nutrient uptake network. This information helps in understanding the adaptation mechanisms of karst plants to habitat soils.

Keywords: golden *Camellia* species; karst plant; calcareous soils; acid soils; plant nutrition; adaptability

Citation: Zhu, X.; Tang, J.; Qin, H.; Bai, K.; Chen, Z.; Zou, R.; Liu, S.; Yang, Q.; Wei, X.; Chai, S. Contrasting Adaptation Mechanisms of Golden *Camellia* Species to Different Soil Habitats Revealed by Nutrient Characteristics. *Agronomy* **2022**, *12*, 1511. https://doi.org/10.3390/agronomy12071511

Academic Editor: Domenico Ronga

Received: 28 April 2022
Accepted: 21 June 2022
Published: 23 June 2022

1. Introduction

Camellia sect. *Chrysantha* Chang, also known as the "the queen of camellias" or "dreaming camellia", is an evergreen shrub or small tree of the Theaceae family that is famous for its golden camellia flowers [1]. These plants primarily grow in Guangxi Province, South China, and North Vietnam [2]. Currently, about 20 species of golden *Camellia* are distributed within China, most of which have a narrow distribution, and all of which are on the List of National Key Protected Wild Plants in China (http://www.gov.cn/zhengce/zhengceku/2021-09/09/content_5636409.htm, accessed date: 8 November 2021). In natural environments, golden *Camellia* species grow in highly specific soil environments, and most species are only native to calcareous soils, except for a few species that grow only in acidic soils [3]. Depending on their preferred type of soil habitat, they can be divided into calcareous-soil golden *Camellia* (CSC) and acid-soil golden *Camellia* (ASC). A recent cultivation trial showed that CSC also grows normally in low-Ca environments, while ASC is less well adapted to high-Ca environments [4]. The adaptation of these plants to environments with different Ca ion concentrations may be related to significant variations in photosynthetic and physiological indices, such as their contents of chlorophyll, proline, soluble sugars, or flavonoids [4,5]. However, our knowledge of the mechanisms by which golden *Camellia* species adapt to different soil habitats, especially in the field, is still poor.

Nutrients play key roles in the growth and development of plants, and their profiles can effectively reflect the adaptation mechanisms of plants to specific soil environments [6,7]. C, N, and P are basic elements required by all organisms and are the core elements used to evaluate the nutritional status of plants [8], while certain nutrients, such as K, Ca, Fe, Mg, and Mn, are sources of energy and regulators of many life activities, and thus are considered essential elements in most plants [9]. Nutrients in plants may interact with each other in unexpected ways to maintain the nutritional balance; for example, a deficiency in a single element can lead to the enhanced or reduced uptake of other nutrients by the plant [10]. These nutrients are also frequently used as important indicators for evaluating soil fertility [11,12].

A considerable number of previous studies have investigated plant–soil nutrient characteristics at different scales, i.e., at regional, ecosystem, and species levels [13–15], providing valuable information on the nutrient interactions between plants and soils. Several plants typical of karstic and non-karstic regions have been found to have significantly different leaf Ca contents and Ca storage forms [14]. Cui et al. [15] observed that leaves from a karst forest in Xishuangbanna, Yunnan, and Nonggang, Guangxi, were generally rich in Ca and Mg due to the influence of carbonate rocks. Qi et al. [16] investigated plants of the same genus from different geological backgrounds and revealed that the leaf Ca content of *Primula* growing in karst soils was significantly higher than that of *Primula* in Danxia soils, suggesting that soil type has an important influence on the enrichment of plant nutrients. In golden *Camellia* species, previous studies have documented the nutrient uptake characteristics and habitat soil physicochemical properties of only a few species [17,18]. Nevertheless, the nutrient uptake characteristics of most golden *Camellia* species have not been reported. Moreover, the lack of knowledge about the nutrient interrelationships between plants and soils limits the successful conservation of this group of plants.

As such, in this study, the nutrient profiles of habitat soils, roots, stems, and leaves were analyzed for 14 golden *Camellia* species from calcareous or acidic soils. We aim to explore: (1) What are the differences in nutrients between calcareous and acidic soil habitats? Are these differences in the soils reflected in the concentrations of elements in the plant tissues? (2) What are the differences in nutrient uptake by different golden *Camellia* species? We are particularly interested in the variation in the uptake and storage of Ca by plants. (3) Are there some significant interrelationships in the nutrient exchange between plants and soil? This information will help to reveal the adaptability of golden *Camellia* species to different soil habitats and provide a scientific basis for their conservation.

2. Materials and Methods

2.1. Plant Materials

The study areas were in Guangxi Zhuang Autonomous Region, China. The 10 species of CSC were *C. impressinervis* (CIM), *C. perpetua* (CPE), *C. longzhouensis* (CLO), *C. pingguoensis var. terminalis* (CPT), *C. flavida* (CFL), *C. huana* (CHU), *C. pubipetala* (CPU), *C. limonia* (CLI), *C. grandis* (CGR), and *C. pingguoensis* (CIM), and the four species of ASC were *C. tunghinensis* (CTU), *C. nitidissima* (CNI), *C. euphlebia* (CEU), and *C. parvipetala* (CPA). Dr. Shengfeng Chai undertook the formal identification of the plant material used in this study. Voucher specimens of this material were deposited in the herbarium of Guangxi Institute of Botany, (voucher number: SFC2017001-SFC2017014). The soil and plant samples were taken from pristine habitats, and the collection sites are shown in Figure 1. The vegetation cover at the collection sites on the northern edge of the tropics was mainly limestone evergreen forest and limestone montane seasonal rainforest, and those in the southern subtropics were evergreen broad-leaved forests with mountain gullies and streams 120–350 m above sea level, as well as limestone karst slope foothills, crested trough valleys, and depressional valley zones.

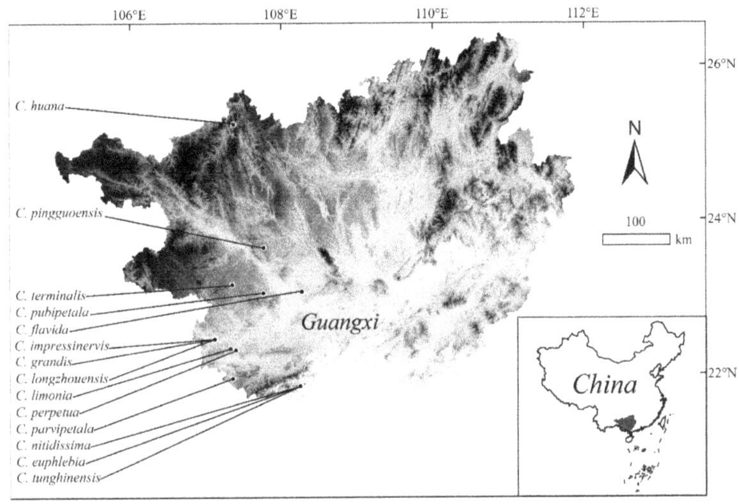

Figure 1. Collection sites of 14 species of golden *Camellia*. The map was downloaded from Geospatial Data Cloud (https://www.gscloud.cn, accessed date: 12 October 2021).

2.2. Sample Collection and Measurements

One representative population of each species of golden *Camellia* at the collection sites was selected, and three adult plants at a similar growth stage in that population were separately sampled as three replicates. First, about 1 kg of soil was collected from the surface layer (0–20 cm) near the roots of the plant. The soils were then taken to the laboratory, where they were naturally dried and impurities were removed, and they were then ground, sieved, and stored in hermetic bags. In addition, 0.5–1.0 cm of the lateral roots, stems of 1–3 annuals, and leaves of 1 annual were collected from the plants accordingly. These plant samples were taken back to the laboratory in separate clean envelopes, washed, dried, crushed, and then sealed in bags for storage.

A total of 42 soil samples and 126 plant samples were taken. For soil samples, the pH was determined by the glass electrode method; the OM content was determined by the high-temperature external heat potassium dichromate oxidation–volumetric method; the total N and P were determined by a graphite digestion–automatic chemical interruption analyzer; the total K was determined by a graphite digestion–flame photometer; and the total Ca, Mg, Fe, and Mn were determined by microwave digestion–flame atomic absorption spectrophotometry. For the plant samples, the contents of nutrients (N, P, K, Ca, Mg, Fe, and Mn) were determined separately for the root, stem, and leaf samples using the same method as described above. A total of 30 different indices were measured in the soil and plant samples. To effectively quantify the absorption capacity of the plants for each nutrient in the soil, the biological absorption coefficient (BAC) for each nutrient was calculated with reference to the equation of de la Fuente et al. [19]:

$$BAC_E = [(C_{PR} + C_{PS} + C_{PL})/3]/C_S \tag{1}$$

where C_{PR}, C_{PS}, and C_{PL} are the element's (E) mean concentrations in the root, stem, and leaf parts of the plants expressed in g kg^{-1} d.w., and C_S is the element's mean concentration in soil expressed in g kg^{-1}. Typically, BAC values in the range of 0.01–0.1, 0.1–1, and 1–10 indicate weak, moderate, and strong absorption, respectively.

2.3. Statistical Analysis

R v4.1.1 [20] was used for statistical analysis. First, all indices were tested for normality with the Shapiro–Wilk test at ≥ 0.9; when necessary, the indices were subjected to mathematic transformations (e.g., logarithm, sine, or cosine). Then, a *t*-test was performed to compare the differences between the indices of the soil, roots, stems, and leaves of CSC and ASC. Differences in the nutritional characteristics (N, P, K, Ca, Mg, Fe, and Mn in roots, stems, and leaves) between species were compared using ANOVA followed by Duncan's multiple comparison test ($p < 0.05$). The nutrient characteristics in the roots, stems, and leaves were clustered using the "ward.D" method [21]. Analyses of the correlations between 30 indices were tested by Pearson's correlation coefficient. Significant correlation coefficients were extracted using a threshold of $p < 0.05$, followed by the construction of relationship network plots using Cytoscape v3.8.2 [22]. Heat maps were drawn using TBtools v1.098669 [23].

3. Results

3.1. Differences in Habitat Soil between CSC and ASC

The soil pH preferred by CSC species ranged from 6.61 (CIM) to 7.53 (CPI), while the soil pH of ASC species ranged from 3.81 (CNI) to 5.86 (CPA) (Table 1). The mean soil organic matter (OM) content for calcareous soil (8.05%) was significantly higher ($p < 0.05$) than that for acidic soil (5.95%). The content of each nutrient was ranked as Fe > Mg > K > Ca > N > Mn > P for calcareous soils, and Fe > K > Mg > N > Ca > P > Mn for acidic soils. Except for the K content, all other nutrients were present at higher concentrations in calcareous soils than in acidic soils.

Table 1. The pH, organic matter (OM), and major nutrient contents of the habitat soils of golden *Camellia* species.

Species	Soil-pH	Soil-OM (%)	Soil-N ($g\ kg^{-1}$)	Soil-P ($g\ kg^{-1}$)	Soil-K ($g\ kg^{-1}$)	Soil-Ca ($g\ kg^{-1}$)	Soil-Mg ($g\ kg^{-1}$)	Soil-Fe ($g\ kg^{-1}$)	Soil-Mn ($g\ kg^{-1}$)
CIM (n = 3)	6.61 ± 0.48 b	6.30 ± 1.76 cd	4.03 ± 1.64 bc	1.88 ± 0.21 ab	9.87 ± 1.91 a	2.54 ± 0.94 de	6.35 ± 1.44 cde	105.83 ± 8.89 a	2.19 ± 0.15 c
CPE (n = 3)	7.00 ± 0.51 ab	5.58 ± 0.72 cd	1.56 ± 0.37 ef	0.54 ± 0.06 e	1.00 ± 0.30 h	3.31 ± 0.96 bcd	7.31 ± 0.13 cd	90.00 ± 1.99 b	1.20 ± 0.05 de
CLO (n = 3)	6.93 ± 0.25 ab	9.17 ± 0.89 b	2.96 ± 0.22 cde	2.03 ± 0.27 a	2.35 ± 0.10 gh	5.12 ± 1.01 ab	9.93 ± 0.46 b	86.39 ± 5.66 bc	2.70 ± 0.37 bc
CPT (n = 3)	6.90 ± 0.35 ab	12.31 ± 2.69 a	6.69 ± 1.64 a	0.74 ± 0.04 de	2.33 ± 0.29 gh	1.95 ± 0.76 e	7.95 ± 0.20 c	80.20 ± 3.29 bcd	1.32 ± 0.19 d
CFL (n = 3)	7.21 ± 0.21 ab	7.29 ± 0.83 c	2.21 ± 0.50 cdef	1.78 ± 0.20 bc	6.65 ± 0.97 bcd	3.53 ± 1.26 bc	5.75 ± 0.44 de	88.44 ± 1.79 bc	4.31 ± 1.16 a
CHU (n = 3)	7.19 ± 0.37 ab	6.95 ± 1.66 c	2.68 ± 0.53 cdef	1.06 ± 0.04 d	8.03 ± 1.09 b	4.50 ± 1.48 ab	3.83 ± 1.36 f	52.17 ± 10.46 f	3.50 ± 0.80 ab
CPU (n = 3)	7.28 ± 0.13 ab	9.87 ± 1.60 b	3.06 ± 0.50 cde	1.69 ± 0.33 bc	4.63 ± 0.56 def	5.96 ± 1.49 a	13.87 ± 1.12 a	88.26 ± 6.91 bc	2.91 ± 0.23 bc
CLI (n = 3)	6.93 ± 0.17 ab	6.46 ± 1.21 cd	3.85 ± 1.42 bcd	0.85 ± 0.07 de	4.5 ± 0.1 ef	3.49 ± 0.57 bc	3.68 ± 0.08 f	82.72 ± 2.76 bc	2.75 ± 0.46 bc
CGR (n = 3)	6.98 ± 0.59 ab	6.62 ± 0.67 c	5.05 ± 1.00 b	1.07 ± 0.33 d	3.87 ± 0.11 fg	5.88 ± 1.62 a	7.89 ± 0.92 c	70.09 ± 10.52 de	2.81 ± 0.74 bc
CPI (n = 3)	7.53 ± 0.28 a	9.94 ± 1.10 b	2.17 ± 1.16 def	1.50 ± 0.20 c	3.10 ± 0.89 fg	5.38 ± 1.36 a	9.75 ± 1.23 b	77.25 ± 2.16 cd	2.36 ± 0.60 c
CSC (n = 30)	**7.06 ± 0.25 ****	**8.05 ± 2.15 ****	**3.43 ± 1.54 ****	**1.31 ± 0.53 ****	**4.63 ± 2.78 ****	**4.16 ± 1.41 ****	**7.63 ± 3.06 ****	**82.13 ± 14.10 ****	**2.61 ± 0.93 ****
CTU (n = 3)	5.00 ± 0.37 d	8.01 ± 0.14 bc	2.11 ± 0.35 def	0.76 ± 0.02 de	7.1 ± 1.31 bc	0.60 ± 0.12 e	6.62 ± 1.12 cde	50.71 ± 2.30 f	0.50 ± 0.10 de
CNI (n = 3)	3.81 ± 0.23 e	5.59 ± 1.44 cd	3.24 ± 0.94 cde	0.64 ± 0.03 e	6.07 ± 0.61 cde	0.40 ± 0.16 e	3.85 ± 1.11 f	51.20 ± 4.18 f	0.33 ± 0.07 e
CEU (n = 3)	5.21 ± 0.57 d	6.20 ± 0.95 cd	2.18 ± 0.27 def	0.73 ± 0.06 de	9.83 ± 1.31 a	0.48 ± 0.10 e	6.34 ± 1.87 cde	61.19 ± 2.89 ef	0.63 ± 0.23 de
CPR (n = 3)	5.86 ± 0.15 c	3.97 ± 1.06 d	1.07 ± 0.18 f	0.56 ± 0.10 e	7.47 ± 1.98 bc	1.47 ± 0.33 e	5.15 ± 1.30 ef	29.99 ± 7.55 g	0.73 ± 0.41 de
ASC (n = 12)	**4.97 ± 0.86 ****	**5.94 ± 1.69 ****	**2.15 ± 0.89 ****	**0.67 ± 0.09 ****	**7.62 ± 1.59 ****	**0.74 ± 0.50 ****	**5.49 ± 1.26 ****	**48.27 ± 13.11 ****	**0.55 ± 0.17 ****

Data are the mean ± standard deviation. Different letters after the data in the same column indicate significant differences ($p < 0.05$) after Duncan's multiple comparison test. The bold text represents the average values of each index of CSC and ASC, and ** indicates extremely significant differences ($p < 0.01$) between CSC and ASC according to the *t*-test.

3.2. Differences in Major Nutrients Contents and Biological Absorption Coefficients between CSC and ASC

Eleven nutrient indices (mainly P, K, and Ca) of the plants differed significantly ($p < 0.05$) between CSC and ASC, eight of which were observed in the aerial parts (stems and leaves) (Table 2). P and Ca were significantly higher in the roots, stems, and leaves of CSC than in ASC, while the opposite was true for K. Fe and Mn differed significantly ($p < 0.05$) only in the stems. In terms of the nutrient uptake capacity, all species had the highest BAC_{Ca} and the lowest BAC_{Fe} (Figure 2). Among the species, CEU had the highest cumulative BAC values for the seven elements, while CPU had the lowest. CPE showed the highest BAC_N and BAC_K. As a whole, CSC showed lower BAC values of most elements compared with ASC.

Table 2. Nutrient differences in the roots, stems, and leaves of CSC and ASC.

Indices (g kg^{-1})	CSC (n = 30)	ASC (n = 12)
Root-N	4.46 ± 2.31 NS	5.72 ± 2.79 NS
Root-P	0.89 ± 0.52 **	0.53 ± 0.10 **
Root-K	2.45 ± 1.48 *	3.46 ± 1.593 *
Root-Ca	17.60 ± 9.13 *	11.16 ± 11.31 *
Root-Mg	2.30 ± 1.13 NS	2.62 ± 1.09 NS
Root-Fe	0.39 ± 0.30 NS	0.31 ± 0.11 NS
Root-Mn	0.75 ± 0.70 NS	0.71 ± 0.54 NS
Stem-N	4.39 ± 2.10 NS	4.01 ± 1.15 NS
Stem-P	0.93 ± 0.39 **	0.62 ± 0.08 **
Stem-K	2.76 ± 1.33 *	4.32 ± 1.94 *
Stem-Ca	22.98 ± 10.33 **	6.62 ± 2.03 **
Stem-Mg	3.06 ± 1.55 NS	2.46 ± 1.42 NS
Stem-Fe	0.30 ± 0.20 *	0.43 ± 0.24 *
Stem-Mn	0.98 ± 0.87 **	0.32 ± 0.08 **
Leaf-N	8.65 ± 5.63 NS	7.12 ± 3.85 NS
Leaf-P	0.84 ± 0.30 **	0.61 ± 0.14 **
Leaf-K	6.46 ± 2.63 **	10.62 ± 4.32 **
Leaf-Ca	20.59 ± 10.14 *	13.45 ± 7.22 *
Leaf-Mg	2.26 ± 0.84 NS	2.40 ± 1.41 NS
Leaf-Fe	0.28 ± 0.16 NS	0.29 ± 0.11 NS
Leaf-Mn	1.49 ± 1.42 NS	0.77 ± 0.37 NS

Data are the mean ± standard deviation. * and ** indicate significant differences between CSC and ASC at the p = 0.05 and p = 0.01 levels, respectively, and NS indicates no significant difference.

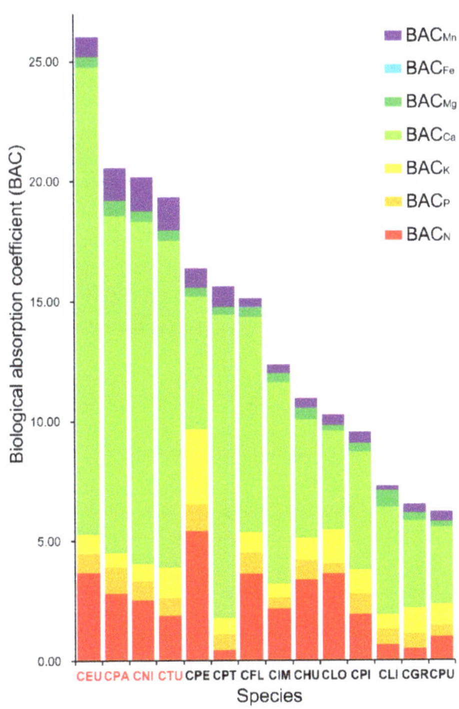

Figure 2. Biological absorption coefficient of each nutrient in golden *Camellia* species. The right subscripts of BAC in the legend represent the different elements. The black and red species labels represent CSC and ASC, respectively.

3.3. Nutrient Characteristics of Golden Camellia Species

The one-way analysis of variance (ANOVA) of the nutrient contents of the roots, stems, and leaves of 14 species of golden *Camellia* showed extremely significant ($p < 0.01$) differences in all 20 indices except for the leaf Fe content ($p < 0.05$) (Figure 3). In most species, the accumulation of N in the leaves was much higher than that in the stems and roots, with the leaf N content being the highest in CLO, at 18.01 g kg^{-1}. We found that two species (CFL and CPI) showed a strong preference for P, as both had considerably higher P contents in their roots, stems, and leaves than the other species. The K content formed a similar pattern to N, with a higher content in the leaves than the stems and roots in most species. However, the Ca content showed greater divergence between species, with most of the CSC plants exhibiting a decrease in Ca in the order of stems, leaves, and roots, while ASC exhibited a decrease in the order of leaves, roots, and stems. Mg mainly accumulated in the stems of most species, including CPT, CGR, CLO, and CLI, and in the leaves in a few species, such as CTU and CPE. All species displayed higher Fe contents in the aerial parts than the roots, except for CPU. The distribution of Mn in golden *Camellia* species showed remarkable variability; for example, Mn was mainly concentrated in the roots of CPI, CPT, and CPA; in the stems of CLO, CGR, and CIM; and in the leaves of CHU, CFL, CPU, CPE, and CTU. Cluster analysis showed that CTU, CTE, and CNI had similar elemental uptake characteristics in the roots, stems, and leaves, while CPA was close to other CSCs (especially CPT). In most species, the concentration of Ca was much higher than that of other nutrients, indicating that Ca is an important plant nutrient in golden *Camellia* species.

3.4. Plant–Soil Nutrient Relationships of Golden Camellia Species

A total of 135 significant ($p < 0.05$) relationships were detected between 30 soil and plant indices (Figure 4). Among them, the highest number of significant relationships (16) was found between soil pH and other indices, followed by soil Ca (15), indicating that the pH and Ca content of soil have the most important effects on nutrient absorption by golden *Camellia* plants. There were 13 and 11 significant relationships between the soil P and Mn and other indices, respectively, such as between the soil P and root P ($R = 0.54$), stem P ($R = 0.75$) and leaf P ($R = 0.60$), and soil Mn and leaf Mn ($R = 0.49$) and stem Mn ($R = 0.32$), further indicating the significant effects of the soil habitat on nutrient absorption. When the relationships among the plant indices were assessed, stem Mg had the highest number of significant relationships (13) with the other indices, while the leaf Fe had the lowest, i.e., one significantly negative relationship with leaf Mn ($R = -0.31$), indicating that stem Mg is susceptible to the influence of other nutrients, while the opposite is true for leaf Fe. There were strong positive relationships between the three essential nutrients N, P, and K in different tissues of golden *Camellia* species; for example, leaf N and stem N ($R = 0.86$), leaf N and root N ($R = 0.78$), and stem N and root N ($R = 0.68$), suggesting that the amount of absorption of these essential elements by other tissues of the plant is reflected in the leaves. In contrast, there were negative associations between the Mg contents of different tissues; for example, leaf Mg and stem Mg ($R = -0.32$). In addition, we found significant synergistic effects (for example, the correlation coefficients for the relationship between Ca and Mn were as high as 0.71, 0.73, and 0.70 in the roots, stems, and leaves, respectively) and antagonistic effects (for example, root N and root Ca ($R = -0.44$) and root Mn ($R = -0.44$), stem Fe and stem Mn ($R = -0.42$), and stem K and stem Mg ($R = -0.43$)) on the absorption of several elements.

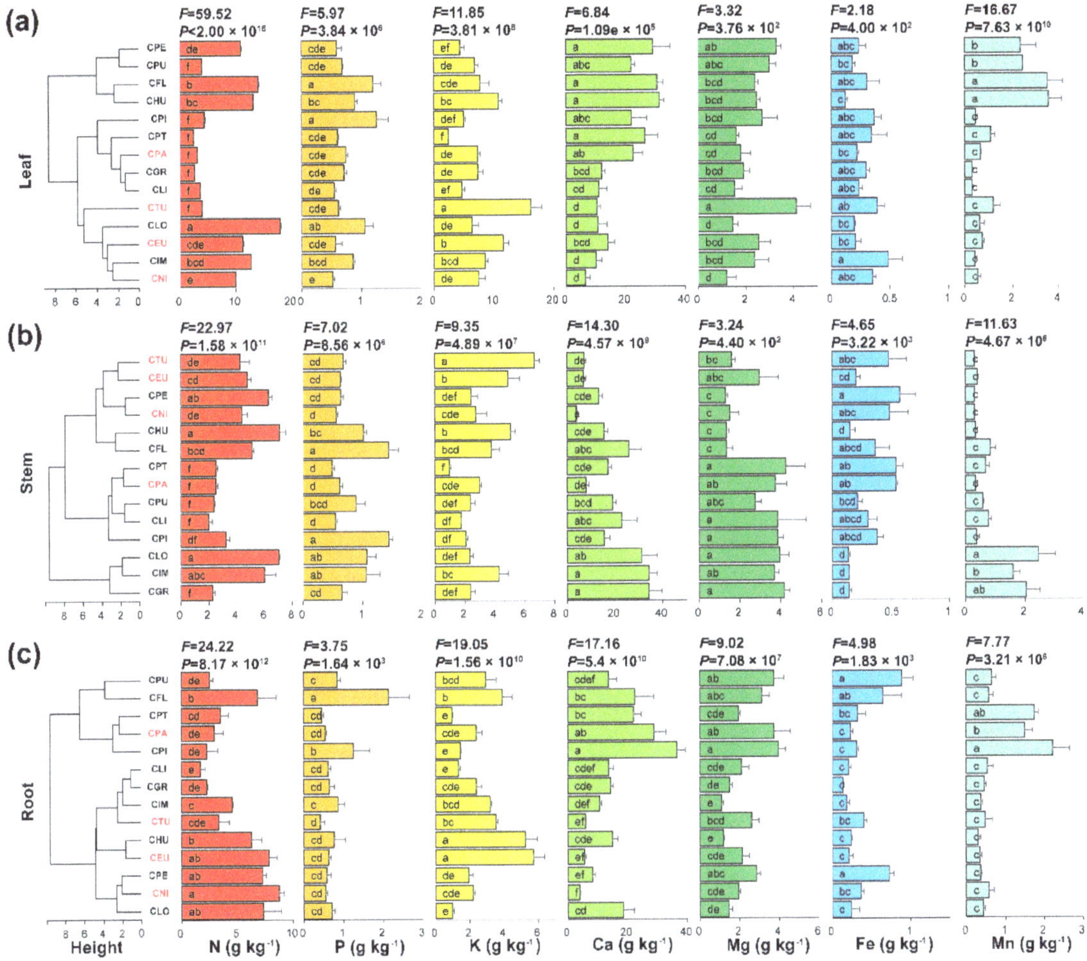

Figure 3. Comparison of the N, P, K, Ca, Mg, Fe, and Mn contents among leaves (**a**), stems (**b**), and roots (**c**) of 14 golden *Camellia* species. The black and red species labels at the ends of the clustering trees represent CSC and ASC, respectively. F and P are the statistics and significance of the ANOVA, respectively. The lines in the bar chart represent the standard deviation. Different letters indicate significant differences ($p < 0.05$) according to Duncan's multiple comparison test.

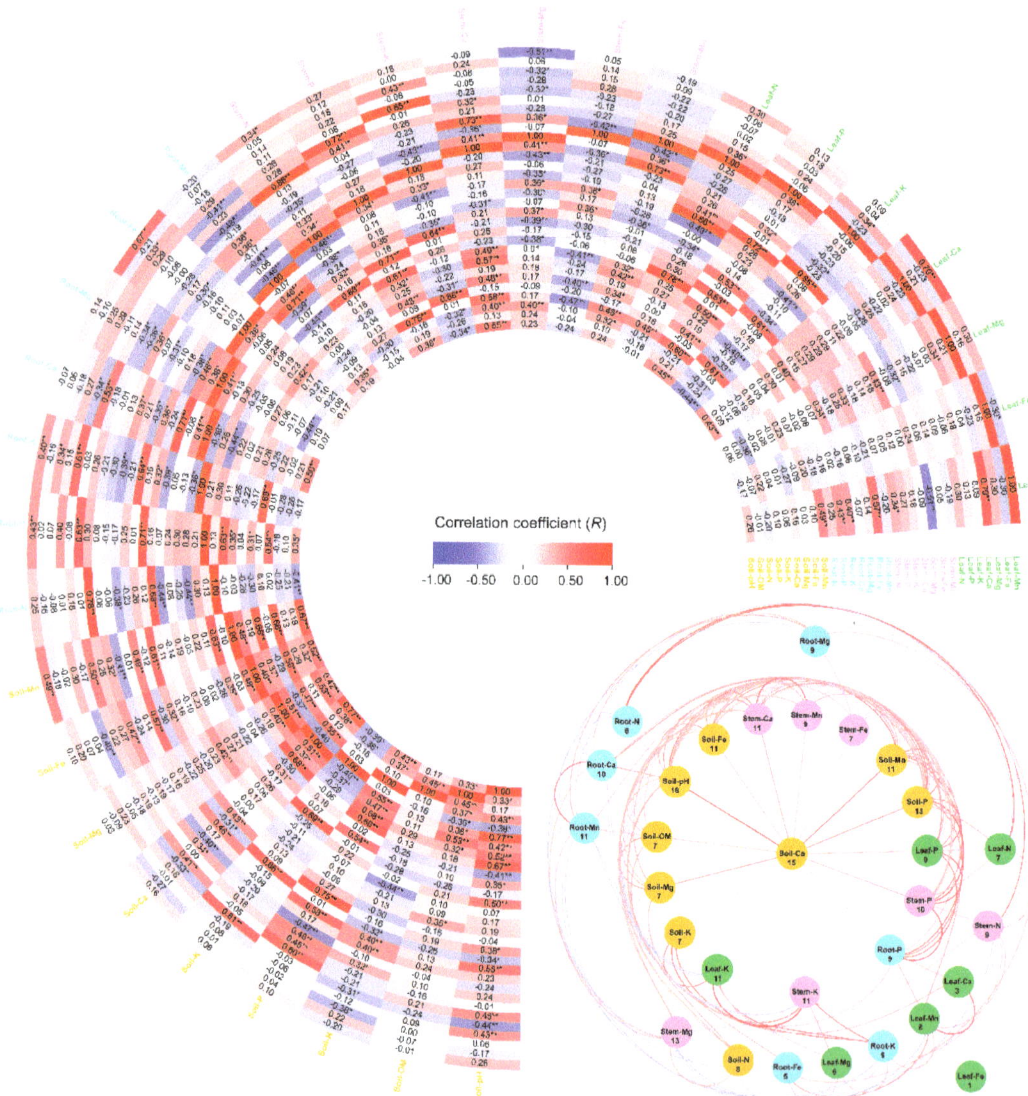

Figure 4. Plant–soil nutrient relationships of golden *Camellia* species. The heat map shows the correlation coefficients between 30 indices of soil and plants. * and ** indicate significant correlation coefficients at the $p = 0.05$ and $p = 0.01$ levels. Networks were constructed by extracting significant correlation coefficients from the heat map ($p < 0.05$). The circles indicate indices, their colors correspond to the indexes on the heat map, and the values underneath each index indicate the number of connected nodes. The color of the lines between the circles represents the correlation coefficient.

4. Discussion

Previous studies highlighted the higher pH and Ca content of calcareous soils in comparison with acidic soils [14,24]. Our results support the previous studies and, in addition, further confirmed that calcareous soils were enriched with higher contents of OM and most nutrients, such as N (3.45 g kg^{-1}), P (1.31 g kg^{-1}), and Mg (7.63 g kg^{-1}) (Table 1). This may be attributed to the abundance of Ca in the bedrock of karstic calcareous soils.

Elemental Ca is reported to have adsorption and precipitation effects on other nutrients, thus allowing for better nutrient fixation in the soil [25]. The soil bedrock of acidic soils is mostly sand shale with a low Ca content, which, together with the acidic soil environment, exacerbates Ca and Mg leaching [26] and promotes the decomposition of OM [27], which may lead to a further decrease in soil fertility. In addition, different topographies, soil textures, weathering environments, and other factors may also lead to differences in soil nutrients [28]. For example, both the N content and P content of CSC soils were higher than those previously observed in karst limestone soils in northwestern Guangxi [29]. Our results suggest that CSC colonized a relatively more fertile habitat than ASC.

The differences in the nutrient elements in soils may directly lead to different distributions of the corresponding nutrient elements in plants [30]. Thus, the ability of plants to store soil nutrients can serve as an adaptive response to soil nutrient variability [31]. The differences in Ca, P, Mn, and K enrichment between CSC and ASC may be a direct result of differences in the amount of nutrients in the habitat soils (Tables 1 and 2), suggesting that golden *Camellia* species are more sensitive to variations in these elements in the soil. Conversely, although the contents of several other nutrients, such as N, Fe, and Mg, also differed considerably between limestone and acidic soils, significant differences in enrichment were not observed in the nutrient tissues of most species (Tables 1 and 2), suggesting that the uptake of this class of nutrients by golden *Camellia* species is weakly regulated by the soil environment. In most species, the accumulation of N, P, and K elements by leaves is dominant, compared with stems and roots (Figure 3). This may be due to leaves being a vital organ for plants to carry out vigorous metabolic functions requiring many basic elements [32,33]. However, the accumulation of several nutrients, especially Ca and Mn, showed significant differences between tissues and species (Figure 3). In regions of high habitat heterogeneity, plants adapt to specific conditions by adjusting their nutrient and water uptake, biomass, spatial distribution characteristics, and morphological structure [34]. Different golden *Camellia* species may be better adapted to karst or non-karst areas with high habitat heterogeneity by regulating nutrient accumulation between different tissues. For instance, the difference in Mn accumulation between CSC and ASC was mainly observed in the stem, while the difference in K accumulation was mainly observed in the leaves (Table 2). Thus, a key insight is to focus on integrating the nutritional contents of different tissues when studying the plants' nutritional characteristics in these highly heterogeneous habitats [35].

The nutrient uptake of plants in karst or non-karst habitats is thought to be related to their calcicole or calcifuge behavior [24]. Therefore, we are particularly interested in the Ca absorption characteristics of golden *Camellia* species. The Ca uptake efficiency and accumulation of golden *Camellia* species were much higher than those of other nutrients (Figures 2 and 3), implying that soil Ca is one of the most important elements affecting growth and development. Depending on their Ca requirements, plants can be classified as calciphile, calcicole, calcifuge, sub-calcifuge, or neutral [36]. Calcicoles, in particular, are characterized by their ability to grow normally in high-Ca soils and are rarely found in acidic soils, while calcifuges grow well in acidic soils but are harmed by a slight increase in the soil Ca content. Calcicoles generally have a greater capacity for Ca uptake and storage than calcifuges [14–16,35]. In fact, CSC and ASC are calcicole and calcifuge plants, respectively. The study of calcicole or calcifuge behavior in plants remains a hot topic; however, it has rarely been discussed at the genus or species scales. Luo et al. [37] found that most typical calcicoles in karst forests were characterized by low P and K and high Ca and Mg, and most are P-limited plants ($N/P > 16$). However, CSC exhibited high P and similar Mg levels to ASC, except for the K and Ca uptake characteristics consistent with typical calcicoles. Furthermore, the N/P values in the roots, stems, and leaves of CSC were all lower than 14, indicating that they were N-limited, rather than P-limited [37]. These elemental uptake characteristics may reveal a unique mechanism for the adaptation of golden *Camellia* species to different soil habitats.

In terms of the uptake efficiency of different nutrients, golden *Camellia* species had a low BAC for Fe, the most abundant element in the soil, but a high BAC for other, less abundant nutrients (Figure 2). Similar nutrient absorption characteristics have been reported in previous studies of plants, such as several dominant species of *Burretiodendron hsienmu*, *Litsea dilleniifolia,* and *Cephalomappa sinensis* in the tropical and subtropical karst forest regions of Guangxi [38], as well as the endemic limestone species *Triadica rotundifolia* [39]. Interestingly, ASC has a higher BAC than CSC for most nutrients. This suggests that the more efficient nutrient uptake efficiency may allow ASC to adapt to more infertile acidic soil habitats.

The soil pH and soil Ca content have significant effects on plant nutrient uptake, growth, and development by influencing the physical, chemical, and biological properties of the soil [39,40]. We further confirmed that these two important soil indices have the most extensive effects on plants by constructing a network of relationships between the major nutrient contents of *Camellia* spp. and soil (Figure 4). The plant index with the highest correlation with both the soil pH and soil Ca was Stem-Ca, indicating that the Ca content in the stems of golden *Camellia* species would be most affected if there were changes in the soil pH or Ca. Therefore, it is necessary to focus on the soil pH and soil Ca content in the future cultivation or conservation of golden *Camellia* species. Additionally, we identified a few significant relationships between the nutrient elements. The relationships between some nutrients were mentioned in previous reports; for example, the synergistic effects of N and P [40]. However, the interactions between some nutrients in this study were inconsistent with previous reports. For instance, in *Brassica napus*, an elevated Mn content had a significant inhibitory effect on the absorption of both Ca and Fe [41]; however, in golden *Camellia* species, Mn showed a significant synergistic effect with Ca. On the one hand, this may be due to the unique biological properties distinguishing the species. On the other hand, synergistic or antagonistic interactions between nutrients in plant–soil systems may be closely related to their concentrations [42]. The absorption of K and Mg, as an example, was synergistic when the soil K content was low, and antagonistic when the soil K content was high [43]. It is worth noting that K and Mg showed antagonistic effects in this study, suggesting that the K content of the habitat soil was more than adequate for golden *Camellia* species. Nevertheless, the nutrient interrelationships between plants and soils need to be further verified in conjunction with physiological experiments.

5. Conclusions

By investigating the nutrient characteristics of the habitat soils and plant tissues of 14 species of golden *Camellia*, the following conclusions were drawn: (1) neutral to weakly alkaline calcareous soil habitats are more fertile than acidic soil habitats. (2) CSC exhibited stronger Ca, P, and Mn and weaker K storage than ASC, yet ASC had higher nutrient uptake efficiency, suggesting their contrasting mechanisms of adaptation to habitat soils. (3) A complex soil–plant nutrient exchange network was found, in which soil Ca and soil pH play important roles. Overall, this study provides a scientific basis for the conservation of the germplasm resources of this rare species by revealing the adaptation of plants of the golden *Camellia* species to different types of soils from the perspective of plant nutrition.

Author Contributions: Conceptualization, X.W. and S.C.; Data curation, X.Z. and J.T.; Formal analysis, X.Z. and S.C.; Funding acquisition, X.W. and S.C.; Investigation, J.T., H.Q., Z.C., R.Z., S.L., Q.Y. and S.C.; Project administration, R.Z.; Resources, S.L., Q.Y. and X.W.; Software, X.Z.; Visualization, X.Z.; Writing—original draft, X.Z.; Writing—review and editing, K.B. and S.C. All authors have read and agreed to the published version of the manuscript.

Funding: This research was supported by the National Natural Science Foundation of China (32060248, 31660092, 31860169), the key research and development program of Guangxi (GuikeAB2119601, the Central Guidance on Local Science and Technology Development Fund (ZY21195035), and Yunfu City 2021 Chinese medicine (southern medicine) industry talent project (Yunke [2022] No.16).

Institutional Review Board Statement: Not applicable.

Informed Consent Statement: Not applicable.

Data Availability Statement: Not applicable.

Acknowledgments: We thank the Golden Camellia National Nature Reserve Management Center, the Administration of Nonggang National Nature Reserve of Guangxi, and the Guangxi Longhushan Nature Reserve Management Office for their help in sample collection.

Conflicts of Interest: The collection of plants and soils of endangered golden *Camellia* species in this study was granted by the Golden Camellia National Nature Reserve Management center, the Administration of Nonggang National Nature Reserve of Guangxi, and the Guangxi Longhushan Nature Reserve Management Office, and the collection of these materials complied with the Regulations of the People's Republic of China on the Protection of Wild Plants and the IUCN Policy Statement on Research Involving Species at Risk of Extinction. The authors declare that they have no competing interests.

References

1. Zhou, X.; Li, J.; Zhu, Y.; Ni, S.; Chen, J.; Feng, X.; Zhang, Y.; Li, S.; Zhu, H.; Wen, Y. De Novo Assembly of the *Camellia nitidissima* Transcriptome Reveals Key Genes of Flower Pigment Biosynthesis. *Front. Plant Sci.* **2017**, *8*, 1545. [CrossRef] [PubMed]
2. Wei, S.-J.; Lu, Y.-B.; Ye, Q.-Q.; Tang, S.-Q. Population Genetic Structure and Phylogeography of *Camellia flavida* (Theaceae) Based on Chloroplast and Nuclear DNA Sequences. *Front. Plant Sci.* **2017**, *8*, 718. [CrossRef] [PubMed]
3. Su, Z.; Mo, X. Geographical Distribution of *Camellia* Section *chrysantha* from China. *Guihaia* **1988**, *8*, 77–83.
4. Chai, S.; Fu, R.; Zou, R.; Tang, J.; Shi, Y.; Wei, J. Effects of Different Calcium Ion Concentrations on Photosynthetic and Physiological Indexes of Calcicole-Type and Calcifuge-Type Golden *Camellia*. *Guihaia* **2021**, *41*, 167–176.
5. Liu, C.; Huang, Y.; Liang, Y. Adaptive Mechanism Exploration of *Camellia limonia* in Karst High Calcium Environment by Integrative Analysis of Metabolomics and Metagenomics. *Trop. Plant Biol.* **2022**, *15*, 22–39. [CrossRef]
6. Pan, F.; Qian, Q.; Liang, Y.; Wang, K.; Zhang, W. Spatial Variations in Fine Root Turnover, Biomass, and Necromass of Two Vegetation Types in a Karst Ecosystem, Southwestern China. *Forests* **2022**, *13*, 611. [CrossRef]
7. Wu, S.; Liang, K.; Fan, Y.; Su, T.; Hu, S.; Chen, Y.; Yu, S.; Shi, J. Nutrient Characteristics and Physiological Adaptation of *Phyllostachys glauca* at Different Habitats in Limestone Mountains. *For. Res.* **2022**, *35*, 180–186.
8. Pavlovic, J.; Kostic, L.; Bosnic, P.; Kirkby, E.A.; Nikolic, M. Interactions of Silicon with Essential and Beneficial Elements in Plants. *Front. Plant Sci.* **2021**, *12*, 697592. [CrossRef]
9. James, D.O.; Maaji, S.P.; Gimba, C.G.; Tsaku, J.E.; Kevin, E.U. Comparative Study of Essential Elements in Parts of *Moringa oleifera* Plant as Promedicine. *J. Appl. Sci. Environ. Manag.* **2019**, *23*, 1491. [CrossRef]
10. Fan, X.; Zhou, X.; Chen, H.; Tang, M.; Xie, X. Cross-Talks between Macro- and Micronutrient Uptake and Signaling in Plants. *Front. Plant Sci.* **2021**, *12*, 663477. [CrossRef]
11. Huang, F.; Wei, X.; Zhu, T.; Luo, Z.; Cao, J. Insights into Distribution of Soil Available Heavy Metals in Karst Area and Its Influencing Factors in Guilin, Southwest China. *Forests* **2021**, *12*, 609. [CrossRef]
12. Sousa, J.S.; Neves, J.C.L.; Martinez, H.E.P.; Alvarez, V.H.V. Relationship between Coffee Leaf Analysis and Soil Chemical Analysis. *Rev. Bras. Ciência Solo* **2018**, *42*, e0170109. [CrossRef]
13. Ji, H.; Wen, J.; Du, B.; Sun, N.; Berg, B.; Liu, C. Comparison of the Nutrient Resorption Stoichiometry of *Quercus variabilis* Blume Growing in Two Sites Contrasting in Soil Phosphorus Content. *Ann. For. Sci.* **2018**, *75*, 59. [CrossRef]
14. Cao, J.; Zhu, M.; Huang, F.; Lu, Q. Comparison Study on Calcium Forms in Plant Leaves under Different Geological Backgrounds—A Case Study in Maolan, Guizhou Province. *Bull. Mineral. Petrol. Geochem.* **2011**, *30*, 251–260.
15. Cui, P.; Shen, Z.; Fu, P.; Bai, K.; Jiang, Y.; Cao, K. Comparison of foliar element contents of plants from natural forests with different substrates in southern China. *Acta Ecol. Sin.* **2020**, *40*, 9148–9163.
16. Qi, Q.; Hao, Z.; Tao, J.; Kang, M. Diversity of Calcium Speciation in Leaves of *Primulina* Species (Gesneriaceae). *Biodivers. Sci.* **2013**, *21*, 715–722.
17. Qin, Y.; Shi, P.; Yu, Y.; Wang, Y.; Jiang, Y.; Nong, Y. Enrichment Characteristics of Metal Elements in Leaves from Four Species of Yellow Flower *Camellia*. *Guihaia* **2016**, *36*, 1416–1421.
18. Zhao, J.; Liu, T.; Zhang, D.; Wu, H.; Zhang, T.; Dong, D.; Liao, N. Bacterial Community Composition in the Rhizosphere Soil of Three *Camellia chrysantha* Cultivars Under Different Growing Conditions in China. *J. Soil Sci. Plant Nutr.* **2021**, *21*, 2689–2701. [CrossRef]
19. de la Fuente, V.; Rufo, L.; Rodríguez, N.; Amils, R.; Zuluaga, J. Metal Accumulation Screening of the Río Tinto Flora (Huelva, Spain). *Biol. Trace Element Res.* **2010**, *134*, 318–341. [CrossRef]
20. R Core Team. *R: A Language and Environment for Statistical Computing*; R Core Team: Vienna, Austria, 2015.
21. Dolnicar, S.; Leisch, F. Using Graphical Statistics to Better Understand Market Segmentation Solutions. *Int. J. Mark. Res.* **2014**, *56*, 207–230. [CrossRef]
22. Smoot, M.E.; Ono, K.; Ruscheinski, J.; Wang, P.-L.; Ideker, T. Cytoscape 2.8: New Features for Data Integration and Network Visualization. *Bioinformatics* **2011**, *27*, 431–432. [CrossRef] [PubMed]

23. Chen, C.; Chen, H.; Zhang, Y.; Thomas, H.R.; Frank, M.H.; He, Y.; Xia, R. TBtools: An Integrative Toolkit Developed for Interactive Analyses of Big Biological Data. *Mol. Plant* **2020**, *13*, 1194–1202. [CrossRef] [PubMed]
24. Liao, J.X.; Liang, D.Y.; Jiang, Q.W.; Mo, L.; Pu, G.Z.; Zhang, D. Growth Performance and Element Concentrations Reveal the Calcicole-Calcifuge Behavior of Three *Adiantum* Species. *BMC Plant Biol.* **2020**, *20*, 327. [CrossRef] [PubMed]
25. Zhang, W.; Zhao, J.; Pan, F.; Li, D.; Chen, H.; Wang, K. Changes in Nitrogen and Phosphorus Limitation during Secondary Succession in a Karst Region in Southwest China. *Plant Soil* **2015**, *391*, 77–91. [CrossRef]
26. Zhu, S.; Liu, C.; Tao, F.; Wang, Z.; Piao, H. Difference in Stable Carbon Isotope Composition and Profile Distribution of Soil Organic Matter between Brown Limestone Soil and Yellow Soil in Karst Areas of Guizhou Province. *Acta Pedol. Sin.* **2007**, *44*, 169–173.
27. Krull, E.S.; Skjemstad, J.O. Δ13C and Δ15N Profiles in 14C-Dated Oxisol and Vertisols as a Function of Soil Chemistry and Mineralogy. *Geoderma* **2003**, *112*, 1–29. [CrossRef]
28. Wang, M.; Chen, H.; Zhang, W.; Wang, K. Influencing Factors on Soil Nutrients at Different Scales in a Karst Area. *CATENA* **2019**, *175*, 411–420. [CrossRef]
29. Wang, M.; Chen, H.; Zhang, W.; Wang, K. Soil Nutrients and Stoichiometric Ratios as Affected by Land Use and Lithology at County Scale in a Karst Area, Southwest China. *Sci. Total Environ.* **2018**, *619–620*, 1299–1307. [CrossRef]
30. Zhu, P.; Li, Y.; Qiu, J.; Wang, M.; Zhang, C.; Ling, L.; Shi, Y.; Hu, J.; Zhang, R.; Li, Y.; et al. Correlation between Soil and Leaf Nutrient Element Contents in Citrus Orchards of Yuxi City, Yunnan Province. *J. Fruit Sci.* **2019**, *36*, 1658–1666.
31. Ferré, C.; Caccianiga, M.; Zanzottera, M.; Comolli, R. Soil–Plant Interactions in a Pasture of the Italian Alps. *J. Plant Interact.* **2020**, *15*, 39–49. [CrossRef]
32. van der Ent, A.; Mulligan, D.R.; Repin, R.; Erskine, P.D. Foliar Elemental Profiles in the Ultramafic Flora of Kinabalu Park (Sabah, Malaysia). *Ecol. Res.* **2018**, *33*, 659–674. [CrossRef]
33. Tang, S.; Liu, J.; Lambers, H.; Zhang, L.; Liu, Z.; Lin, Y.; Kuang, Y. Increase in Leaf Organic Acids to Enhance Adaptability of Dominant Plant Species in Karst Habitats. *Ecol. Evol.* **2021**, *11*, 10277–10289. [CrossRef] [PubMed]
34. He, J.; Yan, Y.; Yi, X.; Wang, Y.; Dai, Q. Soil heterogeneity and its interaction with plants in karst areas. *Chin. J. Appl. Ecol.* **2021**, *32*, 2249–2258.
35. Fu, Y.; Yu, L.; Huang, Z.; Peng, Q. Nutrient utilization characteristics of dominant plants of root underground habitat in karst areas. *Ecol. Environ. Sci.* **2020**, *29*, 2337–2345.
36. Hou, X. *Indicator Plants of Acid Soil, Calcareous Soil and Saline Alkali Soil in China*; Science Press: Beijing, China, 1954.
37. Luo, X.; Zhang, G.; Du, X.; Wang, S.; Yang, Y.; Huang, T. Characteristics of Element Contents and Ecological Stoichiometry in Leaves of Common Calcicole Species in Maolan Karst Forest. *Ecol. Environ. Sci.* **2014**, *23*, 1121–1129.
38. Deng, Y.; Jiang, Z.; Lan, F.; Luo, W.; Qin, X. Biogeochimical characteristics of elements in soil-plant system in tropical and subtropical typical karst area in Guangxi. *Ecol. Environ.* **2008**, *17*, 1140–1145.
39. Liu, X.; Qin, X.; Liang, T.; Lin, S. Characteristics of Chemical-element Contents in Leaves and Growing Soils of *Triadica rotundifolia*, A Limestone Endemic Species. *Southwest China J. Agric. Sci.* **2013**, *26*, 1195–1200.
40. Zhang, Y.; Zhou, C.; Lv, W.; Dai, L.; Tang, J.; Zhou, S.; Huang, L.; Li, A.; Zhang, J. Comparative Study of the Stoichiometric Characteristics of Karst and Non-Karst Forests in Guizhou, China. *J. For. Res.* **2019**, *30*, 799–806. [CrossRef]
41. Zeng, Q.; Geng, M.; Zhang, Z.; Zhou, W.; Zhu, D. Effects of Mn toxicity on the Content of Mn, Ca, Fe and the Activities of POD and CAT in *Brassica napus* L. at Seedling Stage. *J. Huazhong Agric. Univ.* **2004**, *23*, 300–303.
42. Narwal, R.P.; Kumar, V.; Singh, J.P. Potassium and Magnesium Relationship in Cowpea (*Vigna unguiculata* (L.) Walp.). *Plant Soil* **1985**, *86*, 129–134. [CrossRef]
43. Ding, Y.-C.; Chang, C.-R.; Luo, W.; Wu, Y.-S.; Ren, X.-L.; Wang, P.; Xu, G.-H. High Potassium Aggravates the Oxidative Stress Inducedy by Magnesium Deflciency in Rice Leaves. *Pedosphere* **2008**, *18*, 316–327. [CrossRef]

agronomy

MDPI

Article

Leaf Functional Traits and Relationships with Soil Properties of *Zanthoxylum planispinum* 'dintanensis' in Plantations of Different Ages

Yanping Song, Yanghua Yu * and Yitong Li

School of Karst Science/State Engineering Technology Institute for Karst Decertification Control, Guizhou Normal University, Guiyang 550001, China
* Correspondence: yuyanghua2003@163.com

Abstract: To explore the changes of leaf functional traits of *Zanthoxylum planispinum* 'dintanensis' with growth and development and its relationship with soil properties, which can clarify the response of the plantation to soil properties and suitable strategy. The research results can provide a scientific basis for plantations management. We explored the response of leaf functional traits to soil by using redundancy analysis in 5–7-, 10–12-, 20–22-, and 28–32-year *Z. planispinum* 'dintanensis' plantations. The results showed that: (1) The coefficients of variation of leaf traits ranged from 0.41% to 39.51%, with mostly medium and low variation, with the lowest variability in leaf water content (0.51–0.85%); The 5–7, 10–12, 20–22-year-old plantations were laid at the "slow investment-return" end of the economic spectrum while 28–32-year plantations were close to "fast investment-return" end. (2) The *Z. planispinum* 'dintanensis' tended to suit the environment via making trade-off and coordination of leaf functional traits. Leaf dry matter content decreased with an increase in leaf carbon/leaf nitrogen ratio, which is the trade-off between nitrogen usage efficiency and nutrient fixation capacity in *Z. planispinum* 'dintanensis'. (3) Redundancy analysis suggested that soil carbon/nitrogen ratio, soil total calcium, soil water content, soil available phosphorus, soil carbon/calcium ratio were highly correlated with leaf functional traits, while soil elemental stoichiometry had a greater reflection on leaf functional traits than their own content.

Keywords: karst; plantation age; stoichiometry; redundancy analysis; suitable strategy

Citation: Song, Y.; Yu, Y.; Li, Y. Leaf Functional Traits and Relationships with Soil Properties of *Zanthoxylum planispinum* 'dintanensis' in Plantations of Different Ages. *Agronomy* **2022**, *12*, 1891. https://doi.org/10.3390/agronomy12081891

Academic Editor: Yanyou Wu

Received: 29 June 2022
Accepted: 10 August 2022
Published: 12 August 2022

1. Introduction

Plant functional traits are morphological, physiological, and phenological characteristics that affect the growth, reproduction, and survival of plants [1,2]. They function as indicators of the mutual feedback between plants and their environments [3]. Plants are composed of organs, such as roots, stems, and leaves of which leaves are both the main site of photosynthesis and an important organ for maintaining hydrological security. Leaf functional traits can reflect the efficiency of resource use, such as light, temperature, and soil [4], and the combination of a series of trade-offs and synergistic relationships among leaf traits is called the leaf economic spectrum, which integrally characterizes the plant's suitability to habitats and resource use strategies [5]. Therefore, exploring leaf functional traits can better elucidate the dynamics of plant growth suitability with the environment.

There is a substantial body of research on the functional traits of mixed-age plantations and the formulation of suitable strategies. Chen et al. found that fine root carbon/nitrogen/phosphorus ratio *Pinus tabulaeformis* and *Robinia pseudoacacia* plantations was inconsistent with increasing plantation age [6]. Wang et al. showed that the stoichiometry of different organs in a *Metasequoia glyptostroboides* plantation was different in mixed-age plantations, and the growth of young and mature trees were limited by nitrogen and phosphorus, respectively [7]. Chang et al. showed that the nutrient content of leaf nitrogen, phosphorus, and potassium in a larch plantation in the Qinling mountain area did not change consistently with plantation age [8]. Fan et al. suggested that there is a close coupling relationship between soil and plant nutrients in a *Eucalyptus* sp. plantation [9]. Lin et al. studied the effects of soil nutrients and water content on leaf nutrients and stoichiometry in a *Leucaena leucocephala* forest at different ages, and they reached the conclusion that available soil nutrients and soil water content were the main factors limiting plant growth [10]. Nelson et al. concluded that leaf functional traits are influenced by the plantation age and habitat characteristics [11]. He et al. found that *Pinus massoniana* showed strong plasticity in its adaptation to the environment as indicated by coupled, coordinated, or combined traits [12]. These studies indicate that the responses of leaf functional traits to plantation ages are different and they have elucidated the level of environmental influence on plantations at different developmental stages. However, the change of the suitable strategy of *Z. planispinum* 'dintanensis' plantation with age is not clear, which is not conducive to the dynamic adjustment of plantation management measures.

Z. planispinum 'dintanensis' is a variety of *Z. planispinum* [13,14], in the family Rutaceae, which is a deciduous shrub or dungarunga. As a valued ecological and economic plant in Guizhou karst area, *Z. planispinum* 'dintanensis' is lithophytic and has the characteristics of calcium preference and drought tolerance, helping to conserve water and soil, and it plays a significant role in mitigating rocky desertification [15]. The pericarp is well known for its strong aroma, strong hemp flavor, and high yield, which makes it a protected geographical trademark, with high development potential for functional products [16,17]. Research has mainly focused on soil nutrients, functional traits, and stoichiometry [14,18,19], leaving the interactions between growth and environment unclear. Moreover, the fruit harvesting reduces nutrient levels, which leads to differences in available nutrients in plantations of different ages. Thus, the driving effect of soil on leaf functional traits in different ages needs to be further investigated. Since the soil background values in this study are approximate, the leaf functional traits are considered to be mainly influenced by the growth and developmental processes of the *Z. planispinum* 'dintanensis', so the leaf trait change drivers are mainly discussed in relation to the soil. This study selected 5–7-, 10–12-, 20–22-, and 28–32-year-old plantations in Bashan village, Huajiang Town, Guanling County, Anshun City, and Guizhou Province as the experimental sites, and adopted redundancy analysis to investigate the relationships between soil factors and leaf functional traits in *Z. planispinum* 'dintanensis'. The aim was to investigate: (1) how the leaf economic spectrum and suitable strategy vary with plantation age; (2) the connections among leaf functional traits, and (3) how the soil water content, soil elements status, and stoichiometry regulate leaf functional traits. The purpose is to reveal the response and suitable strategy of the *Z. planispinum* 'dintanensis' plantation to the environment and to provide a theoretical basis for plantation management.

2. Materials and Methods

2.1. Overview of the Study Area

The study area has a subtropical, humid, monsoon climate, with an average annual precipitation of 1100 mm. The rainfall is abundant but unevenly distributed throughout the year. The altitude is 530–1473 m, with a relative elevation difference of 943 m, and the terrain fluctuates between largely severe and moderate rock desertification [14]. The area is mainly covered by calcareous soil, rich in calcium and magnesium elements. The soil layer is shallow and discontinuous, resulting in a low water and nutrient holding capacity. The carbonate rocks in the study area are typically developed and can represent the characteristics of the regional habitat. Additionally, the research results can provide a reference for similar areas (Figure 1).

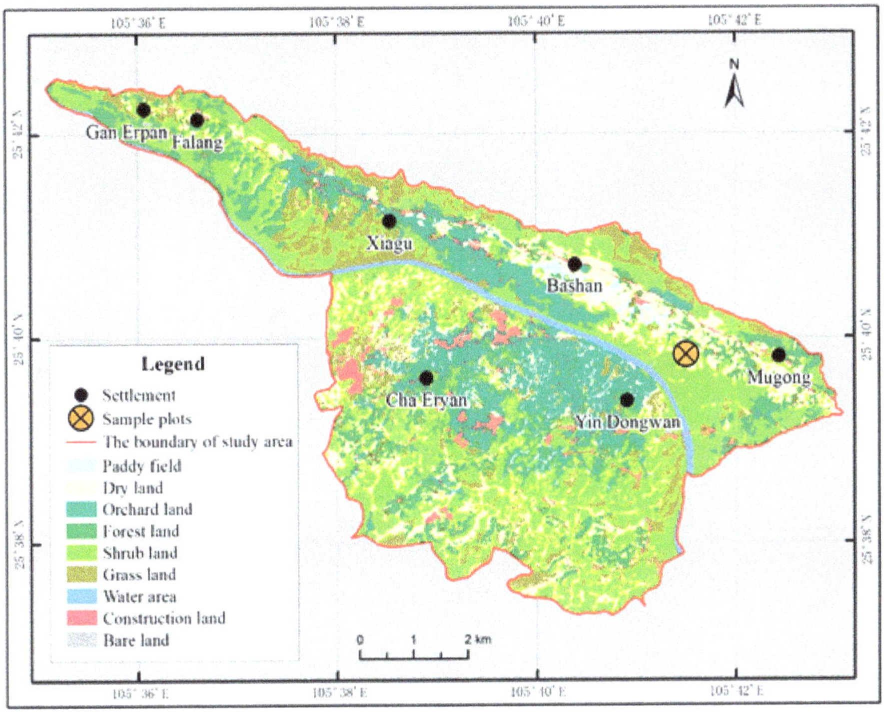

Figure 1. The sample plot distribution map.

Z. planispinum 'dintanensis' is planted mainly in the low-lying soil accumulation area below 700 m. Plantations are established by transplanting seedlings into the field. Management techniques have resulted in relatively stable plantations. *Z. planispinum* 'dintanensis' has been planted in this area for more than 30 years, extending over more than 1000 hectares. The benefits include the control of rocky desertification and sustainable economic growth in the area.

2.2. Sample Plot Setting

The study areas were located in Bashan village, Huajiang Town, Guanling County, Anshun City, and Guizhou Province. During the vigorous growth period in July 2020, *Z. planispinum* 'dintanensis' plantations at sites with similar conditions (middle and lower slope, sunny slope, 5–10° slope, limestone soil, and immediate soil temperature of 25–26.5 °C) were selected in the area centered around 35°39′49.64″ N, 105°41′30.09″ E, with an elevation of 621 m. According to the planting layout, the plantations were divided into 5–7, 10–12, 20–22, and 28–32-year-old groups using space instead of time. Since replanting may happen in the process of stand cultivation, the plantation age is presented as an interval value rather than a specific value. The stands were mainly pure plantations, and there were no other cultivated crops interspersed within the *Z. planispinum* plantation. Due to the fragmentation and high heterogeneity of habitats in karst areas, three 10 m × 10 m sample plots were set up as three replications in plantations of different ages. In total, 12 sample plots were set up with a buffer zone of more than 5 m wide between each sample plot. The height, density, crown width, vegetation coverage, and yield of the trees in each sample plot were measured (Table 1).

Table 1. Basic characteristics of plots.

Age	Tree Height (m)	Density (Plant/ha)	Crown (m)	Coverage (%)	Yield (Plant/kg)
5–7	3.0	1150	3 × 3	100	6–7
10–12	3.0	1150	3 × 3	100	7–8
20–22	3.5	1000	3.5 × 3.5	90	4–5
28–32	4.2	650	4 × 5	70	1–1.5

2.3. Plant Description and Identification

Z. planispinum 'dintanensis' are evergreen shrubs, 2–4.5 m high. The stems and branches have sharp, reddish-brown thorns, and the base of the thorns is wide and flat; Branchlets, shoots, and leaves are glabrous or occasionally pilose. Leaves are pinnately compound. Lobules are usually lanceolate or lanceolate elliptic, opposite or alternate, 4–9 cm long, 1.5–2.5 cm wide, smooth, and glabrous. Cymose panicles axillary or concurrent at the top of lateral branches, with a length of 2–7 cm. The ripe pericarp is mostly olive green, with a number of conspicuously raised round punctured oil glands [20].

Changsheng Wei, the Senior Engineer of Zhenfeng County Forestry Bureau, and Dr. Yanghua Yu, one of the authors of this paper, jointly undertook the formal identification of the plant materials. The voucher specimens were deposited in the publicly available School of Life Sciences, Guizhou Normal University. Deposition code numbers are 990010, 200012, and 200055.

2.4. Sample Collection

Five plants with good growth and consistent tree height and crown width were selected from each sample plot. Four to six mature leaves were selected from each plant along the four directions of east, south, west, and north of each plant. Selected leaves exposed to full sunlight, consistent in size, fully unfolded, and free from obvious symptoms of pathogens, were then placed in cool and dark conditions until further processing in the laboratory. In addition, a mixed sample of 200 g leaves were collected for determining the leaf carbon, nitrogen, and phosphorus contents.

Five soil samples were collected after removing litter and avoiding fertilized areas from each sample plot at a depth from 0 up to 20 cm according to the S-shaped point sampling method [21]. The soil collected from each sample plot was mixed thoroughly after removing plant and animal residues and gravels, and a total of 12 soil samples were obtained. The soil samples were dried naturally in a lab, ground, and passed through 1.0 and 0.25 sieves, respectively.

2.5. Selection of Leaf Functional Traits and Soil Properties Parameters

The fragmented and shallow soil cover in karst regions lead to poor water retention capacity. Leaf thickness can reflect the suitability strategy of plants to water, and leaf water content, leaf water use efficiency, and leaf stable carbon isotope is related to soil water conditions. Carbon, nitrogen, and phosphorus, as important biogenic elements in soil, affect carbon, nitrogen, and phosphorus in leaves, specific leaf area, and leaf dry matter content, then they further affect the photosynthetic capacity of leaves. The process of the soil nitrogen cycle also affects the leaf stable nitrogen isotope. In addition, *Z. planispinum* 'dintanensis' is a calcicole plant, so soil calcium also plays a regulatory role in its growth and development. Therefore, the selected soil factors were soil water content, soil organic carbon, soil total nitrogen, soil total phosphorus, soil total calcium, soil available nitrogen, soil available phosphorus, soil available calcium, soil carbon/nitrogen ratio, soil nitrogen/phosphorus ratio, soil carbon/calcium ratio, and soil calcium/phosphorus ratio—12 in total; leaf table carbon isotope and leaf water use efficiency both indicate the water usage efficiency of plants and the latter was selected as the leaf factor. Thus, the selected leaf factors were leaf thickness, specific leaf area, leaf dry matter content, leaf water content, leaf nitrogen, leaf phosphorus, leaf carbon, leaf stable nitrogen isotope, and leaf water use efficiency. The selected leaf factors and soil factors can further clarify the regulation effects of soil moisture content, element forms, and stoichiometry on leaf functional traits of *Z. planispinum* 'dintanensis'.

2.6. *Index Determination Method*

2.6.1. Leaf Traits Determination

The leaf thickness was measured with Vernier calipers (SF2000, Guilin, China), with an accuracy of 0.01 mm, by taking three measurements from the upper middle and lower leaf regions (avoiding the main leaf veins) and using the average value as the leaf thickness. The leaf area was measured using a leaf area meter (LI-COR 3100C Area Meter, LI-COR, Lincoln, NE, USA). Leaf fresh weight was measured using a balance (JJ124BC, Changshu, China) with an accuracy of 0.0001 g. The leaves were immersed in water for 12 h without light; then, after removing them, the water on the leaf surface was quickly absorbed with absorbent paper, and the leaf saturated fresh weight was obtained. To obtain leaf dry weight, the leaves were dried in an oven for 30 min at 105 °C and then at 70 °C to a constant mass. Leaf nitrogen was determined by Kjeldahl's method; leaf phosphorus was determined by molybdenum antimony anti colorimetry; leaf carbon/nitrogen ratio, leaf carbon/phosphorus ratio, and leaf nitrogen/phosphorus ratio were measured as element mass ratio.

In 2019, the team studied leaf carbon, leaf nitrogen, leaf phosphorus, and their stoichiometric characteristics in the early stage, and the results showed that it could better indicate the nutrient limitation. Therefore, these indicators were still adopted in the more systematic study of leaf functional traits in 2020, but their ages are different, and they can clarify the law of nutrient limitation from a longer time scale. The combination of leaf chemometrics and leaf functional traits is helpful to reflect the process of plant suitability to the environment; leaf stable carbon isotope and leaf stable nitrogen isotope were determined by a stable isotope mass spectrometer (Thermo Scientific MAT253, Bremen, Germany). Specific leaf area, leaf dry matter content, leaf water content, leaf tissue density, and leaf water use efficiency [22] were calculated according to the following formula:

$$\text{Specific leaf area} = \text{Leaf area}/\text{Leaf dry weight} \quad (1)$$

$$\text{Leaf dry matter content} = \text{Leaf dry weight}/\text{Leaf saturated fresh weight} \quad (2)$$

$$\text{Leaf water content} = (\text{Leaf fresh weight} - \text{Leaf dry weight})/\text{Leaf fresh weight} \times 100\% \quad (3)$$

$$\text{Leaf tissue density} = \text{Leaf dry weight}/(\text{Leaf area} \times \text{leaf thickness}) \quad (4)$$

$$\text{Leaf water use efficiency} = C_a/1.6(\delta^{13}C_p - \delta^{13}C_a + b)/(b - a) \quad (5)$$

$$\delta^{13}C_a = -6.429 - 0.006 \exp\,[0.0217(t - 1740)] \quad (6)$$

In Formulas (5) and (6), C_a stands for the CO_2 concentration (approximately 0.038%) in atmosphere; $\delta^{13}C_p$ is the abundance of stable carbon isotope in sample leaves; $\delta^{13}C_a$ is the abundance of stable carbon isotopes in the atmosphere (‰); a is the stable carbon isotope fractionation value produced by diffusion (approximately 4.4‰); b is the stable carbon isotope fractionation value produced by carboxylation reaction (approximately 27‰); t is the year when the sample was taken. In this study, the sample was taken in 2020, and t = 2020 was substituted into Formula (6) to calculate $\delta^{13}C_a$, which was −9.04.

2.6.2. Determination of Soil Physical and Chemical Properties

Soil water content was determined by an electrode method, referring to Bao et al. to determine the soil chemical properties [23] in which soil organic carbon was determined by the potassium dichromate oxidation external heating method; soil total nitrogen and soil available nitrogen were determined by the Kjeldahl method and alkaline hydrolysis diffusion method, respectively. Soil total phosphorus and soil available phosphorus were determined by the acid melting method and hydrochloric acid sulfuric acid extraction, respectively. Soil total calcium and soil available calcium were determined by the diethylenetriamine penta-acetic acid extraction method. Soil carbon/total nitrogen ratio, soil nitrogen/phosphorus ratio, soil carbon/calcium ratio, and soil calcium/phosphorus ratio were measured as the element mass ratio (Table 2).

Table 2. Descriptive statistics of soil factors.

Age (a)	Soil Water Content (%)	Soil Total Nitrogen (g/kg)	Soil Organic Carbon (g/kg)	Soil Total Phosphorus (g/kg)	Soil Total Calcium (g/kg)	Soil Available Nitrogen (mg/kg)
5–7	22.47 ± 0.85 a	2.62 ± 0.34 a	23.65 ± 4.31 a	0.43 ± 0.11 b	8.3 ± 2.69 ab	175 ± 14.14 a
10–12	26.23 ± 1.09 a	2.50 ± 0.30 a	15.3 ± 0.85 b	0.80 ± 0.20 ab	12.75 ± 6.33 a	162 ± 5.66 a
20–22	24.97 ± 4.01 a	2.00 ± 0.52 a	15.05 ± 2.47 b	1.11 ± 0.24 a	2.5 ± 1.20 b	222.5 ± 110.71 a
28–32	23.23 ± 4.81 a	2.12 ± 0.43 a	16.5 ± 2.26 ab	0.77 ± 0.18 ab	3.9 ± 1.27 ab	145 ± 29.70 a
Age (a)	**Soil Available Phosphorus (mg/kg)**	**Soil Available Calcium (mg/kg)**	**Soil Carbon/ Nitrogen Ratio**	**Soil Nitrogen/Phosphorus Ratio**	**Soil Carbon/ Calcium Ratio**	**Soil Calcium/Phosphorus Ratio**
5–7	32.70 ± 5.80 a	1109.00 ± 185.26 a	9.00 ± 0.48 a	6.37 ± 2.35 a	2.92 ± 0.42 ab	20.65 ± 11.31 a
10–12	20.2 ± 5.37 a	989.50 ± 156.27 a	6.16 ± 0.41 b	3.27 ± 1.19 ab	1.36 ± 0.62 b	17.50 ± 12.38 a
20–22	36.65 ± 9.55 a	1018.50 ± 412.24 a	7.625 ± 0.76 ab	1.80 ± 0.08 b	7.39 ± 4.02 a	2.14 ± 1.07 a
28–32	33.65 ± 7.28 a	1145.00 ± 35.36 a	7.86 ± 0.53 a	2.76 ± 0.10 ab	4.37 ± 0.85 ab	5.02 ± 0.47 a

Different letters indicate significant differences among plantation ages at the 0.05 significance level.

2.7. Data Processing

Preliminary data sorting was done in Microsoft Excel 2013. The Kolmogorov–Smirnov method in SPSS 20.0 was used to test the normality of the leaf and soil data. For data that were normally distributed, one-way ANOVA and least significant difference was applied. For non-normally distributed data, Dunett's T_3 method was used. Data were presented in the form of mean ± standard deviation and mean and standard deviation of leaf traits based on plot level. Pearson correlation analysis was performed on the functional traits of each leaf using the "corrplot" packages in R software. Canoco 4.5 was used for redundancy analysis of leaf functional traits and soil factors to investigate the magnitude of the effect of soil factors on leaf functional traits [24]. Coefficients of variation of leaf functional traits = standard deviation/mean × 100%. Sample sizes of three were used to determine the mean, standard deviation, and coefficient of variation. The sample sizes to determine Pearson correlation and redundant analysis was 12.

3. Results

3.1. The Variation Characteristics of Z. planispinum 'dintanensis' Suitability Strategy and Its Economic Spectrum at Different Plantation Ages

As shown in Table 3, the coefficients of variation of leaf functional characters of *Z. planispinum* 'dintanensis' at different ages was 0.41–39.51%, with mainly medium and low coefficients of variation, indicating that the leaf traits had stable variation characteristics during plant development. The coefficients of variation of leaf dry matter content (2.18–4.56%), leaf water content (0.51–0.85%), leaf carbon (2.65–7.61%), and leaf stable carbon isotope (0.41–3.14%) was low. Specific leaf area, leaf water content, and leaf stable nitrogen isotope showed significant differences at different plantation ages. Specific leaf area reached the highest level in the 28–32-year plantation (132.22 cm^2/g), which was significantly higher than that of other three plantation age groups. Leaf water content was 63.46–66.00% and significantly lower in the 5–7- and 10–12-year groups than in the other two age groups. Leaf stable nitrogen isotope reached the highest in the 10–12-year group (3.20‰), suggesting a more open soil nitrogen cycle. Leaf thickness, leaf dry matter content, leaf tissue density, leaf nitrogen, leaf phosphorus, leaf carbon, leaf carbon/nitrogen ratio, leaf carbon/phosphorus ratio, leaf nitrogen/phosphorus ratio, and leaf stable carbon isotope showed no significant differences in all age groups, among which leaf thickness was 0.34–0.37 mm, leaf dry matter content was 31.00–33.00%, and did not change significantly with age. There was no obvious change pattern found in leaf tissue density with age.

Leaf nitrogen, leaf phosphorus, and leaf carbon increased first and then decreased with plantation age. Leaf carbon/phosphorus and leaf nitrogen/phosphorus ratios were rank ordered as 10–12 > 5–7 > 28–32 > 20–22-year group, while leaf carbon/nitrogen ratio was rank ordered as 20–22 > 28–32 > 5–7 > 10–12-year group. Leaf stable carbon isotope and leaf water use efficiency were higher in the late stage of growth, indicating a higher water usage efficiency at this stage.

Table 3. Mean, standard deviation, and coefficient of variation of leaf functional traits of *Z. planispinum* 'dintanensis' at different plantation ages.

Indicator	5–7 a	10–12 a	20–22 a	28–32 a
Leaf thickness (mm)	0.38 ± 0.01 a (3.17%)	0.37 ± 0.03 a (7.65%)	0.36 ± 0.05 a (13.94%)	0.34 ± 0.09 a (27.43%)
Specific leaf area (cm^2/g)	87.37 ± 1.31 b (1.50%)	89.70 ± 6.29 b (7.02%)	91.88 ± 20.65 b (22.47%)	132.22 ± 2.81 a (2.12%)
Leaf dry matter content (%)	32.50 ± 0.71 a (2.18%)	33.00 ± 1.41 a (4.29%)	32.00 ± 1.41 a (4.42%)	31.00 ± 1.41 a (4.56%)
Leaf water content (%)	64.59 ± 0.35 b (0.55%)	63.46 ± 0.54 b (0.85%)	66.00 ± 0.45 a (0.69%)	65.95 ± 0.33 a (0.51%)
Leaf tissue density (g/cm^3)	0.33 ± 0.01 a (2.18%)	0.33 ± 0.03 a (8.58%)	0.32 ± 0.02 a (6.73%)	0.31 ± 0.03 a (9.13%)
Leaf nitrogen (g/kg)	23.70 ± 0.57 a (2.39)	22.25 ± 0.50 a (2.22%)	21.00 ± 1.98 a (9.43%)	22.10 ± 2.97 a (13.44%)
Leaf phosphorus (g/kg)	1.64 ± 0.35 a (21.19%)	1.02 ± 0.23 a (22.19%)	1.67 ± 0.40 a (24.21%)	1.63 ± 0.25 a (15.62%)
Leaf carbon (g/kg)	419.00 ± 18.38 a (4.39%)	373.00 ± 9.90 a (2.65%)	399.50 ± 30.41 a (7.61%)	415.00 ± 12.73 a (3.07%)
Leaf carbon/nitrogen ratio	17.70 ± 1.20 a (6.75%)	16.76 ± 0.07 a (0.42%)	19.04 ± 0.35 a (1.86%)	18.99 ± 3.13 a (16.46%)
Leaf carbon/phosphorus ratio	263.38 ± 67.05 a (25.46%)	373.81 ± 73.22 a (19.59%)	244.91 ± 41.02 a (16.75%)	257.13 ± 32.35 a (12.58%)
Leaf nitrogen/phosphorus ratio	14.79 ± 2.79 a (18.84%)	22.31 ± 4.47 a (20.03%)	12.85 ± 1.92 a (14.92%)	13.87 ± 3.99 a (28.75%)
Leaf stable nitrogen isotope (‰)	0.86 ± 0.02 b (2.17%)	3.20 ± 0.15 a (4.80%)	2.17 ± 0.86 ab (39.51%)	1.80 ± 0.72 ab (39.82%)
Leaf stable carbon (‰)	−28.31 ± 0.30 a (1.07%)	−28.10 ± 0.12 a (0.41%)	−27.84 ± 1.12 a (4.03%)	−27.90 ± 0.88 a (3.14%)
Leaf water use efficiency (umol/mol)	81.21 ± 3.18 a (3.92%)	83.48 ± 1.22 a (1.46%)	86.15 ± 11.81 a (13.71%)	85.54 ± 9.20 a (10.76%)

Different letters indicate significant differences among plantation ages at the 0.05 significance level; Values in brackets indicate the coefficients of variation.

Among many functional traits, specific leaf area, leaf thickness, leaf tissue density, and leaf tissue density are the best variables in the classification axis of plant resource utilization [25,26]. Based on these results, considering that the habitat characteristics of plantations are changing rapidly and that the plantation age in this study varies greatly, we attempted to elucidate the changes in ecological resource strategies of *Z. planispinum* 'dintanensis' to obtain light, temperature, water, air, heat, and soil with plantation age by using leaf economic spectrum. Our analyses demonstrated a lower specific leaf area and a higher leaf tissue density, leaf thickness, and leaf dry matter content, in 5–7, 10–12, and 20–22-year-old plantations, which puts them at the "slow investment-return" end of the leaf economic spectrum, while the lower leaf tissue density, leaf thickness, and leaf dry matter content, and higher specific leaf area, put the 28–32-year-old plantation at the "fast investment-return" end of the same spectrum (Figure 2).

3.2. *Trade-Off and Synergistic Relationship between Leaf Functional Traits of Z. planispinum 'dintanensis'*

Figure 3 shows the relationship among leaf functional traits. Leaf phosphorus showed a significant positive correlation with leaf carbon and leaf water content. Leaf dry matter content decreased with an increase in leaf carbon/nitrogen ratio, which is the trade-off between plant nitrogen usage efficiency and nutrient fixation capacity. Leaf carbon/phosphorus ratio showed a very significant positive correlation with leaf nitrogen/phosphorus ratio, and they both decreased with increasing leaf water content. Leaf stable nitrogen isotope had a reverse effect with leaf phosphorus and leaf carbon, indicating that the nitrogen cycle was affected by the contents of carbon and phosphorus in leaves. There were no significant correlations found among other leaf functional traits. Correlation analysis indicated that *Z. planispinum* 'dintanensis' suits its environment through trade-off and coordination of leaf functional traits.

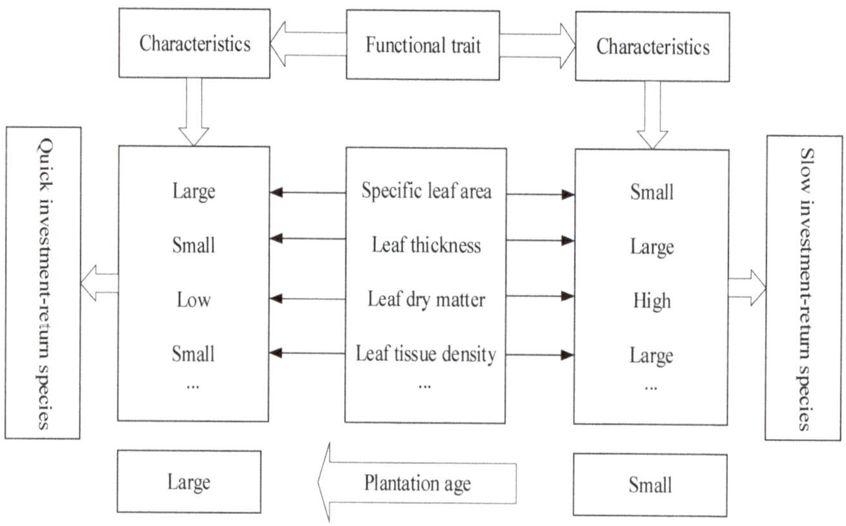

Figure 2. Conceptual illustration of leaf economics spectrum [27,28].

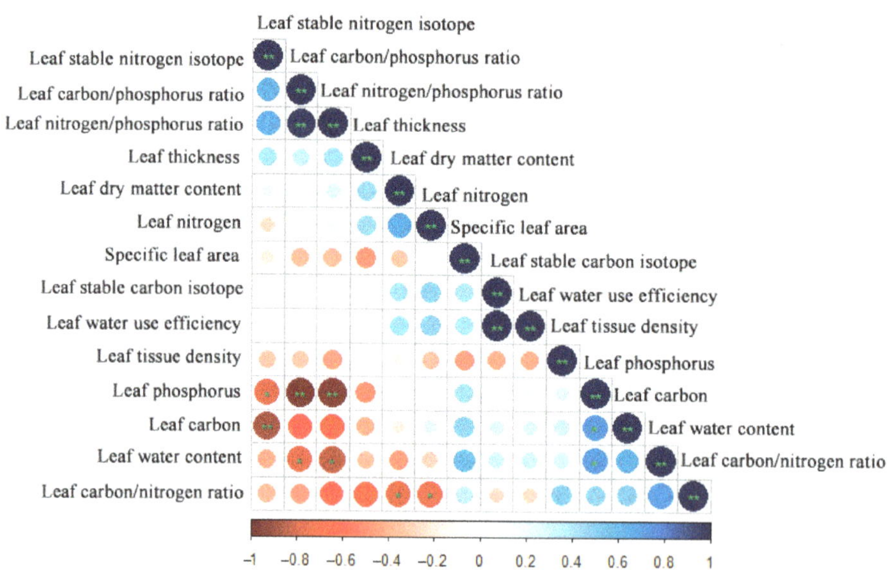

Figure 3. Correlation among leaf functional traits of *Z. planispinum* 'dintanensis' at different plantation ages. Circles with asterisks indicate significant correlations at the 0.05 level (*) and the 0.01 level (**).

3.3. The Relationships between Soil Factors and Leaf Functional Traits of *Z. planispinum* 'dintanensis'

In our previous study, based on the principle of collinearity, the minimum data set was adopted to reduce the soil indicators from 25 to 12, based on which redundancy analysis was performed for leaf functional traits and soil factors. From Table 4, it can be seen that the eigenvalues of redundancy analysis on the first and second axis were 0.434 and 0.240, the accumulated explanation rate was 67.4%, and the total explanation rate was 90.9% on the first four axes. The variation fell mainly on the first and second axis, suggesting a good connection between soil factors and leaf functional traits.

Table 4. Eigenvalues of the ordination axes and the cumulative percentage variance of functional trait–environment relation explained by ordination axes.

Item	Ordination Axes of RDA				
	Axis 1	Axis 2	Axis 3	Axis 4	Sum of All Canonical
Eigenvalues	0.434	0.240	0.161	0.075	1
Cumulative percentage variance of functional traits/%	43.4	67.4	83.4	90.9	_

"_" indicates that there is no data.

In Figure 4, taking soil factors as explanatory variables (blue arrows), and leaf functional traits as response variables to soil factors (red arrows), the relationship between leaf functional traits and soil factors were analyzed by linear constrained redundancy analysis sorting. The length of the line connecting the soil factor arrows represents the magnitude of its effect on leaf functional traits, and the size of the angle between the line connecting the soil and leaf functional traits arrows indicates the level of correlation between them. Smaller acute angles indicate greater positive correlations. While the larger obtuse angles indicate negative correlations, an angle close to 90° indicates the lack of significant correlation.

According to Table 5 and Figure 4, the effects of soil factors on the leaf traits of *Z. planispinum* 'dintanensis' showed a significant effect of soil carbon/nitrogen ratio > soil total calcium > soil organic carbon > soil available phosphorus > soil carbon/calcium ratio, while the other soil factors had less of an effect. The soil carbon/nitrogen ratio was more positively correlated with leaf water use efficiency, leaf nitrogen, leaf carbon, and leaf phosphorus, and negatively correlated with leaf thickness and leaf stable nitrogen isotope. Soil total calcium positively correlated with leaf thickness and leaf dry matter content, and was significantly negatively correlated with leaf carbon. All leaf traits were negatively correlated with soil water content except leaf stable nitrogen isotope, which showed a positively correlation. Leaf water content, specific leaf area, and leaf phosphorus more positively correlated to soil available phosphorus. Soil carbon/calcium ratio was positively correlated with leaf water content, specific leaf area, leaf phosphorus, leaf carbon, and leaf water use efficiency, and negatively correlated with leaf dry matter content, leaf thickness, and leaf stable nitrogen isotope. These results suggested soil carbon/nitrogen ratio, soil total calcium, soil water content, soil available phosphorus, and soil carbon/calcium ratio were highly correlated with leaf functional traits, while soil elemental stoichiometry had a greater correlation on leaf functional traits than their own contents.

Table 5. Soil factor explained variance and significance test.

Soil Factor	Explained Variance (%)	p
Soil carbon/nitrogen ratio	12.06	0.004
Soil total calcium	11.04	0.008
Soil water content	10.52	0.012
Soil available phosphorus	10.27	0.024
Soil carbon/calcium ratio	9.20	0.038

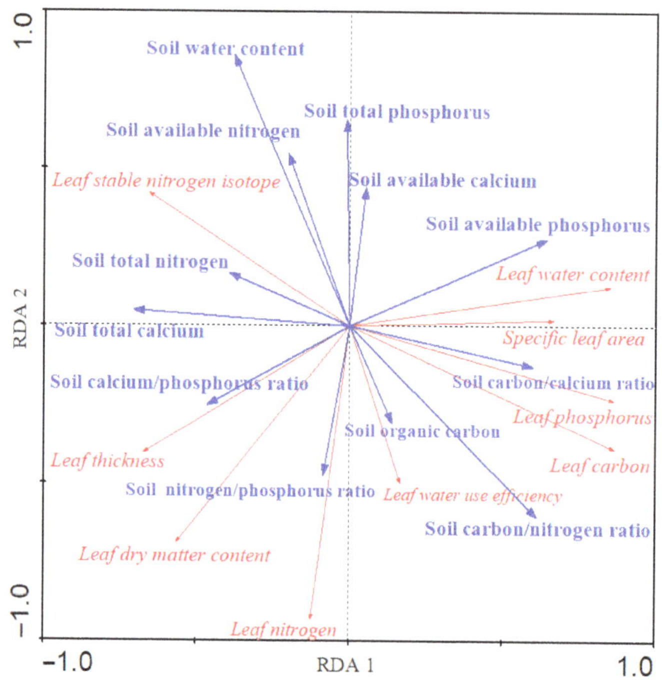

Figure 4. Redundancy analysis of the relationship between leaf functional traits of *Z. planispinum* 'dintanensis' and soil factors.

4. Discussion

4.1. The Leaf Economic Spectrum and Suitability Strategy of Z. planispinum 'dintanensis' Varies with Plantation Age

Soil nitrogen, phosphorus, and other biogenic elements in the study area are deficient [29]. Coupled with the challenges of geological and seasonal drought [30], karst areas gradually degenerate into arid and barren habitats. This study found that most leaf traits did not differ significantly with plantation age and that coefficients of variation were low, indicating that *Z. planispinum* 'dintanensis' plantation gradually develop stable functional traits to suit the habitat during growth. The low coefficient of variation may be a result of the formation of specific traits into which plants devote their resources during long-term suitability for unfavorable habitats [31,32]. The arid and barren habitat in the study area leads to a preference for species with lower trait variability, and thus *Z. planispinum* 'dintanensis' plantations are highly suitable in this area, which is basically consistent with the research results of He et al. [33] in the karst mountains of Guilin. Although the *Z. planispinum* 'dintanensis' is planted artificially, its age is long, and it has formed a long-term coupling relationship with the habitat, which provides strong support for studying its suitability. Some studies have shown that leaf dry matter content is a relatively stable variable on the resource acquisition axis [34], which is similar to the results of this study (the coefficient of variation was 2.18–4.56%). However, this study found that the coefficient of variation of leaf water content (0.51–0.85%) is lower than that of leaf dry matter content. This is because plants growing in arid environments are easily limited by water availability. Their water content is relatively low [35], so the fluctuation of leaf water content is small. Specific leaf area was strongly correlated with photosynthetic capacity and was significantly higher in 28–32 year-old plantations than that of the other three plantation age groups, which was attributed to the reduced density of the stand (Table 1), which allowed the leaves to receive more light and thus enhanced photosynthetic capacity [4,36]. This paper only explores the change of leaf traits with plantation age, and future studies will investigate the change of phenotypic trait plasticity by using plantation age and environment as dual factors to isolate their contribution to phenotypic trait plasticity.

This paper focuses on selecting typical phenotypic traits to characterize the economics of leaf utilization of ecological resources, which is the scientific basis for screening key agronomic traits. The current study showed that the 5–7-, 10–12-, and 20–22-year-old groups stood at the "slow investment-return" end of the leaf economic spectrum, with a tendency toward weak photosynthesis, small specific leaf area, and a longer life. When the specific leaf area is low, the products of photosynthesis are mostly invested to increase the length or resistance of the water diffusion pathway [37]. In addition, the leaf dry matter content, leaf thickness, and leaf tissue density increased correspondingly for the purpose of reducing water loss induced by transpiration, thus increasing the water usage efficiency and the plant suitability to drought and aridity [38,39]. The plantation at the slow end is a conservative strategy, which usually includes a low respiration rate and low leaf turnover rate to prevent carbon loss, with enhancement of the plant resistance to adversity [40]. In the 28–32-year group with growth decline, the assimilation efficiency of leaf carbon by increasing specific leaf area, while leaf tissue density, leaf thickness and leaf dry matter content are reduced, the ecological strategy of "fast investment-return" is adopted to prioritize the nutritional growth. This stage has high productivity and water transport efficiency and a large capacity to obtain and utilize nutrients and fix carbon. However, the ability of the tissue to tolerate drought stress is weak, resulting in poor suitability to a low resource environment [35]. It may be that the growth decline observed in the 28–32-year group resulted in a weaker capacity to use limited environmental resources. In this study, indirect traits, such as chemical traits, were not included in the construction of the leaf economic spectrum. However, in the future, traits, such as metrological and physiological traits, will be included to comprehensively assess the economics of utilizing the resources of *Z. planispinum* 'dintanensis'.

4.2. The Trade-Off and Synergic Relationship among Leaf Functional Traits of *Z. planispinum* 'dintanensis'

In the process of suitability the environment, plants are comprehensively affected by physiological, phylogenetic, environmental, and other factors, resulting in correlation between plant functional traits [27]. Leaf traits manifest a trade-off or synergistic relationship. In this study, leaf carbon was significantly and positively correlated with leaf phosphorus due to the fact that leaf phosphorus characterizes the ability of plants to assimilate carbon dioxide, leaf phosphorus participates in the photosynthetic process of leaves by affecting chlorophyll and protein content in plants [41], and leaf carbon, the main product of photosynthesis, increases with the increase of leaf phosphorus. This study also found that leaf nitrogen/phosphorus ratio had a very significant negative correlation with leaf phosphorus but not with leaf nitrogen, indicating that phosphorus might impose more restrictions to plant growth than nitrogen, which was consistent with the study of Güsewell et al. [42]. Moreover, plants need phosphorus-rich RNA to support protein synthesis during rapid growth, resulting in a faster increase in phosphorus than nitrogen [43], and creating a very significant negative correlation between leaf nitrogen/phosphorus ratio and leaf phosphorus. The leaf dry matter content was observed to decrease with a leaf carbon/nitrogen ratio increase, which indicated the trade-off relationship between plant nitrogen usage efficiency and nutrient fixation (the plant's ability to fix nutrients in the body and use them through growth, physiological and other processes) in our study, because leaf dry matter content is an indicator of nutrient fixation capacity [44], and the leaf carbon/nitrogen ratio is an indicator token of nitrogen utilization efficiency, a higher value indicates a higher nitrogen utilization. In our research, leaf stable nitrogen isotope was negatively correlated with leaf phosphorus and leaf carbon. At present, the mechanism of leaf nutrient elements affecting leaf stable nitrogen isotope is unclear. However, it may be that these elements have a direct or an indirect influence on nitrogen physiological metabolism.

4.3. The Regulation Effect of Soil Factors on the Leaf Functional Traits of Z. planispinum 'dintanensis'

Redundancy analysis showed that leaf functional traits were correlated with multiple soil factors, among which soil carbon/nitrogen ratio, soil total calcium, soil water content, soil available phosphorus, and soil carbon/calcium ratio were most significant. The soil carbon/nitrogen ratio influences the available nutrient supply capacity by affecting the nutrient mineralization rate, thus acting on leaf functional traits [45]. Water supply to plants was dependent on precipitation and soil moisture storage. In spite of the rich precipitation in the karst area, water easily leaks through cracks, resulting in poor soil water retention, and water deficit will inhibit photosynthesis and plant growth [46]. In the research area, soil water content and leaf nitrogen were negatively correlated, which was inconsistent with Cao's research [47]. This is because water deficiency in plants increases the nitrogen allocation to leaves, increases the osmotic pressure in cells, and reduces water loss by reducing stomatal conductance, thus reinforcing water loss protection mechanisms [48]. Calcium can improve the suitability of calcicole plants to arid habitats, yet, it may cause adverse impacts on normal plant growth and development when it exceeds a threshold. In the research area, soil total calcium was positively correlated with leaf thickness and leaf dry matter content, and negatively correlated to leaf carbon, indicating that increased calcium can enhance a plant's capacity to resist adversity. However, in excess it can cause reduced carbon storage, thus reducing plant primary productivity. Zhang et al. [49] indicated that the accumulation of Ca^{2+} in a rocky desertification area led to high soil alkalinity, thus promoting leaf nitrogen content and net photosynthesis rate. The content of phosphorus-related nutrients in soil directly affects the leaf phosphorus absorption capacity. The available soil phosphorus and leaf phosphorus in *Z. planispinum* 'dintanensis' plantations were significantly positively correlated, which is consistent with the conclusion of Fu et al. that plant leaf phosphorus in karst areas is restricted by the availability of soil phosphorus [50], and it also supports the view that soil phosphorus is an important driving factor of leaf phosphorus [51]. Phosphorus is a major element for protein and amino acid synthesis, and phosphorus limitation leads to lower photosynthetic rates and carbon fixation capacity of plants. The leaf nitrogen/phosphorus ratio was less than 14 in the 28–32-year-old plantation, which is limited by nitrogen [52] but not by phosphorus; therefore, soil available phosphorus positively correlated to specific leaf area and leaf carbon, with a relatively high value.

5. Conclusions

(1) The coefficients of variation of leaf functional traits of *Z. planispinum* 'dintanensis' ranged from 0.41% to 39.51%, mostly with medium and low variation. The 5–7, 10–12, and 20–22-year-old plantations were laid at the "slow investment-return" end of the economic spectrum while 28–32-year plantations were close to the "fast investment-return" end.

(2) *Z. planispinum* 'dintanensis' tended to suit the karst environment via making trade-off and coordinating leaf functional traits.

(3) Soil carbon/nitrogen ratio, soil total calcium, soil water content, soil available phosphorus, soil carbon/calcium ratio were highly correlated with leaf functional traits, while soil elemental stoichiometry had a greater reflection on leaf functional traits than their own content.

Author Contributions: Conceptualization, Y.Y. and Y.S.; methodology, Y.S.; software, Y.L.; validation, Y.Y.; formal analysis, Y.S.; investigation, Y.S. and Y.L.; resources, Y.Y.; data curation, Y.Y. and Y.L.; writing—original draft preparation, Y.S.; writing—review and editing, Y.Y. and Y.S.; visualization, supervision, project administration, and funding acquisition, Y.Y. All authors have read and agreed to the published version of the manuscript.

Funding: This research was funded by the Guizhou Province Science and Technology Support Plan Project (Qian-ke-he Zhicheng [2022] Yiban 103), the funder is department of science and technology in Guizhou.

Institutional Review Board Statement: Not applicable.

Informed Consent Statement: Not applicable.

Data Availability Statement: Data sharing not applicable.

Acknowledgments: This work was funded by the Guizhou Province Science and Technology Support Plan Project (Qian-ke-he Zhicheng [2022] Yiban 103).

Conflicts of Interest: The authors declare no conflict of interest.

References

1. Wigley, B.J.; Charles-Dominique, T.; Hempson, G.P.; Stevens, N.; TeBeest, M.; Archibald, S.; Bond, W.J.; Bunney, K.; Coetsee, C.; Donaldson, J.; et al. handbook for the standardised sampling of plant functional traits in disturbance-prone ecosystems, with a focus on open ecosystems. *Aust. J. Bot.* **2020**, *68*, 473–531. [CrossRef]
2. Ameztegui, A.; Paquette, A.; Shipley, B.; Heym, M.; Messier, C.; Grave, D. Shade tolerance and the functional trait: Demography relationship in temperate and boreal forests. *Funct. Ecol.* **2017**, *31*, 821–830. [CrossRef]
3. Pezner, A.K.; Pivovarof, A.L.; Sun, W.; Sharifi, M.R.; Rundel, P.W.; Seibt, U. Plant functional traits predict the drought response of native California plant species. *Int. J. Plant Sci.* **2020**, *181*, 256–265. [CrossRef]
4. Poorter, H.; Niinemets, U.; Poorter, L.; Wright, I.J.; Villar, R. Causes and consequences of variation in leaf mass per area (LMA): A meta-analysis. *New Phytol.* **2009**, *182*, 565–588. [CrossRef]
5. Anderegg, L.D.L.; Berner, L.T.; Badgley, G.; Sethi, M.L.; Law, B.E.; Hillerislambers, J.; Penuelas, J. Within-species patterns challenge our understanding of the leaf economics spectrum. *Ecol. Lett.* **2018**, *21*, 734–744. [CrossRef]
6. Chen, L.L.; Deng, Q.; Yuan, Z.Y.; Mu, X.M.; Kallenbach, R.L. Age-related C:N:P stoichiometry in two plantation forests in the Loess Plateau of China. *Ecol. Eng.* **2018**, *120*, 14–22. [CrossRef]
7. Wang, J.N.; Wang, J.Y.; Wang, L.; Zhang, H.; Guo, Z.W.; Wang, G.G.; Smith, W.K.; Wu, T.G. Does stoichiometric homeostasis differ among tree organs and with tree age? *For. Ecol. Manag.* **2019**, *453*, 117673. [CrossRef]
8. Chang, Y.J.; Li, N.W.; Wang, W.; Liu, X.J.; Du, F.F.; Yao, D.R. Nutrients resorption and stoichiometry characteristics of different-aged plantations of *Larix Kaempferi* in the Qingling Mountains, central China. *PLoS ONE* **2017**, *12*, e0189424. [CrossRef]
9. Fan, H.B.; Wu, J.P.; Liu, W.F.; Yuan, Y.H.; Hu, L.; Cai, Q.K. Linkages of plant and soil C:N:P stoichiometry and their relationships to forest growth in subtropical plantations. *Plant Soil* **2015**, *392*, 127–138. [CrossRef]
10. Lin, Y.M.; Chen, A.M.; Yan, S.W.; Rafay, L.; Du, K.; Wang, D.J.; Ge, Y.G.; Li, J. Available soil nutrients and water content affect leaf nutrient concentrations and stoichiometry at different ages of *Leucaena leucocephala* forests in dry-hot valley. *J. Soils Sediments* **2019**, *19*, 511–521. [CrossRef]
11. Nelson, P.R.; McCune, B.; Roland, C.; Stehn, S. Non-parametric methods reveal non-linear functional trait variation of lichens along environmental and fire age gradients. *J. Veg. Sci.* **2015**, *26*, 848–865. [CrossRef]
12. He, B.; Li, Q.; Feng, T.; Xue, X.H.; Li, W.J.; Liu, Y. Variation in leaf functional traits of different-aged *Pinus massoniana* communities and relationships with soil nutrients. *J. Nanjing For. Univ. (Nat. Sci. Ed.)* **2020**, *44*, 181–190. (In Chinese) [CrossRef]
13. Appelhans, M.S.; Reichelt, N.; Groppo, M.; Paetzold, C.; Wen, J. Phylogeny and biogeography of the pantropical genus *Zanthoxylum* and its closest relatives in the proto-Rutaceae group (Rutaceae). *Mol. Phylogenetics Evol.* **2018**, *126*, 31–44. [CrossRef] [PubMed]
14. Yu, Y.H.; Zheng, W.; Zhong, X.P.; Yin, B. Stoichiometric characteristics in *Zanthoxylum planispinum* var. *dintanensis* plantation of different ages. *Agron. J.* **2020**, *113*, 685–695. [CrossRef]
15. Liu, Z.; Yang, R.; Pei, Y.D. Soil erosion resistance characteristics of *Zanthoxylum Bungeanum* and *Lonicera Japonica* forest land in canyon areas of karst plaeau. *Acta Pedol. Sin.* **2019**, *56*, 466–474. [CrossRef]
16. Wei, C.S.; Zuo, Z.L. Analysis and countermeasure research on the cause of the decline of *Zanthoxyhum planispiunum* var. dingtanensis industry. *Guizhou For. Sci. Technol.* **2016**, *144*, 60–64. (In Chinese) [CrossRef]
17. Wang, Y.; Zhang, L.T.; Feng, Y.X.; Guo, S.S.; Pang, X.; Zhang, D.; Geng, Z.F. Insecticidal and repellent efficacy against stored-product insects of oxygenated monoterpenes and 2-dodecanone of the essential oil from *Zanthoxylum planispinum* var. *dintanensis*. *Environ. Sci. Pollut. Res.* **2019**, *26*, 24988–24997. [CrossRef]
18. Qu, S.; Wang, R.; Yang, Y.; Pei, Y.D.; Li, K.F.; Hu, J.D. Change characteristics of soil nutrients during *Zanthoxylum bungeanum* var. dingtanensis grow process in karst plateau. *Non-Wood For. Res.* **2020**, *38*, 183–191. (In Chinese)
19. Li, H.; Yu, Y.H.; Long, J.; Li, J. Responses of leaf functional traits of *Zanthoxylum planispinum* var. dintanensis to premature senescence. *Chin. J. Ecol.* **2021**, *40*, 1695–1704. (In Chinese) [CrossRef]

20. Tu, Y.L.; Wei, C.S.; Zuo, Z.L.; Lu, Y.M. A new *Zanthoxylum* Genus—*Z. planipinum var.dingtanensis* and the research of its species classification. *Guizhou Sci.* **2001**, *19*, 77–801. (In Chinese)
21. Yu, Y.H.; Wu, Y.G.; Song, Y.P.; Li, Y.T. Carbon and nitrogen stable isotope abundance and soil stoichiometry of *Zanthoxylum planispinum* var. dintanensis plantations of different ages. *Agronomy* **2022**, *12*, 1248. [CrossRef]
22. Cernusak, L.A. Gas exchange and water-use efficiency in plant canopies. *Plant Biol.* **2020**, *22*, 52–67. [CrossRef] [PubMed]
23. Bao, S.D. *Soil Agro-Chemistry Analysis*, 3rd ed.; China Agriculture Press: Beijing, China, 2010.
24. Huang, X.L.; Wang, J.; Zhu, Q.; Wu, M.L.; Liu, Y. Effect of soil nutrients on leaf functional traits of different life form plants. *Acta Bot. Boreali-Occident. Sin.* **2018**, *38*, 2293–2302. (In Chinese) [CrossRef]
25. Puglielli, G.; Varone, L. Inherent variation of functional traits in winter and summer leaves of Mediterranean seasonal dimorphic species: Evidence of a 'within leaf cohort' spectrum. *AOB Plant* **2018**, *10*, ply027. [CrossRef]
26. Hecking, M.J.; Zukswert, J.M.; Drake, J.E.; Dovciak, M.; Burton, J.I. Montane temperate-boreal forests retain the leaf economic spectrum despite intraspecific variability. *Front. For. Glob. Change* **2022**, *4*, 754063. [CrossRef]
27. Wright, I.J.; Reich, P.B.; Westoby, M.; Ackerly, D.D.; Baruch, Z.; Bongers, F.; Cavender-Bares, J.; Chapin, T.; Cornelissen, J.H.C.; Diemer, M.; et al. The worldwide leaf economics spectrum. *Nature.* **2004**, *428*, 821–827. [CrossRef]
28. Westoby, Y.M.; Wright, I.J. Land-plant ecology on the basis of functional traits. *Trends Ecol. Evol.* **2006**, *21*, 261268. [CrossRef]
29. Yu, Y.H.; Song, Y.P.; Zhong, X.P.; Li, Y.T.; Ying, B. Growth decline mechanism of *Zanthoxylum planispinum* var. *dintanensis* in the canyon area of Guizhou Karst Plateau. *Agron. J.* **2021**, *113*, 852–862. [CrossRef]
30. Xu, T.R.; Wu, X.C.; Tian, Y.H.; Li, Y.; Zhang, W.; Zhang, C.C. Soil property plays a vital role in vegetation drought recovery in karst region of Southwest China. *J. Geophys. Res.-Biogeosci.* **2021**, *126*, e2021JG006544. [CrossRef]
31. Li, Q.; Hou, J.H.; He, N.P.; Xu, L.; Zhang, Z.H. Changes in leaf stomatal traits of different aged temperate forest stands. *J. For. Res.* **2021**, *32*, 97–936. [CrossRef]
32. Grassein, F.; Till-Bottraud, I.; Lavorel, S. Plant resource-use strategies: The importance of phenotypic plasticity in response to a productivity gradient for two subalpine species. *Ann. Bot.* **2010**, *106*, 637–645. [CrossRef]
33. He, Y.; Yao, Y.P.; Jiang, Y.; Liang, S.C.; Li, Y.J.; Liang, H.H.; Zhao, Q.N.; Huang, Y.P.; Lin, C.J. Intraspecific variation of functional traits of woody species in the dominant *Cyclobalanopsis glauca* community in the karst area of Guilin city, Southwest China. *Acta Ecol.* **2021**, *41*, 8237–8245. (In Chinese) [CrossRef]
34. Windt, C.W.; Nabel, M.; Kochs, J.; Jahnke, S.; Schurr, U. A mobile NMR sensor and relaxometric method to non-destructively monitor water and dry matter content in plants. *Front. Plant Sci.* **2021**, *12*, 617768. [CrossRef]
35. Wei, Y.H.; Liang, W.Z.; Han, L.; Wang, H.Z. Leaf functional traits of *Populus euphratica* and its response to groundwater depths in Tarim extremely arid area. *Acta Ecol. Sin.* **2021**, *41*, 5368–5376. (In Chinese) [CrossRef]
36. Zhang, K.Y.; Yang, D.; Zhang, Y.B.; Ellsworth, D.S.; Xu, K.; Zhang, Y.P.; Chen, Y.J.; He, F.L.; Zhang, J.L. Differentiation in stem and leaf traits among sympatric lianas, scandent shrubs and trees in a subalpine cold temperate forest. *Tree Physiol.* **2021**, *41*, 1992–2003. [CrossRef]
37. Zhu, J.Y.; Cao, Y.J.; He, W.J.; Xu, Q.; Xu, C.Y.; Zhang, X.N. Leaf functional traits differentiation in relation to covering materials of urban tree pits. *BMC Plant Biol.* **2021**, *21*, 556. [CrossRef]
38. Chang, Y.N.; Xu, C.B.; Yang, H.; Zhou, J.X.; Hua, W.P.; Zhang, S.H.; Zhong, Q.L.; Li, B.Y. Leaf structural traits vary with plant size in even-aged stands of *Sapindus mukorossi*. *Front. Plant Sci.* **2021**, *12*, 692484. [CrossRef]
39. Zhu, J.Y.; Zhu, H.; Cao, Y.J.; Li, J.H.; Zhu, Q.Y.; Yao, J.M.; Xu, C.Y. Effect of simulated warming on leaf functional traits of urban greening plants. *BMC Plant Biol.* **2020**, *20*, 139. [CrossRef] [PubMed]
40. Reich, P.B. The world-wide'fast-slow'plant economics spectrum: A traits manifesto. *J. Ecol.* **2014**, *102*, 275–301. [CrossRef]
41. Tian, D.; Du, E.; Jiang, L.; Zeng, W.J.; Zou, A.L.; Feng, C.Y.; Xing, A.J.; Wang, W.; Zheng, C.Y.; Ji, C.J.; et al. Responses of forest ecosystems to increasing N deposition in China: A critical review. *Environ. Pollut.* **2018**, *243*, 75–86. [CrossRef]
42. Güsewell, S. N:P ratios in terrestrial plants: Variation and functional significance. *New Phytol.* **2004**, *164*, 243–266. [CrossRef] [PubMed]
43. Wang, L.L.; Zhao, G.X.; Li, M.; Zhang, M.T.; Zhang, L.F.; Zhang, X.F.; An, L.Z.; Xu, S.J. C:N:P Stoichiometry and leaf traits of halophytes in an arid saline environment, northwest China. *PLoS ONE* **2015**, *10*, e0119935. [CrossRef] [PubMed]
44. Wilson, P.J.; Thompson, K.; Hodgson, J.G. Specific leaf area and leaf dry matter content as alternative predictors of plant strategies. *New Phytol.* **1999**, *143*, 155–162. [CrossRef]
45. Zeng, Q.C.; Rattan, L.; Chen, Y.A.; An, S.S. Soil, leaf and root ecological stoichiometry of *Caragana Korshinskii* on the Loess Plateau of China in relation to plantation age. *PLoS ONE* **2017**, *12*, e0168890. [CrossRef]
46. Zhang, Q.W.; Zhu, S.D.; Jansen, S.; Cao, K.F. Topography strongly affects drought stress and xylem embolism resistance in woody plants from a karst forest in Southwest China. *Funct. Ecol.* **2021**, *35*, 566–577. [CrossRef]
47. Cao, J.J.; Wang, X.Y.; Adamowski, J.F.; Biswas, A.; Liu, C.F.; Chang, Z.Q.; Feng, Q. Response of leaf stoichiometry of *Oxytropis ochrocephala* to elevation and slope aspect. *Catena* **2020**, *194*, 104772. [CrossRef]
48. Wright, I.J.; Reich, P.B.; Westoby, M. Strategy shifts in leaf physiology, structure and nutrient content between species of high-and low-rainfall and high-and low-nutrient habitats. *Funct. Ecol.* **2001**, *15*, 423–434. [CrossRef]
49. Zhang, S.H.; Zhang, Y.; Xiong, K.N.; Yu, Y.H.; Min, X.Y. Changes of leaf functional traits in karst rocky desertification ecological environment and the driving factors. *Glob. Ecol. Conserv.* **2020**, *24*, e01381. [CrossRef]

50. Fu, P.L.; Zhu, S.D.; Zhang, J.L.; Finnegan, P.M.; Jiang, Y.J.; Lin, H.; Fan, Z.X.; Cao, K.F. The contrasting leaf functional traits between a karst forest and a nearby non-karst forest in south-west China. *Funct. Plant Biol.* **2019**, *46*, 907–915. [CrossRef]
51. Rossatto, D.R.; Carvalho, F.A.; Haridasan, M. Soil and leaf nutrient concentration of tree species support deciduous forests on limestone outcrops as a eutrophic ecosystem. *Acta Bot. Brasílica* **2015**, *29*, 231–238. [CrossRef]
52. Koerselman, W.; Meuleman, A.F.M. The vegetation N:P ratio: A new tool to detect the nature of nutrient limitation. *J. Appl. Ecol.* **1996**, *33*, 1441–1450. [CrossRef]

agronomy

Article

Carbon and Nitrogen Stable Isotope Abundance and Soil Stoichiometry of *Zanthoxylum planispinum* var. *dintanensis* Plantations of Different Ages

Yanghua Yu [1,*], Yingu Wu [2], Yanping Song [1] and Yitong Li [1]

1 School of Karst Science/State Engineering Technology Institute for Karst Decertification Control, Guizhou Normal University, Guiyang 550001, China; syp_2022@163.com (Y.S.); liyitong0711@163.com (Y.L.)
2 School of Geography and Environment Science, Guizhou Normal University, Guiyang 550025, China; wuyingu0918@163.com
* Correspondence: yuyanghua2003@163.com

Abstract: Understanding the relationships between carbon; nitrogen, their stable isotopes $\delta^{13}C$ and $\delta^{15}N$, and soil stoichiometry may further our understanding of the regulatory mechanisms of the soil quality index on the equilibrium on isotopic fractionation. Four plantations of *Zanthoxylum planispinum* var. *dintanensis* (5–7, 10–12, 20–22 and 30–32 years) in the karst plateau gorge area, Guizhou Province, China, were selected to determine the variation characteristics and interactions between leaves, leaf litter, soil carbon (C), soil nitrogen (N) and their isotopes with plantation age, and to explore the relationship between soil stoichiometry and the isotopes $\delta^{13}C$ and $\delta^{15}N$. The results were as follows: (1) the $\delta^{13}C$ in leaves, litter, and soil were −28.04‰ ± 0.59‰, −26.85‰ ± 0.67‰, and −19.39‰ ± 1.37‰, respectively. The contents of $\delta^{15}N$ were 2.01‰ ± 0.99‰, 2.91‰ ± 1.32‰, and 3.29‰ ± 0.69‰, respectively. The contents of $\delta^{13}C$ and $\delta^{15}N$ were ranked in the order, soil > litter > leaf. (2) With increasing plantation age, the soil ^{13}C decreased; the leaf and the litter $\delta^{15}N$ increased first then decreased, and the litter $\delta^{13}C$ and the soil $\delta^{15}N$ did not vary significantly. (3) The litter layer was positively correlated with soil $\delta^{13}C$ and negatively correlated to $\delta^{15}N$. (4) Redundancy analysis showed that the soil microbial biomass carbon (MBC) and the bacteria/fungi (BAC/FUN) were the dominant factors affecting the natural abundance of C and N isotopes

Keywords: age; *Zanthoxylum planispinum* var. *dintanensis*; leaf-litter-soil continuum; carbon and nitrogen stable isotope; soil stoichiometry; rocky desertification area

Citation: Yu, Y.; Wu, Y.; Song, Y.; Li, Y. Carbon and Nitrogen Stable Isotope Abundance and Soil Stoichiometry of *Zanthoxylum planispinum* var. *dintanensis* Plantations of Different Ages. *Agronomy* **2022**, *12*, 1248. https://doi.org/10.3390/agronomy12061248

Academic Editor: Alwyn Williams

Received: 19 April 2022
Accepted: 20 May 2022
Published: 24 May 2022

1. Introduction

The composition of forest carbon (C), nitrogen (N), and their stable isotopes is a result of the continuous recycling of nutrients between plants and their environments, Consequently, $\delta^{13}C$ and $\delta^{15}N$ are comprehensive indicators of the C and the N cycles in terrestrial ecosystems. Their temporal and spatial differentiation, as well as their relationships in the environment, can reveal the source of plant nutrient elements [1,2]. The degree of fractionation is a discrepancy due to different photosynthetic pathways in plants [3], the $\delta^{13}C$ values of plants with C_3, C_4, and CAM are −35–−20‰, −15–−7‰, and −22–−10‰ [4], respectively, which can be used to identify plant photosynthetic types. The effect of atmospheric carbon dioxide (CO_2) in the same photosynthetic pathway produces variations in the $\delta^{13}C$ values within plants [5]. The balance and the cycling laws of multiple elements can be studied through stoichiometry [6], as it analyzes the interaction of multiple elements in an ecosystem, and it clarifies the cycling process and isotope fractionation mechanism of C and N [7,8]. Investigation into the main causes of the differentiation of $\delta^{13}C$ and $\delta^{15}N$ in different forests may help to understand the configuration and the restriction status of nutrient elements in a plantation. This, in turn, may provide theoretical support to explain the mechanism of C and N cycles and isotope fractionation in a karst ecosystem.

In recent years, scholars have studied variations in $\delta^{13}C$ and $\delta^{15}N$ values with plantation age. For example, the study of Wang et al. (2019) showed that $\delta^{13}C$ and $\delta^{15}N$ in leaves of young *Cunninghamia lanceolata* were lower than that of a mature forest [9], which was consistent with the results of studies on *Ulmus pumila* and *Pinus sylvestris* var. *mongolica* plantations [10,11], yet conflicted with the study on *Caragana intermedia* [2]. Zheng et al. (2015) showed that the *C. lanceolata* content of $\delta^{13}C$ was relatively low in 3- and 8-year-old plantations, and the content of $\delta^{15}N$ was significantly higher than that of other stands [12]. These results suggest that there is a dynamic adjustment of $\delta^{13}C$ and $\delta^{15}N$ in different plantations. Factors such as species, ages, and environmental changes affect resource use strategies. Soil is an essential part of a terrestrial ecosystem, and its stoichiometry characteristics are affected by plantation age, forest structure, and habitat [13]. Soil stoichiometric coupling and microbial metabolism could affect nutrient uptake patterns and forest ecological effects, resulting in different C and N isotope fractionation [1,9,14]. Wang et al. (2019) discovered that the alleviation of soil N limitation and the exacerbation of phosphorus limitation could promote the fractionation of plant N isotope [9]. The soil C/N ratio can represent the decomposition rate of organic N and microorganisms; the higher soil C/N ratio, the lower the decomposition rate of organic N; and the opposite is true for microorganisms [15]. In addition, it was found that microbes can optimize their resource utilization strategy according to litter quality, nutrient utilization efficiency, and restriction status, thus affecting the $\delta^{13}C$ and the $\delta^{15}N$ composition [16]. In conclusion, soil stoichiometry and microorganisms are tightly connected to $\delta^{13}C$ and $\delta^{15}N$. Yet, the variation of forest $\delta^{13}C$ and $\delta^{15}N$ with plantation age and the mechanism of soil stoichiometry driving C and N isotope fractionation are still uncertain and need to be further studied.

Zanthoxylum planispinum var. *dintanensis* is a variety of *Zanthoxylum planispinum*, unique to Guizhou province, with characteristics of calcium preference, drought tolerance, and strong adaptability. It plays an important role in rocky desertification control, ecological function improvement, and soil erosion control [17]. *Z. planispinum* var. *dintanensis* is a suitable species for karst restoration. However, due to the continuous cropping obstacle, *Z. planispinum* var. *dintanensis* may experience accelerated aging, a vigorous fruit setting period, and a shortening of the rotation cutting period. At present, the research on *Z. planispinum* var. *dintanensis* has focused on cultivation, management, aging, degradation, improving product yield, and quality [18,19]. Stable isotope technology has not been widely used as a tool to analyze the variation of $\delta^{13}C$ and $\delta^{15}N$ in plantations of different ages and their relationship with soil stoichiometry. Plantation age, availability of soil C and N, and soil microbial communities could be key factors affecting the natural abundance of $\delta^{13}C$ and $\delta^{15}N$. We hypothesized that: (1) the contents of $\delta^{13}C$ and $\delta^{15}N$ are significantly different in young and old plant tissues, and (2) soil $\delta^{13}C$ and $\delta^{15}N$ are negatively correlated with soil C/N ratio. Our study investigated if leaves, leaf litter, soil C, soil N, and stable isotopes varied with plantation age and the internal relationship among leaf litter, soil C, soil N, and stable isotopes. We also studied the driving mechanism of soil stoichiometry to $\delta^{13}C$ and $\delta^{15}N$ fractions, to further understand the C and the N cycles in a karst ecosystem. We elucidated the current status of the nutrient configuration in a plantation, formulated fertilizer application measures, optimized stand structure, and diagnosed degradation mechanisms.

2. Materials and Methods

2.1. Overview of the Study Area

The research area is located in Beipan River Basin, Huajiang Grand Canyon, Guanling, Guizhou, China (E 105°41′30.09″, N 25°39′49.64″). The major characteristics of the research area are as follows. (1) Landform: the area belongs to Karst Plateau Canyon landform, with a broken surface and undulating terrain, and an altitude range of 530–1473 m. (2) A dry and hot climate, primarily a subtropical monsoon climate. The average annual precipitation is approximately 1100 mm; precipitation in May to October accounts for 83% of the annual total. The annual average temperature is 18.4 °C, with the highest and lowest temperatures of 32.4 °C and 6.6 °C, respectively. It is warm and dry in winter and spring, and humid and hot in summer and autumn. (3) Rocky desertification; the soil is mainly limestone and marl, and the exposed area of bedrock is more than 70% [20]. The main vegetation is subtropical evergreen deciduous coniferous broad-leaved mixed forest. *Zanthoxylum planispinum* var. *dintanensis* is the dominant tree species. The plantations were cultivated from transplanted seedlings, with a survival rate of 70% with replanting as necessary; no other dwarf crops were intercropped under the plantation. The management mode was mainly natural renewal with human interference. Generally, *Z. planispinum* var. *dintanensis* plantations grew well, in spite of partial degradations and yield reductions in the over-matured plantations (Figure 1).

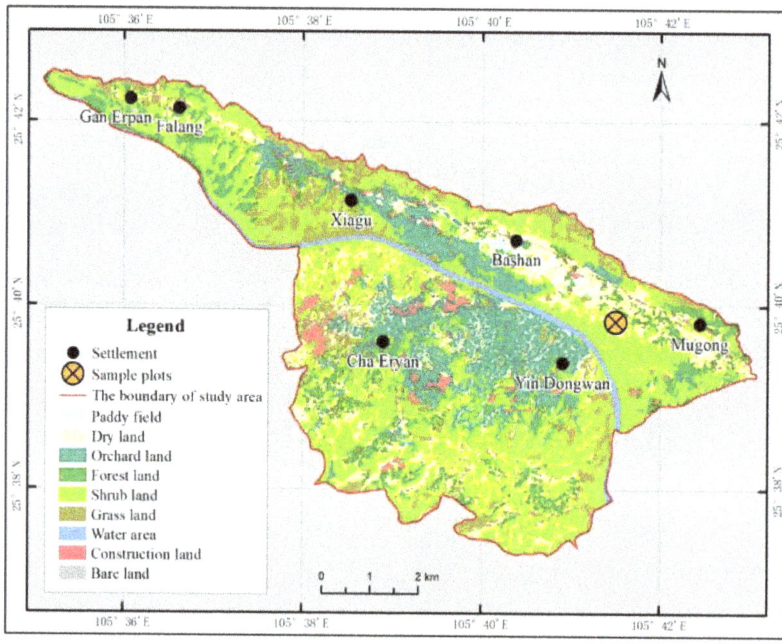

Figure 1. Schematic diagram of the study sample area.

2.2. Experimental Design

2.2.1. Sample Plot Setting

By using space distribution to replace time distribution, the *Z. planispinum* var. *dintanensis* plantations with similar site conditions, such as altitude (621 ± 5 m), slope, aspect, and soil type were selected and divided into four age groups: 5–7 years (initial fruit bearing period), 10–12 years (vigorous fruit bearing period), 20–22 years (end fruit bearing period), and 30–32 years (senescence and death period), which were recorded as yd1–yd4, in turn. Plantation age was recorded as a range rather than specific values because of the replanting regime. Sample plots, 3 10 m × 10 m in size, were set in each plantation age group, with a buffer zone >5 m between each sample plot (Table 1).

Table 1. The basic information for each sample plot.

Plot No	Age	Average Crown Width/m	Height/m	Vegetation Coverage/%	Density/ ikaika (Plant/hm^2)	Yield (Plant/Kg)
YD1	5–7	2.5 × 3	2.7	100	1150	6–7
YD2	10–12	2.5 × 3	2.7	100	1150	7–8
YD3	20–22	3.5 × 3	3.5	90	1000	4–5
YD4	30–32	4 × 5	4	75	650	1–1.5

YD1–YD4: Initial fruit bearing period, vigorous fruit bearing period, ending fruit bearing period, senescence, and death period, respectively.

2.2.2. Sample Collection

In August 2020, five well developed representative plants were selected from each sample plot. The leaves were sampled from the middle part of the canopy, and 30 matured leaves without diseases and pests were picked from east, west, south, and north of the canopy, then mixed and stored in nylon bags [21,22]. In the leaf acquisition area, 31 m × 1 m quadrats were evenly arranged along the diagonal, from which samples were collected from the fully decomposed layer, semi decomposed layer, and undecomposed layer, and then evenly mixed. Five samples were collected from the soil layer at a depth of 0–20 cm along an "S" curve in each sample plot, and then mixed into one sample. Approximately 0.5 kg of a fresh weight soil sample was acquired according to the quartering method from each spot. A total of 12 soil samples were uniformly mixed and brought back to the laboratory.

2.2.3. Sample Determination

The leaf and the litter were transported to the laboratory and washed clean with deionized water. After that, it was dried at 60 °C for 24–28 h and then cooled naturally. Finally, it was crushed with a grinder and screened through 0.2 mm sieves for storage. The visible gravel, roots, and animal and plant debris were removed from the soil. Some of the fresh soil was stored at 4 °C for the purpose of determining the microbial quantity and biomass. The remaining soil was naturally dried, 95% of the samples were ground and passed through a 0.15 mm sieve. The preprocessed samples were sent to the third Marine Research Laboratory of the natural resources sector for determination (Xiamen, China). Soil samples were acidified with 1 mol L^{-1} hydrochloric acid, and carbonate was removed. The reaction time of the acidification process shall not be more than 6 h. The sample was stirred with glass rods every 1 h to ensure complete removal of inorganic carbon in the soil, then stirred and wash with deionized water three to four times to remove excess hydrochloric acid, and dried for reserve. C, N, $\delta^{13}C$, and $\delta^{15}N$ in leaves, leaf litter, and soil were determined by an element analyzer stable isotope mass spectrometer (Flash EA 1112 HT- Delta V Advantages, Thermo Fisher Corporation, Waltham, MA, USA), and the measuring error was <0.05‰. The gas He flow rate was 90 mL·min^{-1}, with a reaction tube temperature of 960 °C and a column temperature of 50 °C. The $\delta^{13}C$ and the $\delta^{15}N$ values are expressed in thousand percentage units (‰), the composition is calculated as follows [1,23]:

$$\delta^{13}C(‰) = \left[\frac{R\left(^{13}C/^{12}C_{sample}\right)}{R(^{13}C/^{12}C_{VPDB})} - 1\right] \times 1000 \quad (1)$$

In the formula, R ($^{13}C/^{12}C_{VPDB}$) represents the C isotope ratio of the international standard VPDB (Vienna Peedee Belemnite), the analysis accuracy of the $\delta^{13}C$ value was ± 0.2 ‰.

$$\delta^{15}N(‰) = \left[\frac{R(^{15}N/^{14}N_{sample})}{R(^{15}N/^{14}N_{air})} - 1\right] \times 1000 \quad (2)$$

In the formula, R ($^{15}N/^{14}N_{air}$) represents the N isotope ratio of N_2 in the atmosphere, the analysis accuracy of the $\delta^{15}N$ value was ± 0.25‰.

Soil organic carbon (SOC) and total nitrogen (TN) were determined by the potassium dichromate oxidation external heating method and the Kjeldahl method, respectively [24]. Soil microbial biomass carbon and nitrogen (MBC and MBN) were measured by chloroform fumigation. Treatment was done by chloroform fumigation followed by extraction with a 0.5 mol·L^{-1}K_2SO_4 solution; MBC was determined by a chloroform fumigation-K_2SO_4 extraction-TOC analyzer, and MBN was determined by chloroform fumigation-K_2SO_4 extraction-potassium persulfate oxidation method, where MBC and MBN calculate using the uniform transformation coefficients 0.45 and 0.54 [25].

Soil bacteria, fungi, and actinomycetes in soil were determined by the beef peptone culture method, the potato glucose agar culture method, and Gao's No.1 [26]. The bacteria and actinomycetes were counted in a 30 °C incubator after 24 h and a 25 °C incubator for 72 h by diluted plate counting; the fungi were cultured in a 28 °C incubator for 96 h after counting, using the inverted plate method.

2.2.4. Statistical Analysis

The data were sorted and analyzed by Microsoft Excel 2010 and SPSS 20.0. The parameter sets were examined for normality; except for N content in leaves, the others parameters were all normally distributed. For N content in leaves, we performed a Kruskal–Wallis test. One-way ANOVA and least significant difference (LSD) were used to test the others parameters. The significant and highly significant levels were $p = 0.05$ and 0.01, respectively. Data in figures and tables are expressed in the form of mean ± standard deviation. Pearson correlation analysis was used to test the correlation between the indicators. Origin 8.6 software was used to depict figures. The relationships between *Z. planispinum* var. *dintanensis* plantation $\delta^{13}C$ and $\delta^{15}N$ and soil stoichiometry were analyzed by Canoco 5 software (Redundancy analysis, RDA).

3. Results

3.1. The Characteristics of Zanthoxylum planispinum var. dintanensis C, N, and Their Stable Isotopes from Different Aged Plantations

There were no significant differences in C and $\delta^{13}C$ content in leaves from the four plantation age groups, which indicate that water use efficiency did not vary significantly with plantation age. Leaf N ranged from 24.95 g·kg^{-1} to 34.75 g·kg^{-1}, showing a decreasing trend with an increase in plantation age. $\delta^{15}N$ in leaves ranged from 0.86‰ to 3.20‰, showing a first increasing and then decreasing trend with plantation age, which suggests that the N use efficiency of leaves in different stands was different. Leaf litter C and N were 413.35–349.65 and 31.2–18.05 g·kg^{-1}, respectively, which were significantly higher in 5–7- and 10–12-year-old plantations than in 20–22- and 30–32-year-old plantations, suggesting that the litter of young leaves decomposes more readily. Leaf litter $\delta^{13}C$ ranged from −25.96‰ to −27.67‰, with no significant difference among the four plantation age groups. The 10–12-year-old plantation had the highest $\delta^{15}N$ (4.15‰ ± 0.92‰), which increased first and then decreased with increasing plantation age. Soil C ranged from 9.1 g·kg^{-1} to 16 g·kg^{-1}, and decreased with increasing plantation age. There were no significant differences in soil N, $\delta^{13}C$, and $\delta^{15}N$ from the four plantation age groups, which suggests that soil nutrient patterns did not change with plantation age (Figure 2).

3.2. Relationship between Zanthoxylum planispinum var. dintanensis Plantation C, N and Their Stable Isotopes

Table 2 shows that there was a highly significant ($p < 0.01$) and positive correlation ($p < 0.05$) between leaf N and soil $\delta^{13}C$ and C, indicating that soil can affect leaf nutrient status. The litter layer was positively related to soil $\delta^{13}C$ and negatively related to soil $\delta^{15}N$. There was a significant positive correlation between leaf litter N with soil $\delta^{13}C$, which suggested there may be promoting or inhibiting effects between litter and soil. Leaf $\delta^{13}C$ was significantly positively correlated to leaf litter $\delta^{15}N$ and negatively correlated to soil $\delta^{15}N$, indicating that leaves, leaf litter, and soil were coupled with each other.

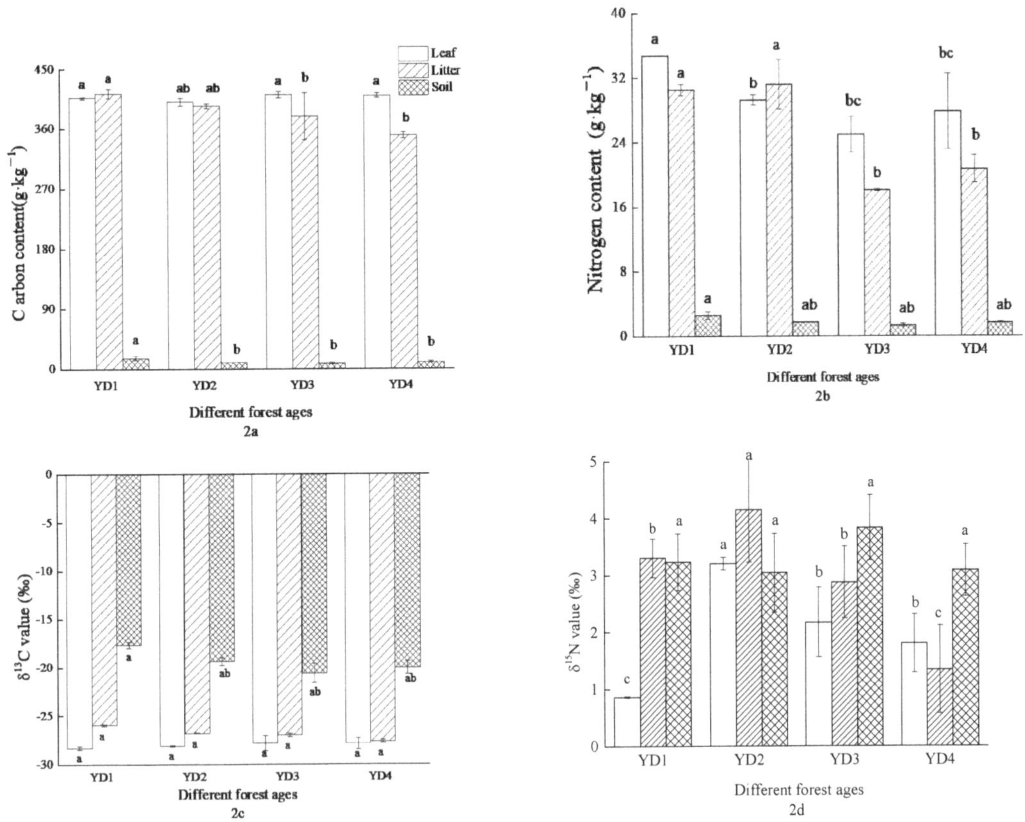

Figure 2. The C, N and their stable isotope characteristics of *Zanthoxylum planispinum* var. *dintanensis* with different ages, in carbon content (**a**), nitrogen content (**b**), $\delta^{13}C$ value (**c**) and $\delta^{15}N$ value (**d**). Data represent mean ± SD (n = 3). The different letters indicate significant differences among the age ($p < 0.05$). YD1, YD2, YD3, and YD4 represent four different forest ages (5–7, 10–12, 20–22, and 30–32), respectively.

Table 2. The C, N and their stable isotope correlation in *Zanthoxylum planispinum* var. *dintanensis* plantation.

Index	C_{leaf}	$\delta^{13}C_{leaf}$	N_{leaf}	$\delta^{15}N_{leaf}$	C_{litter}	$\delta^{13}C_{litter}$	N_{litter}	$\delta^{15}N_{litter}$	C_{soil}	$\delta^{13}C_{soil}$	N_{soil}
$\delta^{13}C_{leaf}$	0.225	1									
N_{leaf}	−0.108	−0.751 *	1								
$\delta^{15}N_{leaf}$	−0.584	0.044	−0.474	1							
C_{litter}	−0.043	0.219	0.260	−0.307	1						
$\delta^{13}C_{litter}$	−0.100	−0.244	0.624	−0.429	0.855 **	1					
N_{litter}	−0.392	−0.389	0.706	−0.021	0.595	0.676	1				
$\delta^{15}N_{litter}$	−0.205	0.210	0.007	0.275	0.727 *	0.575	0.606	1			
C_{soil}	−0.132	−0.545	0.792 *	−0.614	0.400	0.666	0.534	−0.059	1		
$\delta^{13}C_{soil}$	−0.193	−0.704	0.939 **	−0.384	0.271	0.666	0.724 *	0.167	0.747 *	1	
N_{soil}	−0.280	−0.528	0.774 *	−0.484	0.400	0.648	0.628	0.041	0.976 **	0.778 *	1
$\delta^{15}N_{soil}$	−0.180	−0.485	0.054	0.041	−0.286	−0.019	−0.369	−0.484	0.181	−0.045	0.074

Leaf C: Leaf carbon content; Leaf $\delta^{13}C$: Leaf $\delta^{13}C$ vale; Leaf N: Leaf nitrogen content; Leaf $\delta^{15}N$: Leaf $\delta^{15}N$ value; Litter C: Litter carbon content; Litter $\delta^{13}C$: Litter $\delta^{13}C$ value; Litter N: Litter nitrogen content; Litter $\delta^{15}N$: Litter $\delta^{15}N$ value; Soil C: Soil carbon content; Soil $\delta^{13}C$: Soil $\delta^{13}C$ value; Soil N: Soil nitrogen content; Soil $\delta^{15}N$: Soil $\delta^{15}N$ value; * indicates a significant correlation ($p < 0.05$); ** indicates an extremely significant correlation ($p < 0.01$).

3.3. The Impact of Soil Stoichiometry on Plantation C, N, and Their Stable Isotopes

RDA analysis was conducted on soil stoichiometry and plantation components in the different plantation age groups. MBC/MBN was ignored due to its negligible influence. The soil stoichiometry interpreted 90.75% and 4.82% variations on the first and the second axis (Table 3), reflecting the strong connection between soil stoichiometry and some of the plantation components.

Table 3. Redundancy analysis of the component content in the plantation.

Sorting Axis	Axis 1	Axis 2	Axis 3	Axis 4
Explains	90.75	4.82	1.79	0.54
Pseudo-canonical correlation	0.9937	0.9975	0.8307	0.9781
Explained variation (cumulative)	90.75	95.57	97.35	97.89
Explained fitted variation (cumulative)	92.65	97.57	99.36	99.34

The black arrows in Figure 3 indicate parts of the C and N components and their isotopes in the plantation, with the red arrows indicating the soil stoichiometry. According to the two-dimensional diagram of redundancy analysis on plantation components and soil stoichiometry, MBC, BAC/FUN, TN, leaf litter N, soil $\delta^{13}C$, leaf N, leaf litter $\delta^{13}C$, and SOC/TN, MBC, leaf $\delta^{13}C$, and leaf C were positively correlated. BAC/FUN, TN, and leaf C, leaf C and MBN and litter N, soil $\delta^{13}C$, and leaf N were negatively correlated. The angles between MBC and leaf litter $\delta^{15}N$, and BAC/FUN and soil $\delta^{13}C$ were small, showing a strong positive correlation; SOC was negatively correlated with leaf $\delta^{15}N$, but not significantly correlated with other soil stoichiometry. As shown in Table 4, the physical and the chemical variables could be ranked in the order: MBC > BAC/FUN > SOC/TN > MBN > SOC > TN, although no significant influence was found.

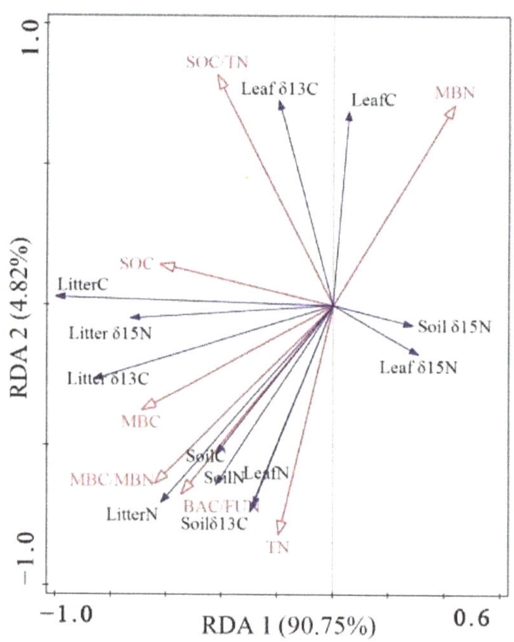

Figure 3. Redundancy analysis of artificial forest C, N, ^{13}C, ^{15}N and soil stoichiometry in the plantation.

Table 4. Importance sequencing and Duncan's test of soil stoichiometry.

Index	Order of Importance	Explains/%	*F*	*p*
MBC	1	44.1	4.7	0.072
BAC/FUN	2	17.6	3.0	0.144
SOC/TN	3	15.1	1.8	0.234
MBN	4	8.6	4.2	0.218
SOC	5	6.7	1.2	0.368
TN	6	5.8	1.1	0.416

SOC: Soil organic carbon; TN: Soil total nitrogen; MBC: Soil microbial biomass carbon; MBN: Soil microbial biomass nitrogen; SOC/TN: Soil C/N ratio; BAC/FUN: Soil bacteria to fungi ratio, the same below.

4. Discussion

4.1. The Abundance Characteristics of $\delta^{13}C$ and $\delta^{15}N$ in Zanthoxylum planispinum var. dintanensis Plantations of Different Plantation Age

The greater the $\delta^{13}C$ value, the higher the water usage efficiency over a long period [27]. The results of our work indicated there was no significant difference in $\delta^{13}C$ among the four plantation age groups, indicating that the water usage efficiency did not vary with plantation age; this was a result of the balance between resource acquisition and spending balance in plantation trees [28]. This may be due to a high competition ability for consumable resources in the vulnerable karst habitat, which minimized age effects. Our results rejected the hypothesis that there was a significant difference in $\delta^{13}C$ between young and old plant tissues. Kieckbusch et al. (2004) [29] showed that there was no significant difference in the $\delta^{13}C$ composition between green leaves and aged leaves, while Lee et al. [30] (2000) showed a significant difference in the $\delta^{13}C$ composition between green leaves and yellow leaves in the two plants. Thus, the observed variation in $\delta^{13}C$ as leaves aged, and its physiological and ecological significance, still need further study. Under the influence of "canopy effect", the closer to the soil surface, the smaller the $\delta^{13}C$ value in plant leaves, and the more obvious the dilution phenomenon [31]. Because the canopy of *Z. planispinum* var. *dintanensis* was relatively small and frequently trimmed, the canopy effect was weak. This meant that it was reasonable to sample middle canopy leaves as representative of the whole canopy. In future research, it will be advantageous to sample leaves from the upper, middle, and lower canopies, to further reveal the mechanisms of $\delta^{13}C$ fractionation. The leaf $\delta^{15}N$ in the 10–12-year-old group was significantly higher than that of the other three age groups, probably due to high N demand during the vigorous fruit bearing period, which stimulated the root system to transfer more N to leaves for photosynthesis. Leaf $\delta^{15}N$ increased first and then decreased with plantation age, which was inconsistent with the research of Wang et al. (2019) [9]. This inconsistency may be due to different N isotope fractionation speeds in the photosynthetic processes of different species.

The $\delta^{13}C$ in soil and litter did not vary significantly with plantation age. This may be because the organic C in soil was derived primarily from litter. C input and output C from litter, combined with the decomposition of soil C, may have determined the characteristics of soil $\delta^{13}C$ [32]. Balesdent et al. (1993) found that soil $\delta^{13}C$ was positively correlated to leaf litter $\delta^{13}C$, although the results were not significant [33], which was consistent with our results, indicating that soil could not fully inherit leaf $\delta^{13}C$, even if C isotope fractionation was not considered in the litter decomposition process. The decomposition rate of litter determines the flow direction of C. Because fresh organic C is more easily decomposed by microorganisms, the soil $\delta^{13}C$ value was the result of mixed new and old C in soil, demonstrating an isotope mixing effect [34–36]. The results of Buchmann et al. (1997) and Farquhar et al. (1989) showed that soil $\delta^{13}C$ generally fell within the range of 1.0–3.0‰; a value higher than 3.0‰ indicates that the organic matter input into the soil may derive from a mixture of C_3 and C_4 plants [3,37]. The average variation of $\delta^{13}C$ in our study was 7.46‰, which is significantly higher than 3.0‰, the reason is that the *Z. planispinum* var. *dintanensis* plantation was initially a mix of coniferous and broad-leaf trees, the soil organic matter would have been influenced by both. With the gradual increase in the atmospheric CO_2

concentration, its $\delta^{13}C$ value decreases continuously, and the amount of CO_2 released into the atmosphere by different ecosystems through respiration varies [5]. It is speculated that the $\delta^{13}C$ value of atmospheric CO_2 changes after agricultural transformation in this area. The variation law of the soil $\delta^{15}N$ presented in this study was inconsistent with that of the Chinese fir forest in Fujian district [9]. The reason was that there were fewer forest plant classes, little difference in litter regression and accumulation, and more intensive human interference in the *Z. planispinum* var. *dintanensis* plantation. Considerable research has shown that soil $\delta^{13}C$ and $\delta^{15}N$ increase with the soil profile [3,12,38,39]. However, this research was conducted in an area with shallow soil cover, and most soil is <20 cm thick and has a high gravel content, so the soil samples were not collected in this study, which limited the understanding of C and N cycles in the soil profile of the karst region. In the future, the soil layer should be divided into finer sublayers, and samples should be collected from different depths; for instance, from 0–2 cm, 2–5 cm, and 5–10 cm soil layers [5]; this will elucidate more clearly the soil C and N cycles and the varying mechanisms of $\delta^{13}C$ and $\delta^{15}N$ fractionations in space.

The $\delta^{13}C$ in *Z. planispinum* var. *dintanensis* leaves was significantly negatively correlated to N content, which is consistent with the results of Tsialtas et al. (2001) [40], yet conflicts with the findings of Zhang et al. (2015) [41]. This finding indicates that N acquisition is different in different environments, which affects the plant leaf $\delta^{13}C$ value. The reason is that leaf N can regulate stomatal density, and higher leaf N content promotes the absorption of CO_2, increases plant photosynthesis rate, and decreases the ratio of intracellular and extracellular CO_2 concentration (C_i/C_a), which leads to an increase in $\delta^{13}C$ [42,43]. Our research area was a barren karst region, where supplementary fertilization is needed for adequate plant growth. Modern agriculture emphasizes supplementation with N and *p*, which leads to greater N leaf uptake, increases stomatal density, and C_i/C_a ratio, thus decreasing $\delta^{13}C$. The results of our study showed that leaf $\delta^{13}C$ was positively correlated to litter $\delta^{15}N$ and negatively correlated to soil $\delta^{15}N$, indicating the coupling relationship between leaves, leaf litter, and soil. The reason was that the C and N cycles in forest ecosystem went through the entire plant-litter-soil continuum. Moreover, the C cycle and N cycles were tightly coupled, and the potential for C fixation was heavily limited by the soil N supply [44,45]. The C and the N cycles in the forest ecosystem were regulated by environmental factors, leading to unique connections between continuums; in addition, nutrient reabsorption and allocation may cause isotope fractionation. However, due to the large number of influencing factors and limited measurement indicators, the reason for the weak inheritance could not be clarified. Further research is required.

4.2. The Driving Mechanism of Soil Stoichiometry to Plantation C and N Isotopes Fractionation

Soil stoichiometry links the chemical cycles in different parts of an ecosystem, reflecting the flowing of elements [46], indirectly regulating forest C and N isotopes fractionation via changing the coupling relationship between soil and microorganism stoichiometry. It is an important index for the evaluation of ecosystem element cycles and internal stability [1,47]. The contents of soil elements can affect these results and restrict the application of stable isotope technology in soil C and N cycles [15,48]. Stevenson et al. (2010) indicated that soil C/N was significantly negatively correlated to $\delta^{15}N$ [49]. The reason was that the biological activity of microbes in soil with different C/N were different, which led to different fractionation speeds and degree in the process of mineralization. Generally, the growth of microorganisms is limited by N content in high C/N soil, thus weakening the $\delta^{15}N$ fractionation in the mineralization process; on the other hand, under low C/N conditions, the growth of microorganisms is limited by C content, thus strengthening the N decomposition in the process of mineralization [7]. The current study showed that there was a weak correlation between soil C/N and $\delta^{15}N$, which was not completely consistent with previous studies [49]. A possible reason could be leaf litter type and quantity are lower in artificial forest, and human interference combined with a high concentration of allelochemicals secreted by *Z. planispinum* var. *dintanensis* inhibited the decomposition of leaf litter and associated microbial activity, leading to a reduced recycling of nutrients. Our research also demonstrated the negative correlation between soil C/N and $\delta^{13}C$. This was attributed to the low C/N soil SOC decomposing faster, more $^{12}CO_2$ being released from the soil, and the remaining soil C library enriched by ^{13}C [50]. Wang et al. (2015) reached a similar conclusion; nonetheless [8], Peri et al. (2012) found that soil C/N did not affect soil $\delta^{13}C$ in their study of the primeval forests in southern Patagonia [32]. A possible reason could be that the climate was different in each research area, leading to different leaf litter types and quantities, and plants adopt different resource utilization and adaptation strategies under different climatic conditions. Soil C and N are indispensable elements for plant survival and it is scientifically feasible to use C/N to determine the composition characteristics of soil $\delta^{13}C$ and $\delta^{15}N$, although it is not the only criterion. In the future, research on coupling with other soil factors should be conducted to comprehensively evaluate soil quality and nutrient status.

As the most active part of soil organic matter [51], biomass can establish good connections with $\delta^{13}C$ and $\delta^{15}N$ through the decomposition of organic matter and microbial activity [52]. Our results showed that soil MBC was positively correlated to soil $\delta^{13}C$, which is related to the isotope fractionation in the process of microorganism decomposition [53]. During the process, ^{12}C enters the released CO_2 preferentially, and the heavier ^{13}C more likely enters the soil microbial biomass before returning to soil organic matter [54]. Relevant research has shown that soil $\delta^{13}C$ was positively correlated with organic C [55]. When the decomposition of organic C speeds up, more $^{12}CO_2$ will be released from the soil system, thus resulting in the enrichment of $\delta^{13}C$ in soil [56]. Inconsistent with these results, organic C did not show significant correlation with $\delta^{13}C$ in our study (Figure 3) indicating that the soil organic C in our research area had no significant influence on C isotope fractionation. This may be because pruning is carried out in winter and in summer to improve the economic value of the *Z. planispinum* var. *dintanensis* plantations, and this reduces the litter return and nutrients; on the other hand, the unique dual structure of karst leads to the aggravation of water loss and soil erosion. In conclusion, litter and microorganisms were important sources of soil nutrients, which should be protected to improve soil quality.

5. Conclusions

Different types of forest have a different natural abundance of $\delta^{13}C$ and $\delta^{15}N$. In *Z. planispinum* var. *dintanensis* plantations, the soil $\delta^{13}C$ value gradually decreased, the $\delta^{15}N$ value of leaves and litter increased first and then decreased, and the remaining indicators did not change; the overall soil showed isotopic enrichment effects. There is a trade-off between plantation indicators between leaf $\delta^{13}C$ and $\delta^{15}N$ only, the parameters of other indicators are synergistic. There is a correlation between soil stoichiometry and *Z. planispinum* var. *dintanensis* plantations; soil MBC and BAC/FUN have a relatively strong driving effect on plantation C and N and their isotopes. Henceforth, attention should be paid to measures that can protect litter and soil microorganisms in order to retain the quality and the nutrient composition of plantation soil, and thus delay its decline.

Author Contributions: Conceptualization, Y.Y. and Y.W.; methodology, Y.Y.; software, Y.W.; validation, Y.Y. and Y.W.; formal analysis, Y.W.; investigation, Y.S. and Y.L.; resources, Y.Y.; data curation, Y.W.; writing—original draft preparation, Y.Y. and Y.W.; writing—review and editing, Y.W., Y.Y. and Y.S.; visualization, Y.L.; supervision, Y.L.; project administration. All authors have read and agreed to the published version of the manuscript.

Funding: Guizhou Province Science and Technology Support Plan Project (Qian-ke-he Zhicheng [2022] Yiban 103).

Informed Consent Statement: Informed consent was obtained from all subjects involved in the study.

Data Availability Statement: No relevant datasets were used. Therefore, no data availability statement exists.

Conflicts of Interest: The authors declare no conflict of interest.

References

1. Zhou, L.; Song, M.H.; Wang, S.Q.; Fan, J.W.; Liu, J.Y.; Zhong, H.P.; Yu, G.R.; Gao, L.P.; Hu, Z.M.; Chen, B.; et al. Patterns of soil ^{15}N and total and their relationships with environmental factors on the Qinghai-Tibetan Plateau. *Soil Sci. Soc. Chin.* **2014**, *24*, 232–242. [CrossRef]
2. Liu, X.Z.; Gao, C.C.; Su, Q.; Zhang, Y.; Song, Y. Altitudinal trends in $\delta^{13}C$ value, stomatal density and nitrogen content of *Pinnus tabuliformis* needles on the southern slope of the middle Qinling Mountains, China. *J. Mount. Sci.* **2016**, *13*, 1066–1077. [CrossRef]
3. Farquhar, G.D.; Ehleringer, J.R.; Hubick, K.Y. Carbon isotope discrimination and photosynthesis. *Annu. Rev. Plant Physiol. Plant Molec. Biol.* **1989**, *40*, 503–537. [CrossRef]
4. Vogel, J.C. Fractionation of the carbon isotopes during photosynthesis. *Architect. Instit. Jpn.* **1980**, *3*, 111–135.
5. Lin, G.H. *Stable Isotope Ecology*; Higher Education Press: Beijing, China, 2013.
6. Fan, H.B.; Wu, J.P.; Liu, W.F.; Yuan, Y.H.; Hu, L.; Cai, Q.K. Linkages of plant and soil C:N:Pstoichimetry and their relationships to forest growth in subtropical plantations. *Plant Soil* **2015**, *39*, 127–138. [CrossRef]
7. Collins, J.G.; Dijkstra, P.; Hart, S.C.; Hungate, B.A.; Flood, N.M.; Schwartz, E. Nitrogen source influences natural abundance ^{15}N of *Escherichia coli*. *FEMS Microbiol. Lett.* **2008**, *282*, 246–250. [CrossRef]
8. Wang, G.A.; Jia, Y.F.; Li, W. Effects of environmental and biotic factors on carbon isotopic fractionation during decomposition of soil organic matter. *Sci. Rep.* **2015**, *5*, 11043. [CrossRef]
9. Wang, L.; Cao, X.Q.; Sun, M.Y.; Li, P.X.; Teng, Z.; Xu, X.N. Foliar carbon and nitrogen stoichiometric trait and isotope discrimination of different *Cunninghamia lanceolata* foliar ages across an altitude gradient. *Ecol. Environ. Sci.* **2019**, *28*, 1776–1784.
10. Tanaka-Oda, A.; Kenzo, T.; Koretsne, S.; Sasaki, H.; Fukuda, K. Ontogenetic changes in water-use efficiency ($\delta^{13}C$) and leaf traits differ among tree species growing in a semiarid region of the Loss Plateau, China. *Forest Ecol. Manag.* **2010**, *259*, 953–9757. [CrossRef]
11. Song, L.N.; Zhu, J.J.; Yan, Q.L.; Li, M.C.; Yu, G.Q. Comparison of intrinsic water use efficiency between different aged *Pinus sylvestris* var. *mongolica* wide windbreaks in semiarid sandy land of northern China. *Agroforest. Syst.* **2015**, *89*, 477–489.
12. Zheng, L.J.; Huang, Z.Q.; He, Z.M.; Wang, X.Y.; Liu, X.M.; Liu, R.Q.; Xiao, H.Y. Influence of forest and foliar ages on the composition of stable carbon and nitrogen isotope of *Cunninghamia lanceolata* in Subtropic China. *Sci. Silv. Sin.* **2015**, *51*, 22–28.
13. Lucas-Borja, M.E.; Hedo, J.; Cerdá, A.; Candel-Pérez, D.; Vinegla, B. Unravelling the importance of forest age stand and forest structure driving microbiological soil properties, enzymatic activities and soil nutrients content in Mediterranean Spanish black pine (*Pinus nigra* Ar. ssp. *Salzmannii*) Forest. *Sci. Total Environ.* **2016**, *562*, 145–154. [CrossRef] [PubMed]
14. Deng, L.; Wang, K.; Tang, Z.S.; Shangguan, Z.P. Soil organic carbon dynamics following natural vegetation restoration: Evidence from stable carbon isotopes ($\delta^{13}C$). *Agric. Ecosyst. Environ.* **2016**, *221*, 235–244. [CrossRef]
15. Zhao, Y.F.; Wang, X.; Ou, Y.S.; Jia, H.X.; Li, J.; Shi, C.M.; Liu, Y. Variations in soil $\delta^{13}C$ with alpine meadow degradation on the eastern Qinghai-Tibet Plateau. *Geoderma* **2019**, *338*, 178–186. [CrossRef]
16. Lorenz, M.; Derrien, D.; Zeller, B.; Udelhoven, T. The linkage of ^{13}C and ^{15}N soil depth gradients with C:N and O:C stoichiometry reveals tree species effects on organic matter turnover in soil. *Biogeochemistry* **2020**, *151*, 203–220. [CrossRef]
17. Yang, L.; Xiong, K.N.; Xiao, S.Z.; Lan, J.C.; Ao, X.H. Carbon sequestration benefits of Zanthoxylum Bungeanum forest in karst rocky desertification areas. *Bull. Soil Water Conserv.* **2016**, *36*, 291–297.
18. Li, A.D.; Yu, L.F.; Wei, X.L. Ecological Water Requirement of *Zanthoxylum bungeanum* Forestland in Typical Karst Valleys. *Soil* **2008**, *40*, 490–494.
19. Qu, S.; Yang, R.; Wang, Y.; Pei, Y.D.; Li, K.F.; Hu, J.D. Change characteristics of soil nutrients during *Zanthoxylum bungeanum* var. *dingtanensis* grow process in karst plateau. *Non-Wood For. Res.* **2020**, *38*, 183–192.
20. Zou, J.; Yu, L.F.; Huang, Z.S. Variation of leaf carbon isotope in plants in different lithological habitats in a karst area. *Forest* **2019**, *10*, 356. [CrossRef]
21. Zeng, D.H.; Chen, G.S.; Chen, F.S.; Zhao, Q.; Ji, X.Y. Foliar nutrients and resorption efficiencies in four *Pinus sylvestris* var. *mongolica* plantations of different ages on sandy soil. *Sci. Silv. Sin.* **2005**, *41*, 21–27.
22. Yan, T.; Lü, X.T.; Yang, K.; Zhu, J.J. Leaf nutrient dynamics and nutrient resorption: A comparison between larch plantations and adjacent secondary forest in northeast, China. *J. Plant Ecol.* **2016**, *9*, 165–173.
23. Farquhar, G.D.; Leary, M.O.; Berry, J. On the relationship between carbon isotope discrimination and the intercellular carbon isotope concentration in leaves. *Austr. J. Plant Physiol.* **1982**, *9*, 121–137. [CrossRef]
24. Bao, S.D. *Soil Agricultural Analysis*, 3rd ed.; China Agricultural Publishing House: Beijing, China, 2000.
25. Chen, Y.L.; Chen, L.Y.; Peng, Y.F.; Ding, J.Z.; Li, F.; Yang, G.B.; Kou, D.; Liu, L.; Fang, K.; Zhang, B.B.; et al. LinKing microbial C:N:P stoichiometry to microbial community and abiotic factors along a 2500-km grassland transect on the Tibetan Plateau. *Glob. Ecol. Biogeogr.* **2016**, *25*, 1416–1427.
26. Lin, X.G. *Principles and Methods of Soil Microbial Research*; Higher Education Press: Beijing, China, 2017.
27. Chen, J.; Chang, S.X.; Anyia, A.O. The physiology and stability of leaf carbon isotope discrimination as a measure of water-use efficiency in barley on the Canadian prairies. *J. Agron. Crop Sci.* **2011**, *197*, 1–11.
28. Heberling, J.M.; Fridley, J.D. Biogeographic constraints on the world-wide leaf economics spectrum. *Glob. Ecol. Biogeogr.* **2012**, *21*, 1137–1146.
29. Kieckbusch, D.K.; Koch, M.S.; Serafy, J.E.; Anderson, W. Trophic linkages among primary producers and consumers in fringing mangroves of subtropical lagoons. *Bull. Mar. Sci.* **2004**, *74*, 271–285.

30. Lee, S. Carbon dynamics of Deep Bay, eastern Pearl River estuary, China. I I: Trophic relationship based on carbon and nitrogen-stable isotopes. *Mar. Ecol. Prog. Ser.* **2000**, *205*, 1–10.
31. Martinelli, L.; Almeida, S.; Brown, I.; Moreira, M.; Victoria, R.; Sternberg, L.; Ferreira, C.; Thomas, W. Stable carbon isotope ratio of tree leaves, boles and fine litter in a tropical forest in Rondonia, Brazil. *Oecologia* **1998**, *114*, 170–179.
32. Peri, P.L.; Ladd, B.; Pepper, D.A.; Bonser, S.P.; Laffan, S.W.; Amelung, W. Carbon ($\delta^{13}C$) and nitrogen ($\delta^{15}N$) stable isotope composition in plant and soil in southern Patagonia's native forests. *Glob. Change Biol.* **2012**, *18*, 311–321. [CrossRef]
33. Balesdent, J.; Cirardin, C.; Mariotti, A. Site-related $\delta^{13}C$ of tree leaves and soil organic matter in a temperate forest. *Ecology* **1993**, *74*, 1713–1721.
34. Högberg, P.; Ekblad, A. Substrate-induced respiration measured in situ in a C_3-plant ecosystem using additions of C_4-sucrose. *Soil Biol. Biochem.* **1996**, *28*, 1131–1138. [CrossRef]
35. Liao, J.D.; Boutton, T.W.; Jastrow, J.D. Organic matter turnover in soil physical fractions following woody plant invasion of grassland: Evidence from natural ^{13}C and ^{15}N. *Soil Biol. Biochem.* **2006**, *38*, 3197–3210. [CrossRef]
36. Huang, Y.H.; Li, Y.L.; Xiao, Y.; Wenigmann, K.O.; Zhou, G.Y.; Zhang, D.Q.; Wenigmann, M.; Tang, X.L.; Liu, J.X. Controls of litter quality on the carbon sink in soils through partitioning the products of decomposing litter in a forest succession series in South China. *Forest Ecol. Manag.* **2011**, *261*, 1170–1177. [CrossRef]
37. Buchmann, N.; Kao, W.Y.; Ehleringer, J. Influence of stand structure on carbon-13 of vegetation, soils, and canopy air within deciduous and evergreen forests in Utah, United States. *Oecologia* **1997**, *110*, 109–119. [CrossRef]
38. Mariotti, A.; Pierre, D.; Vedy, J.C.; Bruckert, S.; Guillenmot, J. The abundance of natural nitrogen 150in the organic matter of soils along an altitudinal gradient. *Catena* **1980**, *7*, 293–300. [CrossRef]
39. Ehleringer, J.R.; Buchmann, N.; Flanagan, L.B. Carbon isotope ratios in belowground carbon cycle processes. *Ecol. Appl.* **2000**, *10*, 412–422. [CrossRef]
40. Tsialtas, J.T.; Handley, L.; Kassioumi, M.T.; Veresoglou, D.S.; Gagianas, A.A. Interspecific variation in potential water- use efficiency and its relation to plant species abundance in a water-limited grassland. *Funct. Ecol.* **2001**, *15*, 605–614. [CrossRef]
41. Zhang, J.; Gu, L.; Cao, Y.; Hao, Y.; He, J.; Li, J.; Li, Y.; Ren, Y.; Wang, F.; Wu, F.; et al. Nitrogen control of ^{13}C enrichment in heterotrophic organs relative to leaves in a landscape-building desert plant species. *Biogeosciences* **2015**, *12*, 15–27. [CrossRef]
42. Macfarlane, C.; Arndt, S.K.; Livesley, S.J.; Edgar, A.C.; White, D.A.; Adams, W.A.; Eamus, D. Estimation of leaf area index in eucalypt forest with vertical foliage, using cover and fullframe fisheye photography. *Forest Ecol. Manag.* **2007**, *242*, 756–763. [CrossRef]
43. Diefendorf, A.F.; Mueller, E.; Wing, S.L.; Koch, P.L.; Freeman, K.H. Global patterns in leaf ^{13}C discrimination and implications for studies of past and future climate. *Proc. Natl. Acad. Sci. USA* **2010**, *107*, 5738–5743. [CrossRef]
44. Li, D.J.; Niu, S.L.; Luo, Y.Q. Global patterns of the dynamics of soil carbon and nitrogen stocks following afforestation: A metanalysis. *New Phytol.* **2012**, *195*, 172–181. [CrossRef] [PubMed]
45. Zechmeister-Boltenstern, S.; Keiblinger, K.M.; Mooshammer, M.; Peuelas, J.; Richter, A.; Sardans, J.; Wanek, W. The application of ecological stoichiometry to plant-microbial-soil organic matter transformation. *Ecol. Monogr.* **2015**, *85*, 133–155. [CrossRef]
46. Yang, Y.; Liu, B.R.; An, S.S. Ecological stoichiometry in leaves, roots, and soil among different plant communities in a desertified region of Northern China. *Catena* **2018**, *166*, 328–338. [CrossRef]
47. Mooshammer, M.; Wanek, W.; Zechmeister, B.S.; Richter, A. Stoichiometric imbalances between terrestrial decomposer communities and their resources: Mechanisms and implications of microbial adaptations to their resources. *Front. Microbiol.* **2014**, *5*, 22. [CrossRef]
48. Wang, C.; Houlton, B.Z.; Liu, D.W.; Hou, J.F.; Cheng, W.X.; Bai, E. Stable isotopic constraints on global soil organic carbon turnover. *Biogeosciences* **2018**, *15*, 987–995. [CrossRef]
49. Stevenson, B.A.; Partt, R.L.; Schipper, L.A.; Baisden, W.T.; Mudge, P. Relationship between soil ^{15}N, C/N and N losses across land uses in New Zealand. *Agric. Ecosyst. Environ.* **2010**, *139*, 736–741. [CrossRef]
50. Sinsabaugh, R.L.; Carreiro, M.M.; Repert, D.A. Allocation of extracellular enzymatic activity in relation to litter composition, N deposition, and mass loss. *Biogeochemistry* **2002**, *60*, 1–24. [CrossRef]
51. Arunachalam, K.; Arunachalam, A.; Melkania, N.P. Influence of soil properties on microbial populations, activity and biomass in humid subtropical mountainous ecosystems of India. *Biol. Fertil. Soil* **1999**, *30*, 217–223. [CrossRef]
52. Nel, J.A.; Craine, J.M.; Cramer, M.D. Correspondence between $\delta^{13}C$ and $\delta^{15}N$ in soils suggests coordinated fraction processes for soil C and N. *Plant Soil* **2018**, *423*, 257–271. [CrossRef]
53. Billings, S.A.; Richter, D.D. Changes in stable isotopic of soil nitrogen and carbon during 40 years of forest development. *Oecologia* **2006**, *148*, 325–333. [CrossRef]
54. De Rouw, A.; Soulileuth, B.; Huon, S. Stable carbon isotope rations in soil and vegetation shift with cultivation practices (Northern Laos). *Agtric. Ecosyst. Environ.* **2015**, *200*, 161–168. [CrossRef]
55. Wynn, J.C.; Bird, M.I. Environmental controls on the stable carbon isotopic composition of soil organ C_3 and C_4 plants, Australia. *Tellus Ser. B Chem. Phys. Meteorol.* **2008**, *60*, 604–621. [CrossRef]
56. Wynn, J. Carbon isotope fractionation during decomposition of organic matter in soil and paleosols: Implication for palaeoecological interpretation of paleosols. *Palaeogeograph. Palaeoclimatol. Palaeoecol.* **2007**, *251*, 437–448. [CrossRef]

agronomy

MDPI

Article

Stoichiometry of Soil, Microorganisms, and Extracellular Enzymes of *Zanthoxylum planispinum* var. *dintanensis* Plantations for Different Allocations

Yitong Li, Yanghua Yu * and Yanping Song

School of Karst Science, State Engineering Technology Institute for Karst Decertification Control, Guizhou Normal University, Guiyang 550001, China; liyitong0711@163.com (Y.L.); syp_2022@163.com (Y.S.)
* Correspondence: yuyanghua2003@163.com; Tel.: +86-138-8505-4791

Abstract: Plantations with different allocation patterns significantly affect soil elements, microorganisms, extracellular enzymes, and their stoichiometric characteristics. Rather than studying them as a continuum, this study used four common allocations of plantations: *Zanthoxylum planispinum* var. *dintanensis* (hereafter *Z. planispinum*) + *Prunus salicina*, *Z. planispinum* + *Sophora tonkinensis*, *Z. planispinum* + *Arachis hypogaea*, and *Z. planispinum* + *Lonicera japonica* plantations, as well as a single-stand *Z. planispinum* plantation as a control. Soil samples from depths of 0–10 and 10–20 cm at the five plantations were used to analyze the element stoichiometry, microorganisms and extracellular enzymes. (1) One-way analysis of variance (ANOVA) showed that the contents of soil organic carbon (C), nitrogen (N), phosphorus (P), and potassium (K) of *Z. planispinum* + *L. japonica* plantation were high, while those of calcium (Ca) and magnesium (Mg) were low compared to the *Z. planispinum* pure plantation; soil microbial and enzyme activities were also relatively high. Stoichiometric analysis showed that soil quality was good and nutrient contents were high compared to the other plantations, indicating that this was the optimal plantation. (2) Two-way ANOVA showed that stoichiometry was more influenced by plantation type than soil depth and their interaction, suggesting that plantation type significantly affected the ecosystem nutrient cycle; soil microbial biomass (MB) C:MBN:MBP was not sensitive to changes in planting, indicating that MBC:MBN:MBP was more stable than soil C:N:P, which can be used to diagnose ecosystem nutrient constraints. (3) Pearson's correlation and standardized major axis analyses showed that there was no significant correlation between soil C:N:P and MBC:MBN:MBP ratios in this study; moreover, MBN:MBP had significant and extremely significant correlations with MBC:MBN and MBC:MBP. Fitting the internal stability model equation of soil nutrient elements and soil MBC, MBN, and MBP failed ($p > 0.05$), and the MBC, MBN, and MBP and their stoichiometric ratios showed an absolute steady state. This showed that, in karst areas with relative nutrient deficiency, soil microorganisms resisted environmental stress and showed a more stable stoichiometric ratio. Overall stoichiometric characteristics indicated that the *Z. planispinum* + *L. japonica* plantation performed best.

Keywords: *Zanthoxylum planispinum* var. *dintanensis*; allocations; plantations; stoichiometry; soil microorganisms; soil extracellular enzymes; internal stability

Citation: Li, Y.; Yu, Y.; Song, Y. Stoichiometry of Soil, Microorganisms, and Extracellular Enzymes of *Zanthoxylum planispinum* var. *dintanensis* Plantations for Different Allocations. *Agronomy* **2022**, *12*, 1709. https://doi.org/10.3390/agronomy12071709

Academic Editor: Yanyou Wu

Received: 18 May 2022
Accepted: 16 July 2022
Published: 19 July 2022

1. Introduction

Soil, as a substrate for plant growth, and a source and sink of water–fertilizer, serves as a medium of continuous material and energy exchanges. The relationship between soil and plant growth can be characterized by the elements, microorganisms, extracellular enzymes of soil, and their interactions [1]. Among them, soil carbon (C), nitrogen (N), phosphorus (P), and other elements are important components of soil fertility, which directly affect plant growth, soil microbial dynamics, litter decomposition, and the accumulation and circulation of soil nutrients [2]. Soil microorganisms mineralize organic matter mainly by secreting extracellular enzymes, which helps absorption of soil nutrients, so as to balance element acquisition and investment [3,4]. Ecological stoichiometry is a discipline that studies ecology from multiple perspectives [5], including the balance of energy and various chemical elements in ecosystems [6], in order to analyze and predict ecosystem structure and dynamics. Soil nutrient stoichiometry can reveal the coupling relationship and effectiveness of soil nutrients. Soil microbial biomass stoichiometry can be used as an important C, N, and P ecosystem flux to reflect soil microbial mineralization or fixation of soil elements [7]. Soil extracellular enzyme stoichiometry is an important indicator to reveal microbial nutrient status and relative resource constraints [8]. Therefore, the study of soil ecological stoichiometry helps to clarify the mechanism of soil nutrient availability, circulation, and balance, and explain soil microbial metabolic activities [9,10]. Studying the continuum of soil, microorganisms, and extracellular enzymes can help to integrate and analyze the circulation rule of soil nutrients in forest ecosystems and explore the balance among different soil components.

Researchers have carried out many studies on soil ecological stoichiometry. Some scholars [11,12] analyzed the soil C:N:P ratio in the Loess Plateau in China and other places, and found that C, N, and P showed significant correlation in varying degrees, which indicates that element stoichiometry is sensitive to variation in contents. Previous studies found that soil element stoichiometry has a close relationship with microbial biomass stoichiometry, and soil microorganisms have strong steady-state behavior, indicating that soil microorganisms adaptively respond to nutrient levels [13–15]. Other scholars [16,17] proposed that the theory of extracellular enzyme stoichiometry can clarify the balance between microbial biomass stoichiometry and soil element composition. Wu et al. [18] clarified that there is a strong covariant relationship between soil, microorganisms, and extracellular enzyme stoichiometry. In conclusion, forest soil stoichiometry emphasizes the correlation between soil elements, microorganisms, and extracellular enzymes. Rather than studying them as a continuum [18,19], most studies focus on elements C, N, and P, and pay less attention to potassium (K), calcium (Ca), and magnesium (Mg). In addition, different plantation types produce different litter and root exudates, resulting in different nutrient transport to the soil, further affecting soil element stoichiometry, microbial dynamics, and nutrient cycles [20,21]. Yet, there have been few studies on the soil stoichiometry of plantations.

Zanthoxylum planispinum var. *dintanensis* (hereafter *Z. planispinum*) has the characteristics of Ca preference and tolerance to drought and rocky soils. It is the most suitable tree species for ecological restoration in many typical rocky desertification areas. It has a balanced variety and rich content of amino acids, proteins, vitamins, mineral elements, and other metabolic substances with outstanding quality advantages. Due to its unique fragrance and numb taste [22], it has become a major seasoning. However, in recent years, some problems, including soil quality degradation, fruit yield, quality reduction, and stand instability, have arisen, probably due to the single-stand structure of plantations. Therefore, clarifying the soil stoichiometry of *Z. planispinum* plantations in different allocations is critical for the improvement of stand stability and delay of any decline. In view of this, this study took four common allocations of plantations in a rocky desertification area of central Guizhou Province, China: *Z. planispinum* + *Prunus salicina*, *Z. planispinum* + *Sophora tonkinensis*, *Z. planispinum* + *Arachis hypogaea*, and *Z. planispinum* + *Lonicera japonica* plantations, as well as a single-stand plantation as a control to study the stoichiometry of soil, microorganisms, and extracellular enzymes. The following issues were addressed: (1) the changing rules of soil ecological stoichiometry characteristics with plantation type; (2) the sensitivity comparison of microbial ecological stoichiometry and soil element stoichiometry indicating nutrient limitation, and (3) the internal relationships among the stoichiometry of elements, microorganisms, and enzymes. The purpose is to understand the interaction rule between soil nutrients and microorganisms in *Z. planispinum* plantations, and to provide a theoretical foundation for optimal stand plantations and soil nutrient regulation.

2. Materials and Methods

2.1. Overview of the Research Site

The study area is located in Beipanjiang Town, Zhenfeng County, Guizhou, and is part of the Beipan River Basin. The habitat is uniquely characterized as (1) a subtropical humid monsoon climate with a dry and hot climate, and an average annual rainfall of about 1100 mm. It is rich in heat resources, with an annual total accumulated temperature of 6542.9 °C, annual average temperature of 18.4 °C, and annual extreme maximum and minimum temperatures of 32.4 and 6.6 °C, respectively; (2) the valley terrain is in the altitude range of 530–1473 m asl and has typical climate characteristics of the region; (3) the habitat has experienced rocky desertification, with low forest coverage. Exposed bedrock is high, with carbonate rocks accounting for 78.45%, and soil is mainly lime soil [23]; and (4) cultivated vegetation includes *Z. planispinum*, *Juglans regia*, and *Tectona grandis*; agricultural crops are mainly *Hylocereus polyrhizus* and maize (*Zea mays*), and shrubs are mainly *L. japonica*.

2.2. Sample Plot Setting and Sample Collection

One sample plot was set up, with basically the same environmental conditions, for each of the five selected stand types to measure geographical factors such as longitude, latitude, altitude, and slope, as well as the community structure indexes of *Z. planispinum*, such as tree age, density, plant height, crown width, and coverage (Table 1). The *Z. planispinum* were planted in five sample plots in 2012. From the end of 2015 to 2016, *Prunus salicina*, *Sophora tonkinensis*, *Arachis hypogaea*, and *Lonicera japonica* were allocated around the *Z. planispinum*, and the allocation time was determined according to plant growth habits. The allocation density was 600–750, 1500–1800, 2500, and 450–600 plants ha^{-1}, respectively. In implementing the policies of returning farmland to forest and reducing maize planting, the ground was already cultivated, which ensured that the soil fertility level of all sample plots was similar before planting *Z. planispinum*. The same fertilization measures were adopted in all sample plots. The specific fertilization measures were to use compound fertilizers based on N, P, and K ($N:P_2O_5:K_2O$ = 15:15:15, total nutrients $\geq$ 45%), fertilizing once in the first 10 days of September and once in the first half of March, both of which used about 0.2 kg of fertilizer per plant, spreading the fertilizer in a ring 20–30 cm away from the main stem of *Z. planispinum* (because *Z. planispinum* fine roots are densely distributed in this area), and covering with 1–2 cm of soil after fertilization to avoid fertilizer loss. After planting *Z. planispinum* for 3 years, fruits were harvested once a year during July–August, producing about 3–4 kg per plant, without other special treatment measures. After allocating *P. salicina* for 3 years, fruits were harvested once a year (a total of twice by the sampling time), producing about 5–6 kg per plant each time. The whole plants of *S. tonkinensis* were harvested once in the third year of allocation. After *A. hypogaea* allocation, the whole plants were harvested once a year, and were harvested a total of four times by the sampling time. *Lonicera japonica* was harvested following the second year of allocation, making a total of three harvested by the sampling time. The soil was mainly sandy and loam, with pH 6.23–8.16; the soil layer was shallow, and the thickness was mainly about 10–18 cm. Soil was mainly yellow and red, plasticity was low, and aggregate structure was poor. The content of gravel in soil was high, and the mass ratio was about 30%.

Table 1. General information of plots.

Plantation Type	Longitude	Latitude	Altitude (m asl)	Slope (°)	Tree Age (a)	Density (m)	Height (m)	Crown Width (m)	Coverage (%)
YD1	105°40′28.33″ E	25°37′57.41″ N	764	10	8	3 × 3	3.5	2 × 2.3	70
YD2	105°40′19.79″ E	25°39′25.75″ N	728	10	8	2 × 2	2.0	1.2 × 1.8	60
TD3	105°38′36.32″ E	25°39′23.64″ N	791	10	8	2 × 2	2.5	2.5 × 2.8	85
YD4	105°38′36.35″ E	25°39′22.29″ N	814	10	8	3.5 × 3	2.5	1.5 × 2.5	70
YD5	105°38′35.64″ E	25°39′23.35″ N	788	10	8	3 × 4	2.2	2.5 × 2.3	65

YD1: *Z. planispinum* + *P. salicina*; YD2: *Z. planispinum* + *S. tonkinensis*; YD3: *Z. planispinum* + *A. hypogaea*; YD4: *Z. planispinum* + *L. japonica*; YD5: *Z. planispinum* pure plantation.

Soil samples were collected in mid-November 2020. The plants were dormant at this time, with weak soil material exchange and low soil microbial activity; thus, soil material composition was more stable. It had been sunny for more than 15 consecutive days before sampling, resulting in relatively low soil variation degree. Due to the high levels of heterogeneous and discontinuous soil in this karst area, three 10 m × 10 m repeated quadrats were set for each sample plot with a sufficient buffer zone between them. Three to five points were selected along an "S" curve in each sample plot. The soil was divided into 0–10 and 10–20 cm sublayers (actual depth was used if it was less than 20 cm). When sampling, we avoided the fertilization area, 10–30 cm around the trunk of *Z. planispinum*, to reduce effects from fertilization, weeding, and other human interference. A total of 30 soil samples were collected from the five sample plots. The fresh samples were divided into two parts after removing the root system, gravel, and animal and plant remains. One part was screened through 2 mm sieves and stored at 4 °C for timely determination of microbial and extracellular enzyme-related indexes; the other part was dried naturally, ground, and screened through a 0.15 mm sieve for soil nutrient analyses.

2.3. *Index Analysis Methods*

2.3.1. Soil Chemical Properties Analysis

Soil organic C (SOC) was determined by the $K_2Cr_2O_7$-external heating method, total N was determined by the semi-micro Kjeldahl method, total P was determined by $HClO_4$–H_2SO_4 digestion–molybdenum antimony anticolorimetry–ultraviolet spectrophotometry, total K was determined by sodium hydroxide melting-flame photometry, and total Ca and total Mg were determined by atomic absorption spectroscopy [24].

2.3.2. Soil Biological Properties Analysis

Soil fungi, bacteria, and actinomycete concentrations were determined by the plate counting method with beef peptone medium, potato glucose agar medium, and Gao's No. 1 medium [25], respectively. Soil microbial biomass C, N, and P were determined by chloroform fumigation extraction [26]. Soil extracellular enzyme activity (EEA) was measured by fluorescence spectrophotometry [27]. The activity of four hydrolases was determined: β-1,4-glucosidase (BG) for C, β-1,4-n-acetylglucosaminidase (NAG), and leucine aminopeptidase (LAP) for N and acid phosphatase (AP) for P. These four enzymes mainly participate in the terminal catalytic reaction, which reflects the metabolic levels of soil C, N, and P. On this basis, the soil extracellular enzyme stoichiometric ratios can be used to characterize the limitations of soil energy and nutrients [28].

2.4. *Data Processing and Analysis*

Soil microbial biomass C, N, and P were calculated according to Wang [29]. Soil EEAs were calculated according to Bell [27], and their stoichiometry calculated according to Sinsabaugh et al. [30]. Data calculation and sorting were carried out by Microsoft Excel 2013 (version 2013, Microsoft, Redmond, WA, USA). All data were analyzed using SPSS 20.0 (version 20.0, IBM SPSS, New York, NY, USA). One-way analysis of variance (ANOVA) with least significant difference was used, taking the planting or soil depth as an independent variable, to test the differences in soil elements, microorganisms, and extracellular enzyme stoichiometry between different plantations of the same depth or between different depths of the same plantation. Two-way ANOVA was used to clarify the effects of planting, soil depth, and their interactions on stoichiometry of soil elements, microorganisms, and extracellular enzymes. Pearson's correlation was used to verify relationships among soil elements, microbes, and extracellular enzyme C:N:P. The relationship between soil elements and microbial biomass was examined using standardized major axis (SMA) estimation and was evaluated using the software SMATR 2.0 (version 2.0, Daniel Falster, Sydney, Australia). Before data analysis, log conversion was performed to improve the distribution and uniformity of variance. Figures were created in Origin 8.6 (version 8.6, OriginLab

Corporation, Northampton, MA, USA), and data in the map and table are presented in the form of mean ± standard deviation.

3. Results

3.1. Soil Elements and Stoichiometry of Z. planispinum Plantations

The SOC and total N contents in the 0–10 cm soil layer were the highest in plot 4 (59.55 and 5.07 g kg^{-1}, respectively), and significantly higher than for plots 2, 3, and 5. Total P was highest in plot 4 (1.73 g kg^{-1}) and lowest in plot 2 (0.99 g kg^{-1}); there was a significant difference between these two plots, but not the other plots. Total K was 5.58–13.00 g kg^{-1}, and significantly lower in plots 1 and 2 than in plots 3–5. Total Ca and total Mg were highest in plot 5, and significantly higher than for the other four plots. Trends for all elements were also similar for the 10–20 cm soil layer. Overall, *Z. planispinum* + *L. japonica* plantation was conducive to accumulation of SOC, total N, total P, and total K on the surface, and the pure plantation was conducive to accumulation of characteristic elements, such as Ca and Mg, in karst areas (Figure 1).

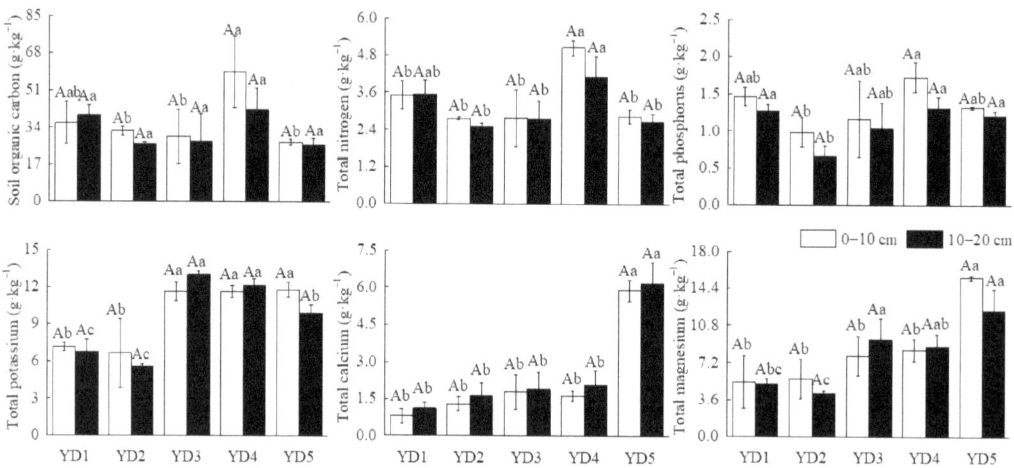

Figure 1. Soil element contents of *Z. planispinum* plantations for different allocations. YD1: *Z. planispinum* + *P. salicina*; YD2: *Z. planispinum* + *S. tonkinensis*; YD3: *Z. planispinum* + *A. hypogaea*; YD4: *Z. planispinum* + *L. japonica*; YD5: *Z. planispinum* pure plantation. The same notation is used in the other figures. Lower case letters, significant differences between different plantation types of the same depth at $p < 0.05$; upper case letters, significant differences between different depths of the same plantation types at $p < 0.05$.

The C:N values were in the range of 9.54–11.7, with no significant differences among the five plots; the values of C:P and N:P were in the order of plot 2 > plot 4 > plot 1 > plot 3 > plot 5, and were significantly greater in plots 2 and 4 than in plot 5, indicating that there was a stable equilibrium relationship between C and N. The values of C:K, N:K, and P:K were in the order of plot 1 > plot 2 > plot 4 > plot 5 > plot 3, indicating that the change trends of C, N, and P contents were basically the same. The C:Ca was highest in plots 1 and 4, and significantly higher than in plot 5; Ca:Mg in plot 5 (0.45) was significantly higher than in other plots. In general, although the variation range of soil element stoichiometry was greater in the 0–10 than the 10–20 cm soil layer, the trends were similar; the effects of plantation type on K, Ca, and Mg were greater than on C, N, and P (Figure 2).

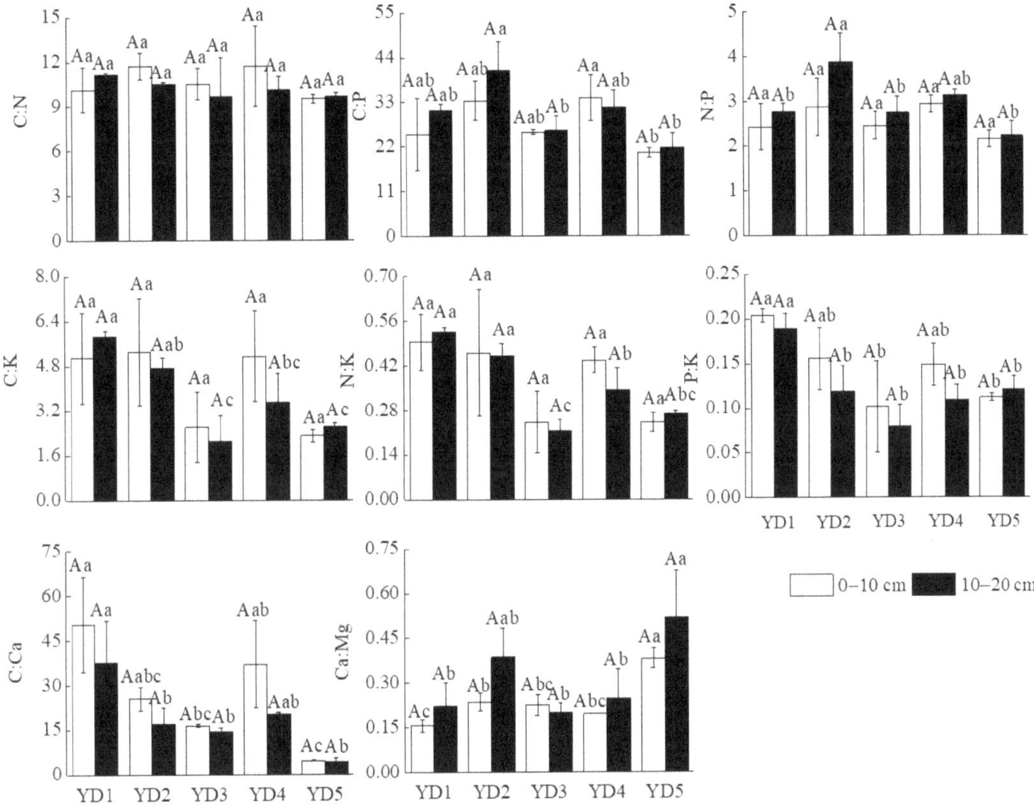

Figure 2. Stoichiometry of soil elements of *Z. planispinum* plantations. C:N, soil C:N ratio; C:P, soil C:P ratio; N:P, soil N:P ratio; C:K, soil C:K ratio; N:K, soil N:K ratio; P:K, soil P:K ratio; C:Ca, soil C:Ca ratio; Ca:Mg, soil Ca:Mg ratio. Lower case letters, significant differences between different plantation types of the same depth at $p < 0.05$; upper case letters, significant differences between different depths of the same plantation types at $p < 0.05$.

3.2. Soil Biological Properties and Stoichiometry of Z. planispinum Plantations

There were no significant differences in the fungi and actinomycete concentrations in both soil layers among the five plots. In the 0–10 cm soil layer, the highest and lowest values of bacterial concentration were in plots 2 and 5, respectively, with no significant differences among other plots. In the 10–20 cm soil layer, the bacteria concentration was highest in plots 1 and 4 and lowest in plot 5. The concentration of soil microorganisms was lowest in plot 5, indicating a stressful soil environment. The effect of plantation type on fungi and actinomycetes was less than on bacteria, and there was no clear effect of soil depth on concentration of soil microorganisms (Figure 3).

In the 0–10 cm soil layer, microbial biomass P (MBP) in plot 4 was significantly higher than in plots 1 and 5, and there was no significant difference in microbial biomass C (MBC) and microbial biomass N (MBN) among the five plots. In the 10–20 cm soil layer, MBC was significantly higher in plot 4 than in plot 1, and there were no significant differences in MBN and MBP among the five plots. The results showed that MBN was more stable than MBC and MBP, and that plantation type had little effect on MBC and MBP. There were no significant differences in MBC:MBN and MBN:MBP between the two soil layers; in the 0–10 cm layer, MBC:MBP in plot 1 (1.95) was significantly higher than in plots 2–4 (1.77,

1.71, and 1.72, respectively), and there were no significant differences in the 10–20 cm layer (Figure 4).

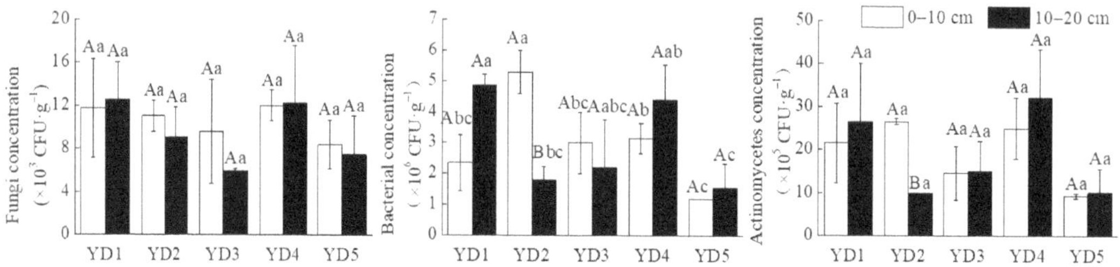

Figure 3. Concentration of soil microorganisms of *Z. planispinum* plantations. Lower case letters, significant differences between different plantation types of the same depth at $p < 0.05$; upper case letters, significant differences between different depths of the same plantation types at $p < 0.05$.

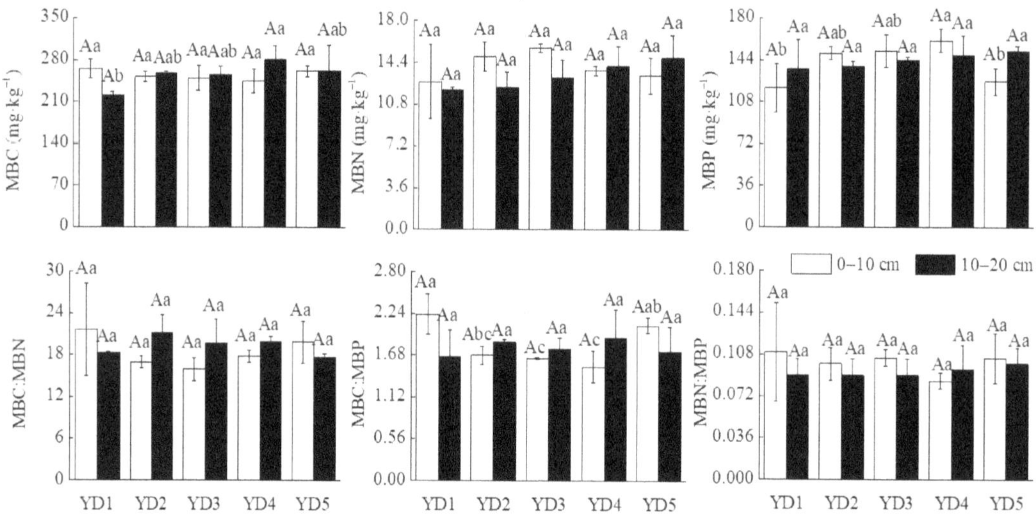

Figure 4. Soil microbial biomass and stoichiometry of *Z. planispinum* plantations. MBC, soil microbial biomass carbon; MBN, soil microbial biomass nitrogen; MBP, soil microbial biomass phosphorus; MBC:MBN, soil microbial biomass carbon to microbial biomass nitrogen ratio; MBC:MBP, soil microbial biomass carbon to microbial biomass phosphorus ratio; MBN:MBP, soil microbial biomass nitrogen to microbial biomass phosphorus ratio. Lower case letters, significant differences between different plantation types of the same depth at $p < 0.05$; upper case letters, significant differences between different depths of the same plantation types at $p < 0.05$.

In the 0–10 cm soil layer, the four EEAs showed no significant differences among the five plots; in the 10–20 cm soil layer, all EEAs were highest in plot 4. Except for plot 1, the EEA decreased significantly with increasing soil depth, with no significant differences among the other plots. The differences in (NAG + LAP):AP were reflected in the 10–20 cm soil layer, being significantly higher for plot 4 than for plots 1–3; there were no significant differences for BG:(NAG + LAP) and BG:AP in the same soil layer in different plots, as well as in different soil layers in the same plot, indicating that soil extracellular enzymes followed a strict proportional relationship (Figure 5).

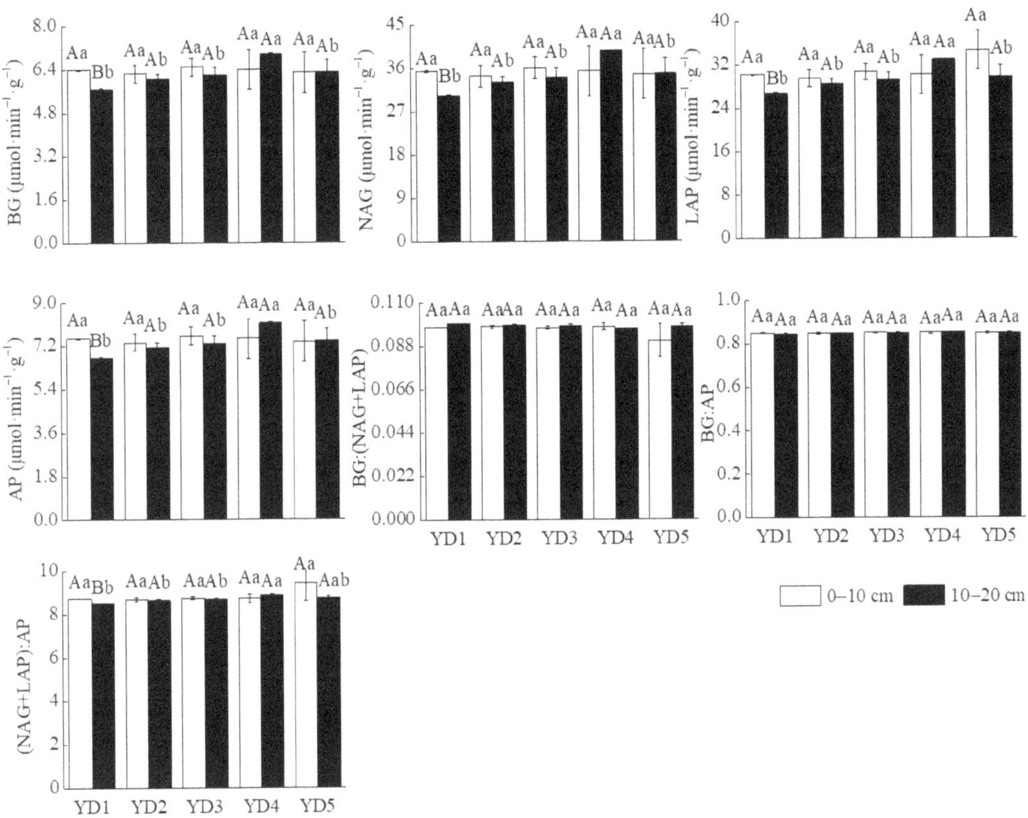

Figure 5. Soil extracellular enzymes activities (EEAs) and their stoichiometry of *Z. planispinum* plantations. BG, β-1,4-glucosidase; NAG, β-1,4-n-acetylglucosaminidase; LAP, leucine aminopeptidase; AP, acid phosphatase; BG:(NAG + LAP), ratio of β-1,4-glucosidase to the sum of β-1,4-n-acetylglucosaminidase and leucine aminopeptidase; BG:AP, β-1,4-glucosidase to acid phosphatase ratio; (NAG + LAP):AP, ratio of the sum of β-1,4-n-acetylglucosaminidase and leucine aminopeptidase to acid phosphatase. Lower case letters, significant differences between different plantation types of the same depth at $p < 0.05$; upper case letters, significant differences between different depths of the same plantation types at $p < 0.05$.

3.3. Effects of Plantation Types and Soil Depth on Soil Properties

Plantation type had significant effects on soil elements and stoichiometry to varying degrees except for C:N. Plantation type had no significant effects on soil biological properties except for bacterial concentration. Soil depth had no significant effect on soil elements and stoichiometry except for total P and Ca:Mg. The interaction between plantation type and soil depth only had significant effects on bacterial concentration. These comprehensive results showed that stoichiometry was more influenced by plantation type than soil depth and their interaction. In terms of stoichiometry, soil microorganisms and extracellular enzymes were more stable than soil elements (Table 2).

3.4. Correlation of Soil Element, Microorganism, and Extracellular Enzyme C:N:P of *Z. planispinum* Plantations

Soil C had high and significant positive correlations with N and P. The absolute value of the negative correlation coefficient between MBN:MBP and MBC:MBN was less than the positive correlation coefficient between MBN:MBP and MBC:MBP, indicating that MBC was more sensitive to the change of MBN than MBP. The correlations among BG:(NAG + LAP),

BG:AP, and (NAG + LAP):AP were all significant ($p < 0.01$), indicating a close relationship among EEAs (Figure 6).

Table 2. Two-way ANOVA for the effects of plantation type and soil depth on soil, microorganisms, and extracellular enzymes and stoichiometry (value in the table are F ratios).

Factors	SOC	TN	TP	TK	TCa	TMg			
A	4.831 *	12.521 ***	5.389 *	30.478 ***	61.275 ***	20.602 ***			
B	1.326	1.678	5.200 *	0.413	1.636	0.669			
A × B	0.750	0.696	0.322	1.423	0.059	1.312			
	C:N	C:P	N:P	C:K	N:K	P:K	C:Ca	Ca:Mg	
A	0.804	6.671 **	5.434 *	6.616 **	9.099 **	9.224 **	6.411 **	8.538 **	
B	0.641	1.336	4.734	0.409	0.157	3.254	2.418	5.177 *	
A × B	0.576	0.697	0.848	0.676	0.383	0.595	0.357	0.950	
	fungi	bacteria	actinomycetes	MBC	MBN	MBP	MBC:MBN	MBC:MBP	MBN:MBP
A	1.570	5.432 *	3.474	0.682	0.925	2.120	0.312	0.890	0.198
B	0.551	0.022	0.033	0.003	1.235	0.106	0.510	0.215	0.798
A × B	0.269	7.235 **	1.369	2.087	1.332	1.676	1.529	2.952	0.319
	BG	NAG	LAP	AP	BG: (NAG + LAP)	BG:AP	(NAG + LAP):AP		
A	1.473	1.472	2.805	1.472	1.478	1.325	1.574		
B	0.470	0.477	3.444	0.485	2.049	0.274	2.147		
A × B	1.393	1.390	2.298	1.406	1.325	1.156	1.383		

A, plantation type; B, soil depth; A × B, the interaction between plantation type and soil depth. SOC, soil organic carbon; TN, total nitrogen; TP, total phosphorus; TK, total potassium; TCa, total calcium; TMg, total magnesium; C:N, soil C:N ratio; C:P, soil C:P ratio; N:P, soil N:P ratio; C:K, soil C:K ratio; N:K, soil N:K ratio; P:K, soil P:K ratio; C:Ca, soil C:Ca ratio; Ca:Mg, soil Ca:Mg ratio; fungi, soil fungi concentration; bacteria, soil bacteria concentration; actinomycetes, soil actinomycetes concentration; MBC, soil microbial biomass carbon; MBN, soil microbial biomass nitrogen; MBP, soil microbial biomass phosphorus; MBC:MBN, soil microbial biomass carbon to microbial biomass nitrogen ratio; MBC:MBP, soil microbial biomass carbon to microbial biomass phosphorus ratio; MBN:MBP, soil microbial biomass nitrogen to microbial biomass phosphorus ratio; BG, β-1,4-glucosidase; NAG, β-1,4-n-acetylglucosaminidase; LAP, leucine aminopeptidase; AP, acid phosphatase; BG:(NAG + LAP), ratio of β-1,4-glucosidase to the sum of β-1,4-n-acetylglucosaminidase and leucine aminopeptidase; BG:AP, β-1,4-glucosidase to acid phosphatase ratio; (NAG + LAP):AP, ratio of the sum of β-1,4-n-acetylglucosaminidase and leucine aminopeptidase to acid phosphatase. *, **, *** indicate significant differences at $p < 0.05$, $p < 0.01$, and $p < 0.001$, respectively.

3.5. Internal Stability Analysis of Soil Elements and Microbial Biomass of *Z. planispinum* Plantations

The ecological stoichiometry internal stability model proposed by Sterner and Elser [31] is expressed in the following equation: $Y = CX^{1/H}$. In the equation, Y represents soil microbial biomass C, N, and P and their stoichiometric ratio; X represents the corresponding soil C, N, and P and their stoichiometric ratio; C is the fitting constant; H is the internal stability index. When the equation fitting result is successful ($p < 0.05$), $H > 4$ shows that Y is in a steady state; $2 < H < 4$ indicates that Y is in a weak steady state; $1.33 < H < 2$ means that Y is in a weakly sensitive state; and $H < 1.33$ indicates that Y is in a sensitive state. When the fitting result of the equation is unsuccessful ($p > 0.05$), Y is in an absolute steady state. Fitting the internal stability model equation of soil nutrient elements and soil microbial biomass C, N, and P was not successful ($p > 0.05$), and the dependent variable showed an absolute steady state [32]. This indicated that soil microorganisms had strong internal stability (Table 3).

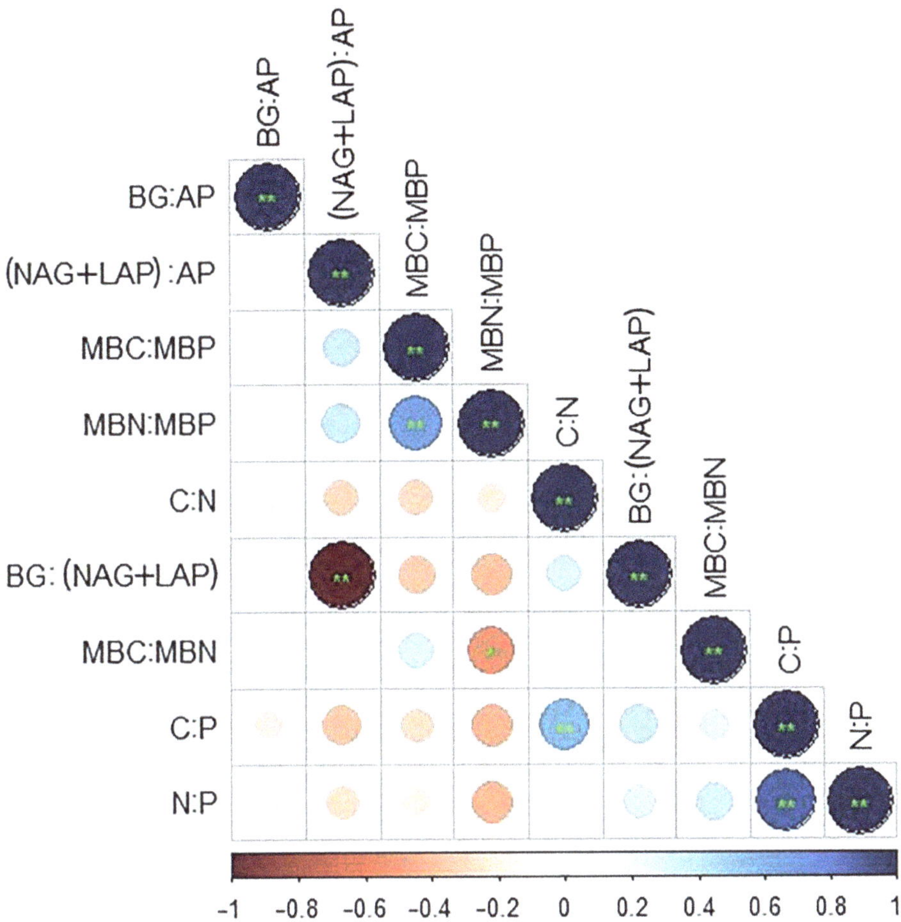

Figure 6. Correlations of soil element, microorganism, and extracellular enzyme C:N:P of *Z. planispinum* plantations. C:N, soil C:N ratio; C:P, soil C:P ratio; N:P, soil N:P ratio; MBC:MBN, soil microbial biomass carbon to microbial biomass nitrogen ratio; MBC:MBP, soil microbial biomass carbon to microbial biomass phosphorus ratio; MBN:MBP, soil microbial biomass nitrogen to microbial biomass phosphorus ratio; BG:(NAG + LAP), ratio of β-1,4-glucosidase to the sum of β-1,4-n-acetylglucosaminidase and leucine aminopeptidase; BG:AP, β-1,4-glucosidase to acid phosphatase ratio; (NAG + LAP):AP, ratio of the sum of β-1,4-n-acetylglucosaminidase and leucine aminopeptidase to acid phosphatase. *, ** indicates significant differences at $p < 0.05$ and $p < 0.01$, respectively.

Table 3. Standardized major axis analysis of C, N, and P concentrations in soil and soil microbial biomass.

Variable		n	r^2	p	Intercept	Slope
X	Y					
SOC	MBC	30	0.003	0.828	2.792	−0.2547
TN	MBN	30	0.005	0.768	1.380	−0.4954
TP	MBP	30	0.001	0.924	2.126	0.3874
C:N	MBC:MBN	30	0.009	0.693	2.405	−1.114
C:P	MBC:MBP	30	0.099	0.176	1.129	−0.6038
N:P	MBN:MBP	30	0.189	0.055	−0.6543	−0.8425

n: sample size.

4. Discussion

4.1. Soil Element Contents and Stoichiometry of Z. planispinum Plantations

The C, N, and P contents were higher for *Z. planispinum* + *L. japonica*/*P. salicina* plantations, indicating that these two plantations were conducive to nutrient accumulation. This is because the roots of *L. japonica* and *P. salicina* are generally well-developed, frequently exchanging materials with soil, where large amounts of nutrients accumulate due to dead root decomposition or root exudates; these things considered, abundant and easily decomposed litter returns nutrients back into soil as fertilizer. In addition, *P. salicina*, a tree with deep roots, can transport nutrients from the deep soil to the surface [33]. The total N and total P contents of the *Z. planispinum* + *S. tonkinensis* plantation were the lowest (2.64 and 0.82 g kg^{-1}, respectively), and because *S. tonkinensis* is a N-fixing plant, this is inconsistent with the conclusion that N-fixing plants help the accumulation of total N [34]. Possible reasons include that (1) the litter of *S. tonkinensis* was less, and the whole plants were harvested, which led to the reduction of nutrients back into soil; and (2) legume plantations had a high absorption of soil P [20], resulting in a relative lack of soil P under the same fertilization measures. Simultaneously, because P deficiency inhibited the secretion of N-fixing enzymes by N-fixing bacteria, it reduced the amount of N fixation [35]. Future management measures will be taken to supplement soil P to improve soil conditions. In this study, total P decreased with deeper soil. Compared with other elements, P was significantly affected by soil depth (Table 2), consistent with the results of Yan et al. [36]. The reason is that P is easily fixed under neutral or alkaline soil conditions, which is altered in the surface soil when there is an abundance of microorganisms and organic matter [37]. In this study, the soil Ca and Mg were in the range of 0.95–6.05 and 4.95–13.85 g kg^{-1}, respectively, which were relatively enriched compared with other subtropical areas. This is because the weathering of carbonate rocks generated this neutral or slightly alkaline soil environment, with Ca and Mg enriched. At the same time, *Z. planispinum* is a calciphile with a strong ability to enrich Ca, which increases the Ca concentration in soil; the soil Ca and Mg contents of *Z. planispinum* plantations for the four allocations were significantly lower than for the pure plantation, indicating that the compound planting mode increased Ca and Mg consumption compared with the pure plantation. Therefore, it is necessary to supplement the corresponding mineral element fertilizers appropriately.

The C:N in the soil tillage layer was inversely proportional to the decomposition rate of organic matter [38]. The soil C:N (10.47) in this area was lower than the Chinese average (14.4), indicating that N was sufficient for microbial growth, and the excess was released into the soil. The lowest value was for the *Z. planispinum* pure plantation (9.59), indicating that soil mineralization rate was highest for the pure plantation. This implies that compound management can reduce mineralization, increase organic matter accumulation, and improve soil C sequestration capacity. The soil C:P is an index of P availability, which reflects the metabolic trend of soil-available P by microorganisms. In this study, the soil C:P (28.9) was less than 200 and lower than the Chinese average (136), indicating that soil microorganisms mineralized SOC to supply the soil-effective P pool, and the soil was mainly limited by C. The soil N:P is an index indicating that the soil is limited by N or P (less than 10 indicates soil N deficiency). In this study, the N:P (2.76) was lower than the Chinese average (9.3), indicating that the soil N limitation was greater than P. The C:P and N:P of *Z. planispinum* + *S. tonkinensis* and *Z. planispinum* + *L. japonica* were significantly higher than for the others, and soil C and N of the *Z. planispinum* + *S. tonkinensis* plantation were low, suggesting that *Z. planispinum* + *S. tonkinensis* and *Z. planispinum* + *L. japonica* promoted the balance of soil C, N, and P; the *Z. planispinum* + *L. japonica* plantation may effectively alleviate the relative lack of C and N. Our results of P saturation are inconsistent with the conclusion that karst areas are generally limited by P [39]. Possible reasons include: (1) intense karstification promoted rock weathering, and with special hydrothermal conditions, the dissolution rate of soil microorganisms was increased, so that more P was dissolved and extracted into soil and (2) weak soil microorganism activity somehow strengthened soil P fixation [37].

4.2. Soil Microbial Properties and Stoichiometry of Z. planispinum Plantations

In this study, the concentrations of soil groups in the five sample plots were in the following order: bacteria > actinomycetes > fungi. This is because the soil was alkaline. This environment was suitable for the growth and reproduction of bacteria and actinomycetes, but unfavorable to the survival of fungi. Therefore, the concentration of bacteria was somewhat higher than that of fungi. The concentrations of bacteria, fungi, and actinomycetes were highest when *Z. planispinum* was mixed with *P. salicina*, *S. tonkinensis*, and *L. japonica*, respectively, followed by that of *Z. planispinum* + *Arachis hypogaea*, and that of the pure plantation was the lowest (Figure 3), indicating that plant species restricted the composition of the soil microbial community. This is because, compared with pure plantations, mixed plantations can increase productivity, improve soil nutrients, and increase soil microbial community abundance and EEAs by increasing the chemical composition of C substrates such as root exudates and litter [40,41]. In this study, MBC:MBN was greater than 10, indicating that the dominant microbial community was fungi, suggesting that local soil nutrients were relatively poor [42].

The MBC:MBN (18.8) was higher for the *Z. planispinum* + *L. japonica* plantation than that of Chinese soil, generally [43] (7.6), and the mean values of MBC:MBP (1.81) and MBN:MBP (0.1) were much lower than that for Chinese soil (70.2 and 6, respectively). This showed that the biological availability of soil N was low; the soil microbial P release supplemented the P pool, ensuring rich P in soil (Figure 1). The relatively poor N in soil limited the MBN fixation. Meanwhile, plants increased investment to P-rich ribosomal RNA in P-rich soil environments [18], resulting in the low N:P ratio (Figure 2).

In this study, the soil enzymatic C:N ratio (0.1) was lower than the mean value of major terrestrial ecosystems (1.41), and the enzymatic C:P (0.85) and N:P (8.8) were higher than the mean of global terrestrial ecosystems (0.62 and 0.44, respectively), indicating that NAG and LAP activities were high and AP activity was low. This is because soil microorganisms secrete more N-decomposing enzymes to relieve N restriction [30] and less P-related enzymes to deal with the P saturation environment. A soil enzymatic C:N ratio of 1:1 indicates that C and N are mineralized at the same rate. We conclude that the mineralization rate of soil N and P was greater than that of soil C in this area. The four enzymes showed higher activities in the *Z. planispinum* + *L. japonica* plantation than in the other modes (Figure 5), and the soil C, N, and P contents in this mode were all the highest (Figure 1). This showed that the higher the nutrient contents, the stronger the EEA, consistent with the conclusions of Zuo et al. [44]. It is speculated that *L. japonica* may persist by regulating soil microorganisms to secrete extracellular enzymes to mineralize more organic matter [45]. In addition, SOC and total N were energy sources for soil enzyme production and secretion: the total N can increase biomass of underground fine roots, promote growth of rhizosphere microorganisms, and finally increase EEAs. In contrast, enzymes are also proteins, and high soil EEA can provide more N to maintain enzyme production. Concerning the effect of soil depth on extracellular enzymes, except for *Z. planispinum* + *L. japonica*, the soil EEA of plantations decreased with increasing soil depth. The reason is that the C input and rhizosphere effect of bottom soil are lower than that of the surface [46]. In contrast, the soil EEA of *Z. planispinum* + *L. japonica* was high in the bottom layer. It may be that the fine roots were deeply distributed, and the root system secreted more enzymes, thus, assisting nutrient absorption. The specific mechanism needs to be further explored.

4.3. Correlation and Internal Stability Analysis of Soil Stoichiometry of Z. planispinum Plantations

The ratio of BG:(NAG + LAP):AP had no significant correlation with soil C:N:P and MBC:MBN:MBP (Figure 6). The reason is that extracellular enzymes are complexes affected by soil properties, vegetation type, and other factors, as well as the fact that EEA is affected by some unmeasured soil biological or non-biological factors [47]. There was no significant correlation between soil C:N:P and MBC:MBN:MBP ratios in this study;

moreover, MBN:MBP had significant and extremely significant correlations with MBC:MBN and MBC:MBP (Figure 6), indicating that soil microorganisms were not affected by soil elements. This is consistent with the results of Cleveland et al. [48] and Hartman et al. [49]. The reason is that soil microorganisms have self-balancing mechanisms which can change their community composition by adjusting their own structure composition and population dynamics to maintain the relative stability of C:N:P [50]. This also indicated that soil microorganisms resist environmental stresses through adjusting their activities, resulting in a stricter C:N:P ratio [13]. This study also showed that soil microbial biomass C:N:P was not sensitive to changes in plantation type (Table 2), indicating that soil microbial biomass C:N:P was more stable than soil C:N:P, which can be used to diagnose ecosystem nutrient constraints.

5. Conclusions

In this study, the soil C:N:P was low, indicating that C and N contents in soil were relatively low, while P content was relatively high, which reflected the low vegetation productivity in this area; moreover, the area was mainly limited by C and N. At the same time, soil microorganisms mineralize soil organic matter to supplement the soil-available P pool. The soil MBC:MBN in this area was high, while MBC:MBP and MBN:MBP were low, indicating that biological availability of soil N was low, and the P released by soil microorganisms supplemented the available P pool. In this study, the low soil enzymatic C:N ratio and the high soil enzymatic C:P and N:P ratios showed that NAG and LAP activities were high and AP activity was low. This is because soil microorganisms secrete more N-decomposing enzymes to relieve N restriction and less P-related enzymes to deal with the P saturation environment. The impact of plantation type on ecological stoichiometry was greater than that of soil depth and their interaction, suggesting that stand type had a more significant impact on ecosystem nutrient cycling. The *Z. planispinum* + *L. japonica* plantation had a large amount of nutrients returned, which improved soil fertility and promoted the growth of soil microorganisms. This study indicated that application of *L. japonica* slowed the nutrient deficit and improved the soil environment. The *Z. planispinum* + *L. japonica* was the best plantation type. Fitting of the internal stability model equation of soil nutrient elements and soil microbial biomass C, N, and P was unsuccessful ($p > 0.05$), and the dependent variable showed an absolute steady state. This indicated that soil microorganisms resisted environmental stress through adjustment of their activities, resulting in a stricter C:N:P ratio. The soil MBC:MBN:MBP was more stable internally than soil C:N:P, which could be applied to diagnose ecosystem nutrient constraints.

Author Contributions: Conceptualization, Y.Y.; formal analysis, Y.L. and Y.Y.; software, Y.L. and Y.S.; investigation, Y.L. and Y.S.; writing—original draft preparation, Y.L.; writing—review and editing, Y.Y. All authors have read and agreed to the published version of the manuscript.

Funding: This research was funded by the Guizhou Province Science and Technology Support Plan Project (Qian-ke-he Zhicheng [2022] Yiban 103).

Institutional Review Board Statement: Not applicable.

Informed Consent Statement: Not applicable.

Data Availability Statement: The data are not publicly available due to privacy.

Acknowledgments: We thank International Science Editing (http://www.internationalscienceediting.com) for editing this manuscript (accessed on 3 May 2022).

Conflicts of Interest: The authors declare no conflict of interest.

References

1. Wang, L.J.; Wang, P.; Sheng, M.Y.; Tian, J. Ecological stoichiometry and environmental influencing factors of soil nutrients in the karst rocky desertification ecosystem, southwest China. *Glob. Ecol. Conserv.* **2018**, *16*, e00449. [CrossRef]
2. Wang, M.M.; Chen, H.S.; Zhang, W.; Wang, K.L. Soil nutrients and stoichiometric ratios as affected by land use and lithology at county scale in a karst area, southwest China. *Sci. Total Environ.* **2018**, *619*, 1299–1307. [CrossRef] [PubMed]

3. Yuan, J.H.; Wang, L.; Chen, H.; Chen, G.L.; Wang, S.Q.; Zhao, X.; Wang, Y. Responses of soil phosphorus pools accompanied with carbon composition and microorganism changes to phosphorus-input reduction in paddy soils. *Pedosphere* **2021**, *31*, 83–93. [CrossRef]
4. Zhang, H.Z.; Shi, L.L.; Wen, D.Z.; Yu, K.L. Soil potential labile but not occluded phosphorus forms increase with forest succession. *Biol. Fertil. Soils* **2016**, *52*, 41–51. [CrossRef]
5. Anderson, T.R.; Boersma, M.; Raubenheimer, D. Stoichiometry: Linking elements to biochemical. *Ecology* **2004**, *85*, 1193–1202. [CrossRef]
6. Elser, J.J.; Fagan, W.F.; Denno, R.F.; Dobberfuhl, D.R.; Folarin, A.; Huberty, A.; Interlandi, S.; Kilham, S.S.; McCauley, E.; Schulz, K.L.; et al. Nutritional constraints in terrestrial and freshwater food webs. *Nature* **2000**, *408*, 578–580. [CrossRef]
7. Ren, C.J.; Zhao, F.Z.; Kang, D.; Yang, G.H.; Han, X.H.; Tong, X.G.; Feng, Y.Z.; Ren, G.X. Linkages of C:N:P stoichiometry and bacterial community in soil following afforestation of former farmland. *For. Ecol. Manag.* **2016**, *376*, 59–66. [CrossRef]
8. Wang, J.P.; Wu, Y.H.; Li, J.J.; He, Q.Q.; Bing, H.J. Soil enzyme stoichiometry is tightly linked to microbial community composition in successional ecosystems after glacier retreat. *Soil Biol. Biochem.* **2021**, *162*, 108429. [CrossRef]
9. Gai, X.; Zhong, Z.K.; Zhang, X.P.; Bian, F.Y.; Yang, C.B. Effects of chicken farming on soil organic carbon fractions and fungal communities in a Lei bamboo (Phyllostachys praecox) forest in subtropical China. *For. Ecol. Manag.* **2021**, *479*, 118603. [CrossRef]
10. Zhong, Z.K.; Li, W.J.; Lu, X.Q.; Gu, Y.Q.; Wu, S.J.; Shen, Z.Y.; Han, X.H.; Yang, G.H.; Ren, C.J. Adaptive pathways of soil microorganisms to stoichiometric imbalances regulate microbial respiration following afforestation in the Loess Plateau, China. *Soil Biol. Biochem.* **2020**, *151*, 108048. [CrossRef]
11. Huang, R.; Lan, T.; Song, X.; Li, J.; Ling, J.; Deng, O.P.; Wang, C.Q.; Gao, X.S.; Li, Q.Q.; Tang, X.Y.; et al. Soil labile organic carbon impacts C:N:P stoichiometry in urban park green spaces depending on vegetation types and time after planting. *Appl. Soil Ecol.* **2021**, *163*, 103926. [CrossRef]
12. Fang, Z.; Li, D.D.; Jiao, F.; Yao, J.; Du, H.T. The Latitudinal Patterns of Leaf and Soil C:N:P Stoichiometry in the Loess Plateau of China. *Front. Plant Sci.* **2019**, *10*, 85. [CrossRef]
13. Hu, N.; Li, H.; Tang, Z.; Li, Z.F.; Li, G.C.; Jiang, Y.; Hu, X.M.; Lou, Y.L. Community size, activity and C:N stoichiometry of soil microorganisms following reforestation in a karst region. *Eur. J. Soil Biol.* **2016**, *73*, 77–83. [CrossRef]
14. Ravindran, A.; Yang, S.S. Effects of vegetation type on microbial biomass carbon and nitrogen in subalpine mountain forest soils. *J. Microbiol. Immunol. Infect.* **2015**, *48*, 362–369. [CrossRef]
15. Wang, Y.; Ren, Z.; Ma, P.P.; Wang, Z.M.; Niu, D.C.; Fu, H.; Elser, J.J. Effects of grassland degradation on ecological stoichiometry of soil ecosystems on the Qinghai-Tibet Plateau. *Sci. Total Environ.* **2020**, *722*, 137910. [CrossRef]
16. Dai, X.L.; Zhou, W.; Liu, G.R.; Liang, G.Q.; He, P.; Liu, Z.B. Soil C/N and pH together as a comprehensive indicator for evaluating the effects of organic substitution management in subtropical paddy fields after application of high-quality amendments. *Geoderma* **2019**, *337*, 1116–1125. [CrossRef]
17. Zechmeister-Boltenstern, S.; Keiblinger, K.M.; Mooshammer, M.; Penuelas, J.; Richter, A.; Sardans, J.; Wanek, W. The application of ecological stoichiometry to plant-microbial-soil organic matter transformations. *Ecol. Monogr.* **2015**, *85*, 133–155. [CrossRef]
18. Wu, X.Z.; Yan, X.; Wang, B.; Liu, R.T.; An, H. Effects of desertification on the C:N:P stoichiometry of soil, microbes, and extracellular enzymes in a desert grassland. *Chin. J. Plant Ecol.* **2018**, *42*, 1022–1032. (In Chinese) [CrossRef]
19. Yin, S.; Wang, C.K.; Jin, Y.; Zhou, Z.H. Changes in soil-microbe-exoenzyme C:N:P stoichiometry along an altitudinal gradient in Mt. Datudingzi, Northeast China. *Chin. J. Plant Ecol.* **2019**, *43*, 999–1009. (In Chinese) [CrossRef]
20. Chen, H.; Chen, M.L.; Li, D.J.; Mao, Q.G.; Zhang, W.; Mo, J.M. Responses of soil phosphorus availability to nitrogen addition in a legume and a non-legume plantation. *Geoderma* **2018**, *322*, 12–18. [CrossRef]
21. Hernández, J.; del Pino, A.; Vance, E.D.; Califra, Á.; Del Giorgio, F.; Martínez, L.; González-Barrios, P. Eucalyptus and Pinus stand density effects on soil carbon sequestration. *For. Ecol. Manag.* **2016**, *368*, 28–38. [CrossRef]
22. Yu, Y.H.; Wang, Z.Y.; Ying, B.; Yang, D.L. *Management Principle and Technology of Ecological Industry in Karst Plateau Canyon*, 1st ed.; China Environmental Science Press: Beijing, China, 2021; pp. 28–30. (In Chinese)
23. Yu, Y.; Song, Y.P.; Zhong, X.P.; Li, Y.T.; Ying, B. Growth decline mechanism of *Zanthoxylum planispinum* var. *dintanensis* in the canyon area of Guizhou karst plateau. *Agron. J.* **2021**, *113*, 852–862. [CrossRef]
24. Bao, S.D. *Soil Agrochemical Analysis*, 3rd ed.; China Agriculture Press: Beijing, China, 2000; pp. 22–173. (In Chinese)
25. Shen, P.; Fan, X.R.; Li, G.H. *Microbiology Experiment*, 3rd ed.; Higher Education Press: Beijing, China, 1999; pp. 92–95. (In Chinese)
26. Vance, E.D.; Brookes, P.C.; Jenkinson, D.S. An extraction method for measuring soil microbial biomass C. *Soil Biol. Biochem.* **1987**, *19*, 703–707. [CrossRef]
27. Bell, C.W.; Fricks, B.E.; Rocca, J.D.; Steinweg, J.M.; McMahon, S.K.; Wallenstein, M.D. High-throughput Fluorometric Measurement of Potential Soil Extracellular Enzyme Activities. *J. Vis. Exp.* **2013**, *81*, e50961. Available online: https://www.jove.com/video/50961 (accessed on 26 October 2021). [CrossRef]
28. Hu, Z.Y.; Li, J.T.; Shi, K.W.; Ren, G.Q.; Dai, Z.C.; Sun, J.F.; Zheng, X.J.; Zhou, Y.W.; Zhang, J.Q.; Li, G.L.; et al. Effects of Canada Goldenrod Invasion on Soil Extracellular Enzyme Activities and Ecoenzymatic Stoichiometry. *Sustainability* **2021**, *13*, 3768. [CrossRef]
29. Wang, L.D.; Yao, T.; Wang, F.L.; Wei, L.Y.; Guo, C.X. Soil microbial and soil enzyme activity in a discontinued farmland by the Lower Shiyang River. *Acta Ecol. Sin.* **2016**, *36*, 4769–4779. [CrossRef]

30. Sinsabaugh, R.L.; Hill, B.H.; Shah, J.J.F. Ecoenzymatic stoichiometry of microbial organic nutrient acquisition in soil and sediment. *Nature* **2009**, *462*, 795–798. [CrossRef]
31. Sterner, R.W.; Elser, J.J. *Ecological Stoichiometry: The Biology of Elements from Molecules to the Biosphere*, 1st ed.; Princeton University Press: Princeton, NJ, USA, 2002; pp. 117–795.
32. Persson, J.; Fink, P.; Goto, A.; Hood, J.M.; Jonas, J.; Kato, S. To be or not to be what you eat: Regulation of stoichiometric homeostasis among autotrophs and heterotrophs. *Oikos* **2010**, *119*, 741–751. [CrossRef]
33. Ding, Y.; Huang, X.; Li, Y.; Liu, H.Y.; Zhang, Q.C.; Liu, X.M.; Xu, J.M.; Di, H.J. Nitrate leaching losses mitigated with intercropping of deep-rooted and shallow-rooted plants. *J. Soils Sediments* **2021**, *21*, 364–375. [CrossRef]
34. Stinca, A.; Chirico, G.B.; Incerti, G.; Bonanomi, G. Regime Shift by an Exotic Nitrogen-Fixing Shrub Mediates Plant Facilitation in Primary Succession. *PLoS ONE* **2015**, *10*, e0123128. [CrossRef] [PubMed]
35. Png, G.K.; Turner, B.L.; Albornoz, F.E.; Hayes, P.E.; Lambers, H.; Laliberte, E. Greater root phosphatase activity in nitrogen-fixing rhizobial but not actinorhizal plants with declining phosphorus availability. *J. Ecol.* **2017**, *105*, 1246–1255. [CrossRef]
36. Yan, Y.J.; Dai, Q.H.; Jin, L.; Wang, X.D. Geometric morphology and soil properties of shallow Karst fissures in an area of karst rocky desertification in SW China. *Catena* **2019**, *174*, 48–58. [CrossRef]
37. Cen, L.P.; Yan, Y.J.; Dai, Q.H.; Jiao, Q.; Hu, G.; Gao, R.X.; Fu, W.B. Occurrence characteristics of organic carbon and phosphorus in fissured soil under different land use types in Karst area. *Acta Ecol. Sin.* **2020**, *40*, 7567–7575. (In Chinese) [CrossRef]
38. Li, Y.Q.; Zhao, X.Y.; Zhang, F.X.; Awada, T.; Wang, S.K.; Zhao, H.L.; Zhang, T.H.; Li, Y.L. Accumulation of soil organic carbon during natural restoration of desertified grassland in China's Horqin Sandy Land. *J. Arid Land* **2015**, *7*, 328–340. [CrossRef]
39. Chen, H.; Li, D.J.; Mao, Q.G.; Xiao, K.C.; Wang, K.L. Resource limitation of soil microbes in karst ecosystems. *Sci. Total Environ.* **2019**, *650*, 241–248. [CrossRef]
40. Wolfe, B.E.; Klironomos, J.N. Breaking new ground: Soil communities and exotic plant invasion. *Bioscience* **2005**, *55*, 477–487. [CrossRef]
41. Rousk, J.; Brookes, P.C.; Baath, E. Contrasting soil pH effects on fungal and bacterial growth suggest functional redundancy in carbon mineralization. *Appl. Environ. Microbiol.* **2009**, *75*, 1589–1596. [CrossRef]
42. Sun, J.; Lang, J.X.; Kong, D.J.; Guo, X.N.; Wei, Y.D.; Zhou, T. Effects of biochar and straw on the C:N:P stoichiometry of soil, microbes, and extracellular enzymes in an aeolian sandy soil. *Acta Pratacult. Sin.* **2021**, *30*, 29–39. (In Chinese) [CrossRef]
43. Xu, X.F.; Thornton, P.E.; Post, W.M. A global analysis of soil microbial biomass carbon, nitrogen and phosphorus in terrestrial ecosystems. *Glob. Ecol. Biogeogr.* **2013**, *22*, 737–749. [CrossRef]
44. Zuo, Y.P.; Zhang, X.Y.; Zeng, H.; Wang, W. Spatiotemporal Dynamics of Soil Extracellular Enzyme Activity and Its Influence on Potential Mineralization Rate of Soil Organic Carbon in Forests of Daxing'an Mountain Range. *Acta Sci. Nat. Univ. Pekin.* **2018**, *54*, 1311–1324. (In Chinese) [CrossRef]
45. Xu, Z.W.; Yu, G.R.; Zhang, X.Y.; He, N.P.; Wang, Q.F.; Wang, S.Z.; Wang, R.L.; Zhao, N.; Jia, Y.L.; Wang, C.Y. Soil enzyme activity and stoichiometry in forest ecosystems along the North-South Transect in eastern China (NSTEC). *Soil Biol. Biochem.* **2017**, *104*, 152–163. [CrossRef]
46. Peng, X.Q.; Wang, W. Stoichiometry of soil extracellular enzyme activity along a climatic transect in temperate grasslands of northern China. *Soil Biol. Biochem.* **2016**, *98*, 74–84. [CrossRef]
47. Xie, M.Y.; Feng, X.X.; Ma, H.F.; Hu, H.; Wang, J.Y.; Guo, Y.X.; Ren, C.J.; Wang, J.; Zhao, F.Z. Characteristics of soil enzyme activities and stoichiometry and its influencing factors in Quercus aliena var. acuteserrata forests in the Qinling Mountains. *Chin. J. Plant Ecol.* **2020**, *44*, 885–894. (In Chinese) [CrossRef]
48. Cleveland, C.C.; Liptzin, D. C:N:P stoichiometry in soil: Is there a "Redfield ratio" for the microbial biomass? *Biogeochemistry* **2007**, *85*, 235–252. [CrossRef]
49. Hartman, W.H.; Richardson, C.J. Differential Nutrient Limitation of Soil Microbial Biomass and Metabolic Quotients (qCO2): Is There a Biological Stoichiometry of Soil Microbes? *PLoS ONE* **2013**, *8*, e57127. [CrossRef]
50. Khan, K.S.; Joergensen, R.G. Stoichiometry of the soil microbial biomass in response to amendments with varying C/N/P/S ratios. *Biol. Fertil. Soils* **2019**, *55*, 265–274. [CrossRef]

agronomy

MDPI

Article

Competition and Niche Differentiation of Water and Nutrients between *Broussonetia papyrifera* and *Platycladus orientalis* under Prolonged Drought Stress

Kai Yao [1,*], Yanqing Wang [1] and Yanyou Wu [2,*]

[1] School of Life Sciences, Guizhou Normal University, Guiyang 550025, China; wangyanqing0129@163.com
[2] State Key Laboratory of Environmental Geochemistry, Institute of Geochemistry, Chinese Academy of Sciences, Guiyang 550081, China
* Correspondence: sc.catcher@hotmail.com (K.Y.); wuyanyou@mail.gyig.ac.cn (Y.W.)

Abstract: Little is known about the competition between and niche differentiation of water and nutrients between angiosperm and gymnosperm tree species under prolonged drought stress, especially in fragile environments. We imposed 21 d drought and competition treatments on *Broussonetia papyrifera* and *Platycladus orientalis* and measured water, N, and P contents, the isotopic composition of N and C, the activity of P assimilation enzymes, and stomatal conductance under solo planting and mixed planting to characterize resource diversity and competition in response to treatments. The N content, δ^{13}C, δ^{15}N, phosphomonoesterase, phosphodiesterase, g_s, and foliage water content were significantly affected by the soil water content. The δ^{15}N content in young leaves showed that N competition between these two plants could be alleviated through niche differentiation, but the changes in the PDE: PME ratio for these two plants indicated that they lost the niche differentiation of the P source under drought stress. Additionally, it was observed that foliage water content, WUE, N contents, and N and P sources were significantly affected by interspecific competition, and *Broussonetia papyrifera* benefited from water competition under moderate drought. Our results indicate that plants have different competition and niche differentiation modes to different nutrients under drought stress, and the effect of interspecific water competition should be seriously considered in mixed forests in semiarid areas.

Keywords: angiosperm; gymnosperm; interspecific competition; water stress; stable isotopic

Citation: Yao, K.; Wang, Y.; Wu, Y. Competition and Niche Differentiation of Water and Nutrients between *Broussonetia papyrifera* and *Platycladus orientalis* under Prolonged Drought Stress. *Agronomy* **2022**, *12*, 1489. https://doi.org/10.3390/agronomy12071489

Academic Editor: Camilla Pandolfi

Received: 1 April 2022
Accepted: 15 June 2022
Published: 22 June 2022

1. Introduction

Global warming may lead to an increase in the duration, intensity, and frequency of drought worldwide [1]. Long-term or extreme drought will lead to plant death or change the composition, structure, and function of plant communities. When studying the impact of environmental change, interspecific relationships and the effects of species competition have been viewed as dominant factors influencing community structure in ecological research. It is generally believed that niche differentiation can mitigate the negative effects of drought stress on plants, but Kunstler et al. [2] found that the degree of trait dissimilarity between species had little influence on reducing competition. Competition determines the interspecific or intraspecific distribution pattern of net available resources [3]. Dominant, highly competitive species in the community can alleviate the risk of drought stress by "plundering resources", but interspecific competition also makes the position of the lowly competitive species deteriorate more. Although a broad body of literature exists regarding the study of plant performance responses to drought combined with competition, less is known about the underlying variation in physiology [4]. Many studies have reported the water source niche differentiation and competition of mixed-species stands under water deficiency, but their results are inconsistent. [5–7]. The fact that resources competition and niche differentiation occur in mixed-species stands can be obtained from in situ studies, but the results are often formed by many ecological factors [8] and species-specific functional traits of plants, so confusing results are easily obtained. To study the interspecific competition more accurately under drought conditions, artificial simulation experiments under controllable laboratory conditions and in situ experiments are both indispensable.

Drought not only affects the water characteristics of plants but also affects their nutrient absorption [9,10]. Soil water deficit significantly directly inhibited the absorption of soil elements with transpiration flow and reduced ion mobility in the soil [11] and limited the decomposition, mineralization, and nutrient cycling of soil elements [12]. Therefore, under drought conditions, the water competition of plants in a community will also affect the nutritional status of plants. Among all the nutrient elements required for plant growth, nitrogen (N) and phosphorus (P) are undoubtedly the most important and widely considered. The N: P ratio can determine the main characteristics and structure of plant communities and reflect the relationship between supply and demand and the circulation of nutrition [13–15], especially in nutrient-limited systems. Inorganic N, which includes nitrate and ammonium, may nonetheless be generally thought to be the dominant source of N for plants [16]. The nitrogen source composition of plants has always been an issue that has received much attention. P is an essential nutrient taken up by plants as the orthophosphate anion (Pi). Phosphate monoesters are the main phosphorus sources for plants, and phosphate diesters also can be absorbed by plants in cold or damp environments [17–19]. Phosphomonoesterase (PME, EC 3.1.3.2) and phosphodiesterase (PDE, EC 3.1.4.1) are rate-limiting enzymes in the assimilation processes of phosphate monoesters and phosphate diesters, respectively [20]; therefore, the PDE: PME ratio can be used for the qualitative analysis of plant P source changes [21].

B. papyrifera and *P. orientalis* are native to eastern Asia [22,23] and frequently coexist in mixed forests. With the main objective of studying the variations in water, nitrogen, and phosphorus nutritional status between *B. papyrifera* and *P. orientalis* under the combined effects of drought and competition, we conducted a simulation experiment in an artificial climate chamber and focused on the following specific questions:

(1) Does the niche differentiation of nitrogen and phosphorus occur between *B. papyrifera* and *P. orientalis* under drought stress?
(2) Does interspecific competition affect the status of the water, nitrogen, and phosphorus nutrition of *B. papyrifera* and *P. orientalis* under drought stress, and who is the winner?

2. Materials and Methods

2.1. Plant Material and Drought Treatment

The angiosperm specie—*Broussonetia papyrifera* L. Vent. and gymnosperm specie—*Platycladus orientalis* L. Franco were selected as the experimental materials. *B. papyrifera* belongs to the Moraceae family and is a large, fast-growing, shallow-root-system, deciduous broad-leaved woody species. *P. orientalis* from the Cupressaceae family is an evergreen tree species with a well-developed root system and a long lifespan. Seeds were sown in wet perlite and germinated at 25 °C in a greenhouse. On the 15th day after seed germination, vigorous seedlings were transplanted into humus soil. The environmental conditions for seedling growth were as follows: a 12 h photoperiod with a day/night temperature of 25/16 °C and a photosynthetic photon flux density of 400 μmol m^{-2} s^{-1} with 60–65% relative humidity. To eliminate individual size differences as much as possible, an experiment was started when the seedlings grew to 30 cm tall (approximately 2.5 months old for *B. papyrifera* and 10 months old for *P. orientalis*).

To study the effect of soil water competition in these two plant species under drought stress, plants were planted alone in a 40 cm × 30 cm box with 18 cm of nutrient soil, and they were also planted with two species together in a 40 cm × 60 cm box with the same thickness of nutrient soil. Figure 1 shows the planting modes and soil sampling points. In order to supplement the water consumed via evapotranspiration, and according to the results of the pre-experiment, 75 mL and 55 mL of water was added to the control group of *B. papyrifera* and *P. orientalis* every day to maintain soil moisture, respectively. The prolonged drought treatment group stopped adding water until the end of the experiment (21 days) to simulate long-term drought without rainfall. Following treatment, soils were sampled once every ten days at the positions shown in Figure 1 to analyze the water content of plants; the parameters of leaf gas exchange were measured once every five days; and the leaves and roots were sampled to determine the water content, carbon, phosphorus, and nitrogen contents and $\delta^{13}C$, $\delta^{15}N$, and PME and PDE activities. Plant tissue was stored at −80 °C until further analysis was performed.

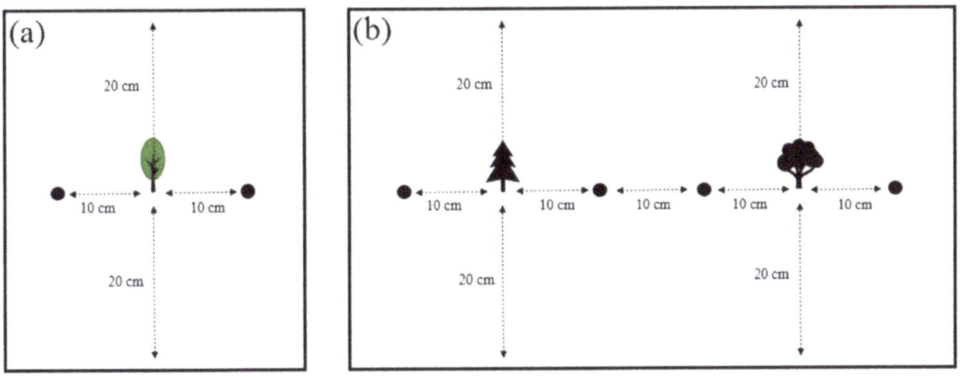

Figure 1. Planting modes and soil sampling points. (**a**) Solo planting; (**b**) mixed planting; • are the soil sampling points.

2.2. Measurement of Leaf Gas Exchange

The latest fully developed leaves in the middle and upper parts of the plant were selected for the determination of leaf gas exchange. The stomatal conductance (g_s) was measured using a Li-6400 portable photosynthesis measurement system (Li-6400, Li-Cor, Lincoln, NE, USA) from 10: 00 to 11: 00 am. The PPFD was 400 μmol m^{-2} s^{-1}, and the temperature was 25 °C. Here, the leaf area of *P. orientalis* was converted from its leaf dry weight using the following formula [24]:

$$S_{ori} = 161 \times M_{ori} \quad (1)$$

where S_{ori} is the leaf area of *P. orientalis* and M_{ori} is the leaf dry weight of *P. orientalis*.

2.3. Nitrogen and Phosphorus Contents

Fresh plant samples were dried at 65 °C for 3 days and ground for N and P content testing, and a 0.3 g sample was added to 10 mL of H_2SO_4. Then, a graphite digestion instrument (SH420, Hanon, Dezhou, China) was used. A total P assay kit was used for sample P content determination, and the test was carried out with a microplate reader (Synergy H1, BioTek, Winooski, VT, USA) according to the instructions. A Kjeldahl apparatus (K1100, Hanon, Dezhou, China) was used for N content determination.

2.4. Carbon ($\delta^{13}C$) and Nitrogen ($\delta^{15}N$) Isotope Compositions

The fresh samples were soaked in 1 mol L^{-1} HCl for 1 h to remove exogenous inorganic carbon. Then, the samples were dried at 65 °C for 3 days and ground to a fine powder. A 2.0 mg fine-powder sample was wrapped in a tin capsule (5 × 8 mm), and the $\delta^{13}C$ and $\delta^{15}N$ values were determined using a continuous-flow isotope ratio mass spectrometer (MAT 253; Thermo Fisher Scientific, Waltham, MA, USA).

2.5. Root PME and PDE Activities

PME activity was assayed according to the method of Tabatabai and Bremmer [25] with some modifications. The reaction mixture containing 0.2 g of fresh root sample was incubated for 10 min at 37 °C with 4 mL of modified universal buffer (MUB), pH 8, and 1 mL of 5 mmol L^{-1} para-nitrophenyl phosphate (pNPP; MFCD00284586, Sigma–Aldrich, St Louis, MO, USA). The reaction was stopped by adding 1 mL of 0.5 mol L^{-1} $CaCl_2$ and 4 mL of 0.5 mol L^{-1} NaOH and immediately centrifuging the mixture for 2 min at 12,000 g. The amount of p-nitrophenol released was measured in the supernatant at 412 nm.

PDE activity was assayed according to the method of Eivazi and Tabatabai [26] with some modifications. The reaction mixture containing 0.2 g of fresh root sample was incubated for 10 min at 37 °C with 4 mL of modified universal buffer (MUB; 0.1 mol L^{-1}) pH 8.0 and 1 mL of 5 mmol L^{-1} bis para-nitrophenyl phosphate (bis-pNPP; CAS645-15-8, Aladdin, Shanghai, China). The reaction was stopped by adding 1 mL of 0.5 mol L^{-1} $CaCl_2$ and 4 mL of 0.5 M NaOH and immediately centrifuging the mixture for 2 min at 12,000 g. The amount of p-nitrophenol released was measured in the supernatant at 412 nm.

2.6. Data Analysis

Five replicates were used for soil and leaf water contents measurement, and three replicates were used for physiology and biochemistry parameter measurement. One-way ANOVA followed by Duncan's multiple range test was performed to explore the differences among the plant parameters under separate treatments. A general linear model was used to explore the interspecific competition effect, with the planting pattern (sole or mixed) as a fixed effect and the plant parameters at every sampling time as dependent variables. If there were significant differences ($p < 0.05$) between the two groups, we judged that a competition effect definitely existed. To explore the difference in the water control strategy between *B. papyrifera* and *P. orientalis*, ANOVA and a general linear model were performed using SPSS 25.0 (SPSS Inc., Chicago, IL, USA). Linear regression analysis was performed to calculate Pearson's correlations between soil water content and N content, P content, $\delta^{13}C$, $\delta^{15}N$, PME, PDE, g_s, and foliage water content, and this test was performed using ORIGIN 95 (OriginLab Inc., Northampton, MA, USA).

3. Results

3.1. Soil Water Content

To study the water competition between B. papyrifera and *P. orientalis* under mixed planting conditions, the change in soil water content was continuously monitored. As shown in Figure 2, at 11 d, soil water contents in the rhizosphere of B. papyrifera were not significantly different between the solo planting and mixed planting conditions within the 3 cm deep soil layer via Duncan's test ($p < 0.05$). Within the 9 cm deep soil layer, the water content was significantly different between the mixed planting inside (MPI) and mixed planting outside (MPO) conditions, but the difference was not significant compared with the solo planting (SP) condition and MPO or MPI, and MPO < SP < MPI. The differences among the three simple points disappeared within the 15 cm deep soil layer. At 21 d, the water content in the SP condition was significantly lower than in the MPI and MPO conditions, and there no significant differences between the MPI and MPO conditions within the 3 cm deep soil layer. Within 9 cm deep soil layers, the water content was significantly different between the MPI and SP conditions, but the difference was not significant compared with the MPO and MPI or SP conditions, and SP < MPO < MPI, and there were no significant differences among the three sample points at the 15 cm depth.

At 11 d, the water contents in the rhizospheres of *P. orientalis* displayed significant differences between the solo planting (SO) and mixed planting inside (MOI) conditions, and MOI < MOO (mixed planting outside) < SO within the 3 cm and 9 cm deep soil layer, and the water content in the MOI condition was significantly lower than that in the MOO and SO conditions, and there no significant differences between the MOO and SO conditions within the 15 cm deep soil layer. At 21 d, the water content in the MOI condition was significantly lower than that in the MOO and SO conditions, and there no significant differences between the MOO and SO conditions within the 3 cm and 15 cm deep soil layer; the water content in the MOI condition was significantly lower than that in the SO condition within the 9 cm deep soil layer.

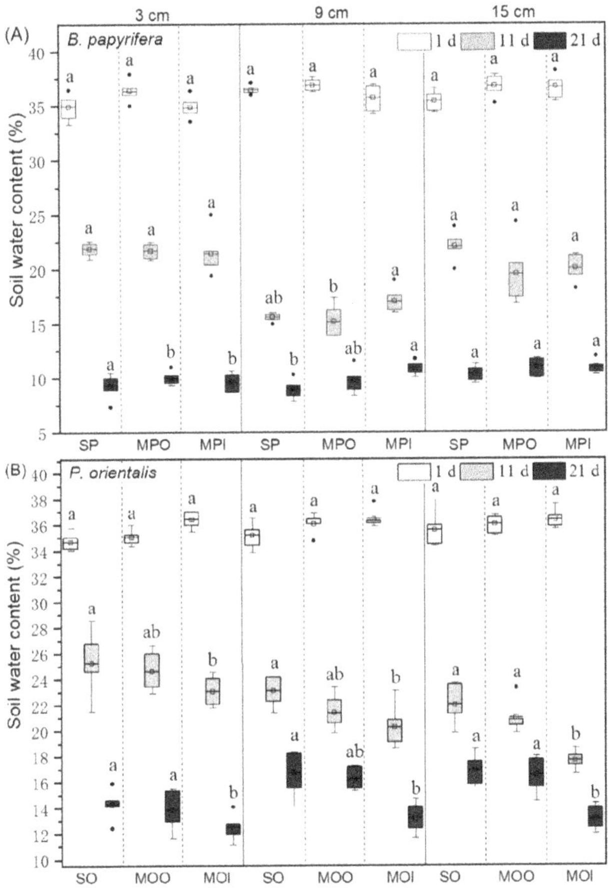

Figure 2. Soil water contents in the rhizosphere of *B. papyrifera* (**A**) and *P. orientalis* (**B**). Letters denote significant differences among soil sampling points ($p < 0.05$). SP—*B. papyrifera* under solo planting, MPO—outside of *B. papyrifera* under mixed planting, MPI—inside of *B. papyrifera* under mixed planting, SO—*P. orientalis* under mixed planting, MOO—outside of *P. orientalis* under mixed planting, MOI—inside of *P. orientalis* under mixed planting.

3.2. Leaf Water Contents, $\delta^{13}C$ Value, and g_s

As shown in Figure 3, the water contents in both plant leaves steadily dropped under prolonged drought conditions. Using Duncan's test ($p < 0.05$), significant differences existed between the solo planting and mixed planting conditions at 11 d and 16 d for *B. papyrifera* and 11 d for *P. orientalis*. Under prolonged drought conditions, the $\delta^{13}C$ values in both plants increased at 11 d.

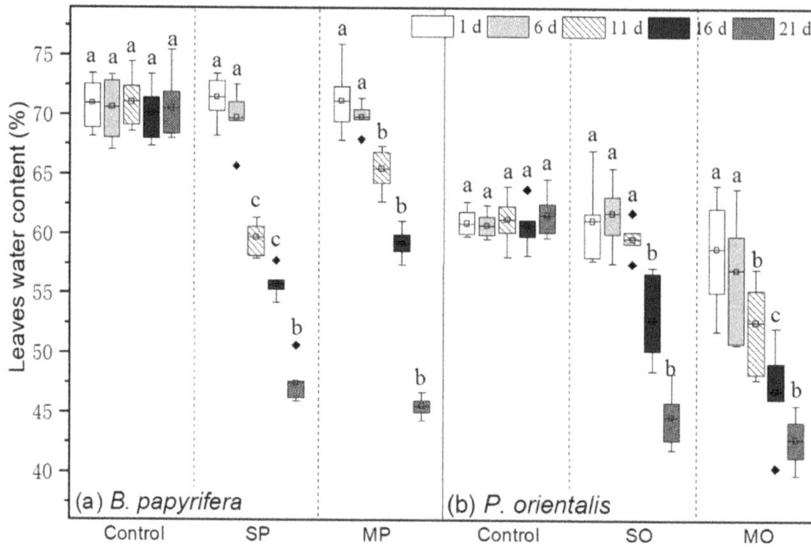

Figure 3. Leaf water contents in *B. papyrifera* and *P. orientalis*. Letters denote significant differences among different planting and treatment modes ($p < 0.05$). SP—*B. papyrifera* under solo planting, MP—*B. papyrifera* under mixed planting, SO—*P. orientalis* under mixed planting, MO—*P. orientalis* under mixed planting.

After 11 d, the $\delta^{13}C$ value in *B. papyrifera* continued to increase, but the $\delta^{13}C$ value in *P. orientalis* was almost maintained at the same level. Significant differences ($p < 0.05$) existed between the solo planting and mixed planting conditions and only occurred on day 16 for *B. papyrifera* (Figure 4A,B). As shown in Figure 4C,D, under prolonged drought conditions, the g_s in *B. papyrifera* under the two planting modes both decreased with the extension of processing time after 6 d. The g_s in *P. orientalis* decreased sharply at 11 d and then declined slightly.

3.3. N Contents and $\delta^{15}N$ Values

The N contents relative to water contents of leaves are represented in Figure 5A,B. Using linear regression analysis, the nitrogen content and water content in *B. papyrifera* leaves showed a significant positive relation under the solo planting and mixed planting conditions (solo: $r = 0.7229$, $p < 0.01$; mixed: $r = 0.7702$, $p < 0.01$), On the contrary, there were significant negative correlations between the nitrogen content and water content in *P. orientalis* leaves under the solo planting and mixed planting conditions (solo: $r = -0.6459$, $p < 0.01$; mixed: $r = -0.7976$, $p < 0.01$).

Under prolonged drought conditions, the $\delta^{15}N$ value in *B. papyrifera* leaves began to decrease sharply from the 11th d, and there was a significant difference ($p < 0.05$) between the mixed planting and solo planting conditions. The $\delta^{15}N$ value in *P. orientalis* leaves decreased slightly from the 16th d, and a significant difference between the two planting modes only existed at 16 d (Figure 5C,D).

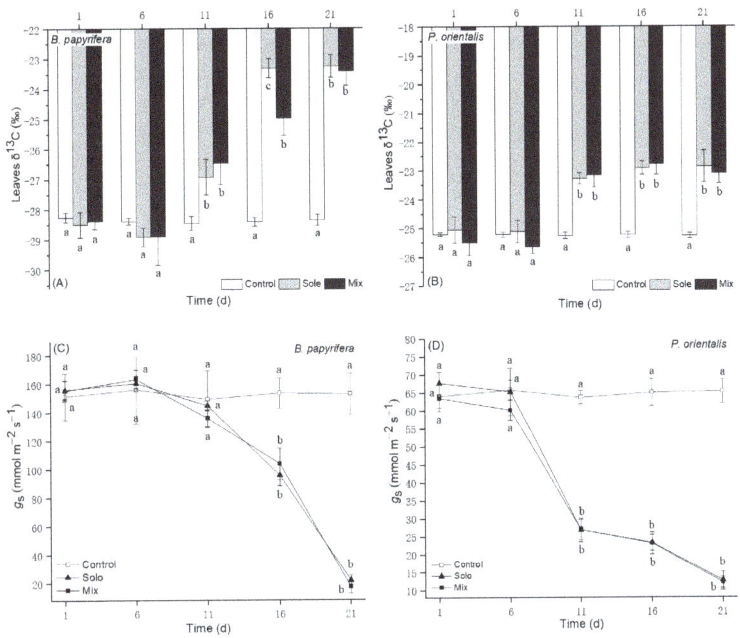

Figure 4. The leaves' $\delta^{13}C$ value in *B. papyrifera* (**A**) and *P. orientalis* (**B**). Leaves' g_s value in *B. papyrifera* (**C**) and *P. orientalis* (**D**). Letters denote significant differences among different planting and treatment modes ($p < 0.05$).

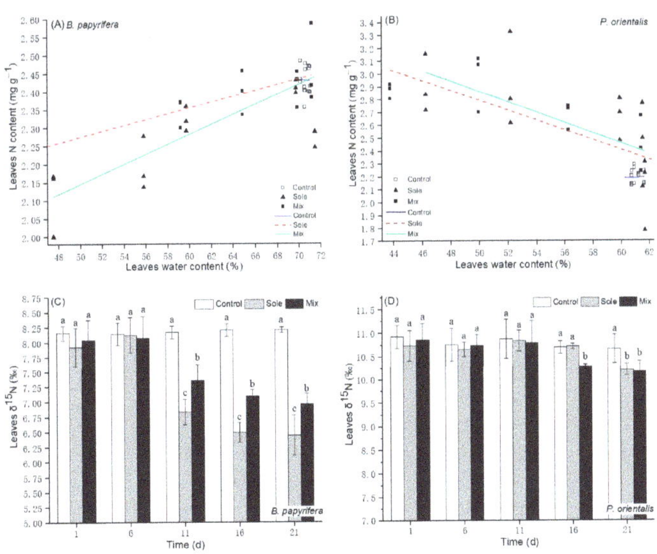

Figure 5. The leaves' N content relative to water content in *B. papyrifera* (**A**) and *P. orientalis* (**B**). Curve fits in the graphs are separate linear regressions. Leaves' $\delta^{15}N$ value in *B. papyrifera* (**C**) and *P. orientalis* (**D**). Letters denote significant differences among different planting and treatment modes ($p < 0.05$).

3.4. P Contents and PDE: PME Ratio

The P contents relative to the water contents in leaves are represented in Figure 6A,B. There was no correlation between the phosphorus content and water content in *B. papyrifera* leaves under the solo and mixed planting conditions (solo: $p = 0.7652$; mixed: $p = 0.1913$), and the same was true for *P. orientalis* (solo: $p= 0.7652$; mixed: $p = 0.1913$). Under prolonged drought, the PDE: PME ratio in *B. papyrifera* roots decreased from 0. 56 on day 1 to 0.24 at 11 d under the solo planting condition, and from 0.54 at 1 d to 0.30 at 11 d under the mixed planting condition. Then, this ratio increased at 21 d and was 0.33 under the solo and mixed planting conditions. The PDE: PME ratio in *P. orientalis* roots decreased from 1.39 at 1d to 0.39 at 11 d under the solo planting condition and from 1.31 at 1 d to 0.43 at 11 d under the mixed planting condition. Then, this ratio increased at 21 d and was 0.78 under the solo and mixed planting conditions. (Figure 6C,D).

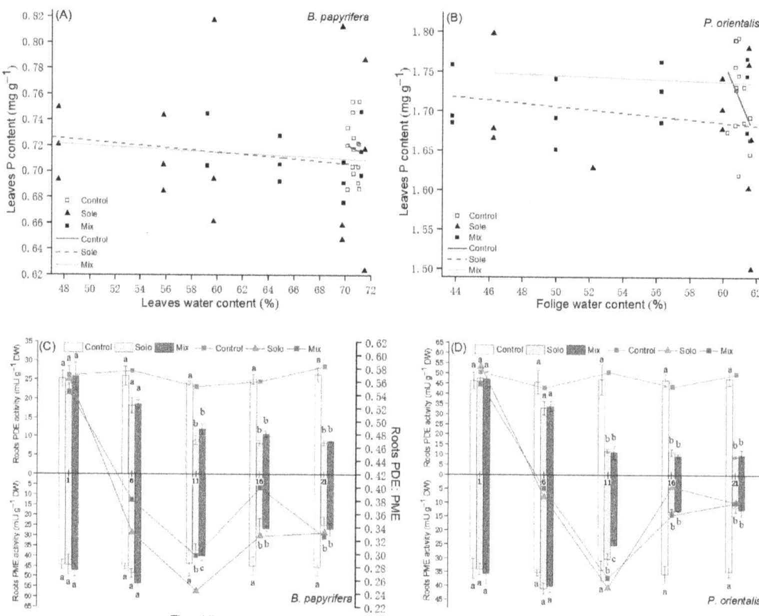

Figure 6. The leaves' P content relative to water content in *B. papyrifera* (**A**) and *P. orientalis* (**B**). Curve fits in the graphs are separate linear regressions. The roots' PDE: PME ratio in *B. papyrifera* (**C**) and *P. orientalis* (**D**). Letters denote significant differences among different planting and treatment modes ($p < 0.05$).

3.5. Correlation and Difference Analysis

As shown in Table 1, linear regression analysis was performed to calculate Pearson's correlations between the soil water content and N content, P content, $\delta^{13}C$, $\delta^{15}N$, PME, PDE, g_s, and foliage water content. For *B. papyrifera*, there was no correlation between soil water contents and P contents; there was a significant positive correlation between soil water contents and N contents; and there was an extremely significant positive correlation between soil water contents and $\delta^{13}C$, $\delta^{15}N$, PME, PDE, g_s, and leaf water contents. For *P. orientalis*, there was no correlation between soil water contents and P contents; a significant positive correlation between soil water contents and $\delta^{15}N$; an extremely significant positive correlation between soil water contents and $\delta^{13}C$, PME, PDE, g_s, and leaves' water contents; and an extremely significant negative correlation between soil water contents and N contents.

Table 1. *p*-value and Pearson's r between soil water content and N content, P content, $\delta^{13}C$, $\delta^{15}N$, PME, PDE, g_s, and leaf water content.

	B. papyrifera *p*-Value	Pearson's r	*P. orientalis* *p*-Value	Pearson's r
$\delta^{13}C$	3.65×10^{-6} ***	0.8648	2.20×10^{-7} ***	0.9064
N content	0.0206 *	0.5405	2.30×10^{-7} ***	−0.9059
$\delta^{15}N$	2.28×10^{-4} ***	0.7634	0.0128 *	0.5739
P content	0. 8149	−0.0594	0.6883	−0.1016
PME	6.24×10^{-5} ***	0.8021	2.50×10^{-7} ***	0.9048
PDE	2.54×10^{-7} ***	0.9046	6.07×10^{-9} ***	0.9411
g_s	2.79×10^{-4} ***	0.7567	3.61×10^{-13} ***	0.9828
Leaf water content	1.89×10^{-5} ***	0.8316	4.71×10^{-7} ***	0.8967

Soil water content is the average value of soil layers and sampling points. *p*-value and Pearson's r were calculated with linear regression analysis (* $p < 0.05$ and *** $p < 0.001$).

As shown in Table 2, a general linear model was used to analyse the difference between solo planting and mixed planting. For *B. Papyrifera*, a significant difference between the two planting methods in leaf water contents emerged at 11 d, and in $\delta^{15}N$, $\delta^{13}C$, N content, PDE and leaf water content emerged at 16 d. For *P. orientalis*, a significant difference between the two planting methods in PME and leaf water contents emerged at 11 d, in $\delta^{15}N$, $\delta^{13}C$ and PDE at 16 d.

Table 2. *p*-values in solo planting and mixed planting conditions in *B. papyrifera* and *P. orientalis*.

	p-Value									
	B. papyrifera					*P. orientalis*				
	1 d	6 d	11 d	16 d	21 d	1 d	6 d	11 d	16 d	21 d
$\delta^{13}C$	0.689	0.964	0.440	**0.012**	0.655	0.689	0.964	0.440	**0.012**	0.655
N content	0.870	0.959	0.136	**0.035**	0.157	0.911	0.935	0.874	0.861	0.822
$\delta^{15}N$	0.675	0.865	0.059	**0.006**	0.072	0.676	0.619	0.878	**0.001**	0.800
P content	0.845	0.799	0.761	0.779	0.791	0.490	0.833	0.588	0.299	0.977
PME	0.426	0.091	0.264	0.568	0.587	0.542	0.611	**0.036**	0.710	0.617
PDE	0.815	0.817	0.058	**0.013**	0.334	0.815	0.817	0.058	**0.013**	0.334
g_s	0.938	0.567	0.340	0.659	0.658	0.205	0.286	0.980	0.954	0.758
Leaf water content	0.873	0.930	**0.010**	**0.040**	0.073	0.585	0.951	**0.010**	0.212	0.695

p-values were calculated with general linear model.

4. Discussion

4.1. Plants' Water Status and Water Source Competition

How much water plants can get from soil is not only related to the distribution and absorption capacity of their roots but is also affected by the water competitiveness of neighboring plants [27]. In this study, the variation in water distribution in rhizosphere soil between the two planting modes was caused by water competition under mixed planting conditions, and *B. papyrifera* benefited from water competition, as shown by Duncan's test. The $\delta^{13}C$ was positively related to WUE under drought stress [28]. The competition effect emerged in *B. papyrifera* at 16 d, and plants under the mixed planting condition had lower WUEs (Figure 4A), which would imply that they were in a better water available environment. There was no difference in the $\delta^{13}C$ value of *P. orientalis* between the two planting modes (Figure 4B), and the slow growth rate and rigid stomatal regulation strategy of *P. orientalis* may have weakened the effect of interspecific competition. Conifers and broad-leaved plants have different strategies in water and nutrition regulation in arid environments because of their significant differences in morphological structures and physiological mechanisms [29,30]. Stomatal behavior can respond to environmental changes and interspecific competition [31]. Under prolonged drought conditions, the g_s values in *P. orientalis* showed that this plant took more rigid stomatal regulation and tried to reduce water loss, but *B. papyrifera* still maintained stomatal conductance before extreme drought occurred (Figure 4C,D). Previous studies observed that seedlings employed a riskier strategy when planted with a high-water-use competitor, and seedlings employed a more conservative strategy when planted with a low-water-use competitor [31]. However, in the current study, there were no significant differences in plants' stomatal behavior between the solo planting and mixed planting conditions. This competitive advantage is related to the two species exhibiting significant differences in their root morphology. *B. papyrifera* has more root branches and fine roots, which is beneficial for enhancing the absorption of soil water; the main root of the *P. orientalis* root is wider and has fewer branches, which is better for the expansion of roots to further ranges and deeper soil layers. Additionally, there is a difference in the xylem structure between angiosperms and gymnosperms [32]. Due to the low hydraulic conductivity of tracheids relative to vessels, the absorption and transportation of water are limited in gymnosperms [33].

4.2. Plants N Status and N Sources Change

B. papyrifera and *P. orientalis* displayed opposite change patterns in the relationship between leaf water and nitrogen content (Figure 5A,B). Under solo planting and mixed planting conditions, there was a significant positive correlation between the leaf nitrogen content and water content in *B. papyrifera*, which was similar to the results of many studies [34–36]; however, there was a significant negative correlation between the leaf nitrogen content and water content of *P. orientalis*. It is speculated that the young leaves of *P. orientalis* obtain N from old leaves to synthesize abundant N-containing osmoregulation substances to maintain osmotic pressure during stomatal closure under water deficiency conditions. Under prolonged drought conditions, the change in $\delta^{15}N$ value in *B. papyrifera* leaves was greater than that in *P. orientalis* leaves (Figure 5C,D), which also implied that *B. papyrifera* obtain more N from the environment and have more flexible N sources, while N in *P. orientalis* was transported from old tissues. In addition, because NO_3^- must be reduced to NH_4^+ before it can be used for plant metabolism, the assimilation event of nitrate is more complex, as mentioned by Evans et al. [37]; if plants use NO_3^- as a nitrogen source, the leaves have relatively high $\delta^{15}N$. From this view, *B. papyrifera* used more ammonium as a nitrogen source under drought stress in this study. The differences in the N source composition and the N pool in young leaves may show that the N competition pressure between these two plants could be alleviated under drought stress.

4.3. Plants P Status and P Sources Change

The availability of N and P is affected by drought in soil through different mechanisms. There was no correlation between the leaf P contents and water contents in *B. papyrifera* and *P. orientalis* (Figure 6A,B), which suggests that there was enough P stored in the plants to supply the growth of young leaves. P stored in roots could be used by plants for growth at times when availability is low [38]. The P content in young *P. orientalis* leaves was much higher than that in *B. papyrifera*, which should be because the growth time of *P. orientalis* of the same size is much longer than that of *B. papyrifera*, so more P accumulates in the P body. A large number of nutrient reserves obtained before drought events would make *P. orientalis* acquire a stronger recovery ability after rainfall [39–41]. Under the control treatment, the PME activity in *B. papyrifera* was higher than that in *P. orientalis*, but the PDE activity in *P. orientalis* was higher than that in *B. papyrifera* (Figure 6C,D), which should be related to the fact that angiosperms are more suitable for mesophytic habits and gymnosperms are more suitable for hydrophytic habits [42] and enable niche differentiation between these two plants in P resource utilization. Under drought stress, the PDE: PME ratio in both plants decreased sharply under prolonged drought, showing that the activity of PDE is heavily dependent on soil moisture; this may be because phosphate diesters are usually degraded rapidly in dry environments and the enzyme activity is driven by relative resource availability [43,44]. The PDE: PME ratio in these two plants underwent almost the same changes under prolonged drought, which indicated that they might have lost the niche differentiation of the P source and that P competition was exacerbated under drought stress.

4.4. Interspecific Competition Analysis

The competition mode of trees is driven by environmental factors and controlled by the dominant limiting factor [45]. In this study, except for the P content in leaves, the physiological and biochemical parameters of plants were significantly affected by soil water contents under prolonged drought conditions (Table 1). This shows that water competition was a decisive factor in interspecific nutritional competition between *B. papyrifera* and *P. orientalis*. Meanwhile, a general linear model was used to analyze the interspecific competition effect (the difference between the solo planting and mixed planting conditions) (Table 2). The results showed that water characteristics, N and P characteristics, and resources were significantly affected by interspecific competition under moderate drought conditions, but the interspecific competition effect did not exist under heavy drought conditions. In the current study, *B. papyrifera* benefited from interspecific competition. Previous studies also indicated that broad-leaved species competition dominates interspecific interactions in growth and water and nutrition utilization in mixed stands [46,47]. However, Liu et al. found that the level of N uptake in *Pinus massoniana* and *Pinus elliottii* were inhibited by the presence of *Michelia maudiae* and *Schima superba*, respectively, but almost no inhibitions occurred when *M. maudiae* was grown together with *P. elliottii* or *S. superba* was grown together with *P. massoniana* [48]. This indicates that angiosperms do not always have the advantage when competing with gymnosperms.

5. Conclusions

The present study indicated that *B. papyrifera* benefits from water and nutrient competition under moderate drought, and the competition mode of *B. papyrifera* and *P. orientalis* is controlled by the dominant limiting factor—soil water content. The N competition between these two plants could be alleviated through niche differentiation, but they lose the niche differentiation of the P source under drought stress conditions. This study illustrated that when coexisting with *B. papyrifera* in mixed stands, the effect of drought stress on *P. orientalis* will be amplified through interspecific competition; this means that the effect of interspecific water competition on mixed forests should be seriously considered in semiarid areas.

Author Contributions: Conceptualization, K.Y. and Y.W. (Yanyou Wu); experimentation, K.Y. and Y.W. (Yanqing Wang); writing, K.Y.; data analysis, Y.W. (Yanqing Wang); review and editing, Y.W. (Yanyou Wu). All authors have read and agreed to the published version of the manuscript.

Funding: This research was funded by the National Natural Science Foundation of China (U1612441) and the Doctoral Research Program of Guizhou Normal University (11904/0517067).

Institutional Review Board Statement: Not applicable.

Informed Consent Statement: Not applicable.

Data Availability Statement: Not applicable.

Acknowledgments: The authors are grateful to staff members from the 3rd Institute of Oceanography, Ministry of Natural Resources for the measurement of the isotopic composition.

Conflicts of Interest: The authors declare no conflict of interest.

References

1. Cramer, W.; Guiot, J.; Fader, M.; Garrabou, J.; Gattuso, J.P.; Iglesias, A.; Lange, M.A.; Lionello, P.; Llasat, M.C.; Paz, S. Climate change and interconnected risks to sustainable development in the Mediterranean. *Nat. Clim. Chang.* **2018**, *8*, 972–980. [CrossRef]
2. Kunstler, G.; Falster, D.; Coomes, D.A.; Hui, F.; Kooyman, R.M.; Laughlin, D.C.; Poorter, L.V.; Vanderwel, M.; Vieilledent, G.; Wright, S.J.; et al. Plant functional traits have globally consistent effects on competition. *Nature* **2016**, *529*, 204–207. [CrossRef] [PubMed]
3. Calama, R.; Conde, M.; Dios-García, J.; Madrigal, G.; Vázquez-Piquéc, J. Linking climate, annual growth and competition in a Mediterranean forest: Pinus pinea in the Spanish Northern Plateau. *Agric. For. Meteorol.* **2019**, *264*, 309–321. [CrossRef]

4. Lorts, C.M.; Lasky, J.R. Competition × drought interactions change phenotypic plasticity and the direction of selection on *Arabidopsis* traits. *New Phytol.* **2020**, *227*, 1060–1072. [CrossRef] [PubMed]
5. Grossiord, C. Having the right neighbors: How tree species diversity modulates drought impacts on forests. *New Phytol.* **2019**, *228*, 42–49. [CrossRef]
6. Jourdan, M.; Lebourgeois, F.; Morin, X. The effect of tree diversity on the resistance and recovery of forest stands in the French Alps may depend on species differences in hydraulic features. *For. Ecol. Manag.* **2019**, *450*, 117486. [CrossRef]
7. Schwarz, J.A.; Bauhus, J. Benefits of mixtures on growth performance of silver fir (*Abies alba*) and european beech (*Fagus sylvatica*) increase with tree size without reducing drought tolerance. *Front. For. Glob. Chang.* **2019**, *2*, 79. [CrossRef]
8. Forrester, D.I.; Bauhus, J. A review of processes behind diversity-productivity relationships in forests. *Curr. For. Rep.* **2016**, *2*, 45–61. [CrossRef]
9. He, M.; Dijkstra, F.A. Drought effect on plant nitrogen and phosphorus: A meta-analysis. *New Phytol.* **2014**, *204*, 924–931. [CrossRef]
10. Sardans, J.; Urbina, I.; Grau, O.; Asensio, D.; Ogaya, R.; Peñuelas, J. Long-term drought decreases ecosystem C and nutrient storage in a Mediterranean holm oak forest. *Environ. Exp. Bot.* **2020**, *114*, 104135. [CrossRef]
11. Suriyagoda, L.D.B.; Ryan, M.H.; Renton, M.; Lambers, H. Plant responses to limited moisture and phosphorus availability: A meta-analysis. *Adv. Agron.* **2014**, *124*, 143–200.
12. Cregger, M.A.; McDowell, N.G.; Pangle, R.E.; Pockman, W.T.; Classen, A.T. The impact of precipitation change on nitrogen cycling in a semi-arid ecosystem. *Funct. Ecol.* **2014**, *28*, 1534–1544. [CrossRef]
13. Agren, G.I. Stoichiometry and nutrition of plant growth in natural communities. *Annu. Rev. Ecol. Evol. Syst.* **2008**, *39*, 153–170. [CrossRef]
14. Le Bauer, D.S.; Treseder, K.K. Nitrogen limitation of net primary productivity in terrestrial ecosystems is globally distributed. *Ecology* **2008**, *89*, 371–379. [CrossRef]
15. Vitousek, P.M.; Porder, S.; Houlton, B.Z.; Chadwick, O.A. Terrestrial phosphorus limitation: Mechanisms, implications, and nitrogen-phosphorus interactions. *Ecol. Appl.* **2010**, *20*, 5–15. [CrossRef]
16. Homyak, P.M.; Slessarev, E.W.; Hagerty, S.; Greene, A.C.; Marchus, K.; Dowdy, K.; Iverson, S.; Schimel, J.P. Amino acids dominate diffusive nitrogen fluxes across soil depths in acidic tussock tundra. *New Phytol.* **2021**, *231*, 2162–2173. [CrossRef]
17. Leake, J.R.; Miles, W. Phosphodiesters as mycorrhizal P sources: I. Phosphodiesterase production and the utilization of DNA as a phosphorus source by the ericoid mycorrhizal fungus *Hymenoscyphus ericae*. *New Phytol.* **1996**, *132*, 435–443. [CrossRef]
18. Makarov, M.I.; Haumaier, L.; Zech, W. The nature and origins of diester phosphates in soils: A ^{31}P-NMR study. *Biol. Fertil. Soils* **2002**, *35*, 136–146.
19. Turner, B.L.; Newman, S. Phosphorus cycling in wetland soils: The importance of phosphate diesters. *J. Environ. Qual.* **2005**, *34*, 1921–1929. [CrossRef]
20. Turner, B.L. Resource partitioning for soil phosphorus: A hypothesis. *J. Ecol.* **2008**, *96*, 698–702. [CrossRef]
21. Rejmánková, E.; Dagmara Sirová, D.; Carlson, E. Patterns of activities of root phosphomonoesterase and phosphodiesterase in wetland plants as a function of macrophyte species and ambient phosphorus regime. *New Phytol.* **2011**, *190*, 968–976. [CrossRef]
22. Huston, M.A. Management strategies for plant invasions: Manipulating productivity, disturbance, and competition. *Divers. Distrib.* **2004**, *10*, 167–178. [CrossRef]
23. Lei, H.P.; Wang, Y.G.; Liang, F.Y.; Su, W.W.; Feng, Y.F.; Guo, X.L.; Wang, N. Composition and variability of essential oils of *Platycladus orientalis* growing in China. *Biochem. Syst. Ecol.* **2010**, *38*, 1000–1006. [CrossRef]
24. Li, J.Y. Studies on drought tolerance of some main used in afforestation in Taihang mountain tree species region (III)—Water parameters. *J. Beijing For. Univ.* **1990**, *13*, 230–239.
25. Tabatabai, M.A.; Bremmer, J.A. Use of p-nitrophenylphosphate for assay of soil phosphatase activity. *Soil Biol. Biochem.* **1969**, *1*, 301–307. [CrossRef]
26. Eivazi, F.; Tabatabai, M.A. Phosphatases in soils. *Soil Biol. Biochem.* **1977**, *9*, 167–172. [CrossRef]
27. Farrior, C.E.; Dybzinski, R.; Levin, S.A.; Pacala, S.W. Competition for water and light in closed-canopy forests: A tractable model of carbon allocation with implications for carbon sinks. *Am. Nat.* **2013**, *181*, 314–330. [CrossRef]
28. Jucker, T.; Grossiord, C.; Bonal, D.; Bouriaud, O.; Gessler, A.; Coomes, D.A. Detecting the fingerprint of drought across Europe's forest: Do carbon isotope ratios and stem growth rates tell similar stories? *For. Ecosyst.* **2017**, *4*, 24. [CrossRef]
29. DeSoto, L.; Cailleret, M.; Sterck, F.; Jansen, S.; Kramer, K.; Robert, E.M.R.; Aakala, T.; Amoroso, M.M.; Bigler, C.; Camarero, J.J.; et al. Low growth resilience to drought is related to future mortality risk in trees. *Nat. Commun.* **2020**, *11*, 545. [CrossRef]
30. Adams, H.D.; Zeppel, M.J.B.; Anderegg, W.R.L.; Hartmann, H.; Landhäusser, S.M.; Tissue, D.T.; Huxman, T.E.; Hudson, P.J.; Franz, T.E.; Allen, C.D.; et al. A multi-species synthesis of physiological mechanisms in drought-induced tree mortality. *Nat. Ecol. Evol.* **2017**, *1*, 1285–1291. [CrossRef]
31. Zenes, N.; Kerr, K.L.; Trugman, A.T.; Anderegg, W.R.L. Competition and drought alter optimal stomatal strategy in tree seedlings. *Front. Plant Sci.* **2020**, *11*, 478. [CrossRef] [PubMed]
32. Brodribb, T.J.; Field, T.S. Leaf hydraulic evolution led a surge in leaf photosynthetic capacity during early angiosperm diversification. *Ecol. Lett.* **2010**, *13*, 175–183. [CrossRef] [PubMed]
33. Piper, F.I.; Hoch, G.; Fajardo, A. Revisiting the relative growth rate hypothesis for gymnosperm and angiosperm species co-occurrence. *Am. J. Bot.* **2019**, *106*, 101–112. [CrossRef] [PubMed]

34. Urbina, I.; Sardans, J.; Beierkuhnlein, C.; Jenntsch, A.; Backhaus, S.; Grant, K.; Peñuelas, J. Shifts in the elemental composition of plants during a very severe drought. *Environ. Exp. Bot.* **2015**, *111*, 63–73. [CrossRef]
35. Mouradi, M.; Bouizgarenb, A.; Farissi, M.; Latracha, L.; Qaddourya, A.; Cherki, G. Seed osmopriming improves plant growth, nodulation, chlorophyll fluorescence and nutrient uptake in alfalfa (*Medicago sativa* L.)—Rhizobia symbiosis under drought stress. *Sci. Hortic.* **2016**, *213*, 232–242. [CrossRef]
36. Mariotte, P.; Cresswell, T.; Johansen, M.P.; Harrison, J.J.; Keitel, C.; Dijkstra, F.A. Plant uptake of nitrogen and phosphorus among grassland species affected by drought along a soil available phosphorus gradient. *Plant Soil.* **2020**, *448*, 121–132. [CrossRef]
37. Evans, R.D.; Bloom, A.J.; Sukrapanna, S.S.; Ehleringer, J.R. Nitrogen isotope composition of tomato (*Lycopersicon esculentum* Mill. cv. T-5) grown under ammonium or nitrate nutrition. *Plant Cell Environ.* **1996**, *19*, 1317–1323. [CrossRef]
38. Peuke, A.D.; Rennenberg, H. Carbon, nitrogen, phosphorus, and Sulphur concentration and partitioning in beech ecotypes (*Fagus sylvatica* L.): Phosphorus most affected by drought. *Trees* **2004**, *18*, 639–648. [CrossRef]
39. Hacke, U.G.; Sperry, J.S.; Wheeler, J.K.; Castro, L. Scaling of angiosperm xylem structure with safety and efficiency. *Tree Physiol.* **2006**, *26*, 689–701. [CrossRef]
40. Gessler, A.; Schaub, M.; McDowell, N.G. The role of nutrients in drought-induced tree mortality and recovery. *New Phytol.* **2017**, *214*, 513–520. [CrossRef]
41. Schönbeck, L.; Gessler, A.; Schaub, M.; Rigling, A.; Hoch, G.; Kahmen, A.; Li, M.H. Soil nutrients and lowered source: Sink ratio mitigate effects of mild but not of extreme drought in trees. *Environ. Exp. Bot.* **2020**, *169*, 103905. [CrossRef]
42. Bond, W.J. The tortoise and the hare: Ecology of angiosperm dominance and gymnosperm persistence. *Bot. J. Linn. Soc.* **1989**, *36*, 227–249. [CrossRef]
43. Bowman, R.A.; Cole, C.V. Transformations of organic phosphorus substrates in soils as evaluated by $NaHCO_3$ extraction. *Soil Sci.* **1978**, *125*, 49–54. [CrossRef]
44. Sinsabaugh, R.L.; Follstad Shah, J.J. Ecoenzymatic stoichiometry and ecological theory. *Annu. Rev. Ecol. Evol. Syst.* **2012**, *43*, 313–343. [CrossRef]
45. Pretzsch, H.; Biber, P. Size-symmetric versus size-asymmetric competition and growth partitioning among trees in forest stands along an ecological gradient in central Europe. *Can. J. For. Res.* **2010**, *40*, 370–384. [CrossRef]
46. Reyer, C.; Lasch, P.; Mohren, G.M.J.; Sterck, F.J. Inter-specific competition in mixed forests of Douglas-fir (*Pseudotsuga menziesii*) and common beech (*Fagus sylvatica*) under climate change—A model-based analysis. *Ann. For. Sci.* **2010**, *67*, 805. [CrossRef]
47. Ferrio, J.P.; Shestakova, T.A.; Castillo, J.; Voltas, J. Oak competition dominates in-terspecific interactions in growth and water-use efficiency in a mixed pine–oak medi-terranean forest. *Forests* **2021**, *12*, 1093. [CrossRef]
48. Liu, Q.Y.; Xu, M.J.; Yuan, Y.; Wang, H.M. Interspecific competition for inorganic nitrogen between canopy trees and underlayer-planted young trees in subtropical pine plantations. *For. Ecol. Manag.* **2021**, *494*, 119331. [CrossRef]

agronomy

MDPI

Article

Karst Soil Patch Heterogeneity with Gravels Promotes Plant Root Development and Nutrient Utilization Associated with Arbuscular Mycorrhizal Fungi

Qing Li [1], Muhammad Umer [1,2], Yun Guo [1], Kaiping Shen [1], Tingting Xia [1], Xinyang Xu [1], Xu Han [1], Wenda Ren [1], Yan Sun [1], Bangli Wu [1], Xiao Liu [3] and Yuejun He [1,2,*]

[1] Forestry College, Research Center of Forest Ecology, Guizhou University, Guiyang 550025, China; lq192508@163.com (Q.L.); umer@gzu.edu.cn (M.U.); zihanyun2013@163.com (Y.G.); skp0825@163.com (K.S.); xtt1268@163.com (T.X.); xxy2592076369@163.com (X.X.); hanxukumo@163.com (X.H.); rwenda2021@163.com (W.R.); suny1360@163.com (Y.S.); wubangli1116@163.com (B.W.)

[2] Institute for Forest Resources & Environment of Guizhou, Guizhou University, Guiyang 550025, China

[3] Forestry Survey and Planning Institute of Guizhou Province, Guizhou University, Guiyang 550003, China; lxmq1005@163.com

* Correspondence: hyj1358@163.com; Tel.: +86-1376-51440-53

Abstract: Arbuscular mycorrhizal (AM) fungi associated with plant roots play an essential role in the belowground ecological process in karst habitats with high spatial and substrate heterogeneity. However, the effects of AM fungi on root morphology and nutrient uptake under different soil patch sizes and gravel content in karst habitats are still unclear. A controlled experiment was conducted using a square device divided into 16 grid patches. This experiment had three treatments, including the mycorrhizal fungal treatment inoculated with (M^+) or without *Glomus etunicatum* Becker & Gerd (M^-), the patch heterogeneity treatment through the homogeneous patch (Homo), heterogeneity-large patch (Hetl) and heterogeneity-small patch (Hets), and substrate heterogeneity treatment through the gravel-free substrate (GF), gravel-low substrate (GL), and gravel-high substrate (GH). Root traits and nutrients of *Bidens pilosa* L were analyzed, and the result showed the AM fungi significantly increased the dry weight, length, surface area, average diameter, volume, tips, branching points, and N, P, and K acquisitions of *B. pilosa* roots, but significantly decreased the specific root length. The Hets with soil and gravel increased the dry weight, length, surface area, tips, branching points, and N, P, and K acquisitions of *B. pilosa* roots compared with Hetl regulated by AM fungi. The GL and GH treatments also increased the dry weight, length, surface area, tips, branching points, and N, P, and K acquisitions of *B. pilosa* roots compared with GF regulated by AM fungi. These results indicate that the *B. pilosa* roots' nutritional acquisition benefits were higher in Hets mixed with gravel for its root morphological development regulated by AM fungi in karst soil. In conclusion, we suggest that soil patch heterogeneity with gravels promotes root morphological development and nutrient utilization to karst plants associated with arbuscular mycorrhizal fungi.

Keywords: arbuscular mycorrhizal fungi; karst; patch and substrate heterogeneity; root morphology; nutrients uptake

Citation: Li, Q.; Umer, M.; Guo, Y.; Shen, K.; Xia, T.; Xu, X.; Han, X.; Ren, W.; Sun, Y.; Wu, B.; et al. Karst Soil Patch Heterogeneity with Gravels Promotes Plant Root Development and Nutrient Utilization Associated with Arbuscular Mycorrhizal Fungi. *Agronomy* **2022**, *12*, 1063. https://doi.org/10.3390/agronomy12051063

Academic Editor: Alwyn Williams

Received: 25 March 2022
Accepted: 26 April 2022
Published: 28 April 2022

1. Introduction

Karst landforms mainly developed through carbonate rocks and are a crucial terrestrial ecosystem with the most fragile ecological landscape [1]. The karst landform in southwest China, having an area of approximately 530,000 km^2, is the largest in the world [2], and is characterized by soil and water loss, low vegetative coverage, bedrock exposure, and rocky terrain desertification [3]. Karst landscape disintegration induces a higher spatial heterogeneity by the shallow and discontinuous soil layer crossing with the exposed rocks through various heterogeneous habitats such as rocky fissures, rocky gullies, soil faces, and rocky pores [4,5]. Additionally, the karst soil's natural habitat composition is usually mixed with different gravels due to the differential weathering of rock through physicochemical disintegration [6], which presents soil resource heterogeneity in terms of mechanical components. Therefore, the karst plants always suffer from the spatial patch heterogeneity induced by different microhabitats, such as the soil face, stone face, and swallet. Meanwhile, the substrate heterogeneity of soil substrate is caused by mechanical composition.

Habitat spatial heterogeneity and substrate heterogeneity affect biodiversity and further change the stability and sustainability of the karst ecosystem [7]. The heterogeneous soil resource heterogeneity has long been theoretically recognized as an essential driver of plant species' coexistence and community diversity [8,9], including soil nutrients and water-induced spatial-scale heterogeneity [10,11], which significantly affects plants and community [12–14]. Liu et al. [15] found that soil resource availability was much more important than soil resource heterogeneity in determining the species diversity of plant communities of karst, and it was supported by the resources availability hypothesis that the species diversity among the community was lower in general under low resource availability because a few species could possibly grow and survive [16]. The spatial heterogeneity in essential resource availability can affect placement and growth of leaves and roots, the growth of whole plants, the intensity of inter-plant competition, and the yield and structure of plant populations [17–20]. For instance, Tamme et al. [20] found that species that segregate along the heterogeneity niche axis grow better in heterogeneous soil, and species' responses to soil heterogeneity would fluctuate depending on the size of the soil patches. Hutchings et al. [21] demonstrated that plants are most advantaged in heterogeneous conditions, and the proportion of roots developed in rich and poor patches closely matches the relative quality of each patch type.

Furthermore, Qian et al. [22] also revealed that different patch sizes significantly affect biomass; plants' plastic responses can effectively utilize nutrient-rich patches at larger spatial scales, in contrast to minor spatial scales. Mi et al. [23] found that plants grown in soils containing rock debris were shorter in heterogeneous substrates, and had smaller basal stem diameters and lower root biomass, than plants grown in rock-free soils. Similarly, Masoni et al. [24] indicated that the increase in soil gravel decreased the biomass, N, and P of durum wheat. Therefore, patch and substrate heterogeneity are crucial for plant growth and community establishment under karst habitat conditions. Physiological and morphological responses of plants to habitat heterogeneity initially occur at plant root systems [25]. Plant roots play an essential role in the underground ecological process and maintain aboveground productivity. Plant root developments also affect nutrient uptake [3,26]. The root system is jointly affected by spatial and resource heterogeneity [23,24]. For example, Wijesinghe et al. [27] showed that plants in nutrient-rich patches produced significantly more root biomass than in nutrient-poor patches when patch size was kept large. Hutchings et al. [21] also discovered that plants can respond with great effectiveness to the heterogeneous substrate, closely matching the mass of roots produced in different patches to the relative quality of each patch. Day et al. [28] illustrated that when plants grow in heterogeneous conditions, the plant yield strongly depends on the patch scale, and plant roots obtained more biomass at the larger patch scale than at the smaller patch scale. Alagna et al. [29] stated that soil–rock ratio has a significant impact on root growth, morphological establishment, the gravel contents determining soil pore status, and the hydrological process [30,31]. The gravel contents further affect root growth [32], compared with gravel-free soil, and gravel forms more pores in the soil, which reduce mechanical resistance, thereby encouraging the extension of roots [31,33]. Therefore, patch heterogeneity and substrate heterogeneity jointly affect plant growth, root development, and nutritional uptakes.

Changes in soil physicochemical properties have a response regulation for a plant with the participation of microorganisms [34], and soil microorganisms are essential drivers of plant growth and the establishment of the community [35–37]. Changes in soil physicochemical properties also affect plant growth, root development, and nutrient uptake in heterogeneous environments. As an example, arbuscular mycorrhizal (AM) fungi can have symbiotic relationships with most land plants to form mycorrhizae, which have slender hyphae and strong branching ability, and regulate the morphological development of the root system and improve nutrient uptakes [38–41]. Facelli et al. [18] found that the arbuscular mycorrhizal symbiotic relationship can strongly influence plant growth in heterogeneous soil by facilitating the pre-emption of limiting resources. Plants preferentially obtain resources from soil patches through plant–AM fungi symbionts instead of root foraging because the metabolic cost of mycelial growth is lower than that of root growth [42,43]. Specifically, AM fungal hyphae can respond to heterogeneous soil in which plant roots cannot obtain more soil resources; thus, AM fungi significantly improved biomass, N, and P of plants [42,44]. Furthermore, the extensive underground mycelial network can redistribute the nitrogen and phosphorus in heterogeneous soil patches and enhance plant ability to obtain optimal growth [45–47]. Liang et al. [48] found that heterogeneous karst habitats maintain high AM fungal diversity. The karst soil contains significant amounts of detritus and gravel in natural heterogeneous karst habitats, resulting in soil substrate heterogeneity and patch heterogeneity. Many studies have reported the effects of AM fungi on plant growth and nutrient acquisition in karst ecosystems. In particular, He and Xia [49,50] suggested that the external root mycelia of AM fungi can obtain soil resources outside the rhizosphere to increase the biomass, N, and P acquisitions of karst plants. Zhang et al. [51] also showed that AM fungi have a positive effect on the growth of karst plants in heterogeneous soil. However, AM fungi jointly regulate the soil, and patch heterogeneity, which affects plant root development, remains unclear. Therefore, we hypothesized that: (1) AM fungi can promote root growth and nutrient uptakes of plants in heterogeneous karst soil; (2) heterogeneous patches promote more root growth and nutrient uptake of plants than homogeneous patches regulated by AM fungi; and (3) soil mixed with gravels promotes more root growth and nutrient uptake of plants than soil substrate through AM fungi.

2. Materials and Methods

2.1. Experimental Treatments

A controlled experiment was conducted using a square microcosm composed of polypropylene plastic (26 cm × 26 cm × 15 cm, length × width × height) divided into 16 small grid cells through a movable grid plate for forming heterogeneous patches by quantitatively filling with growth substrates in each grid (Figure 1). There were 4 round holes of 0.5 cm diameter at the bottom of the device to prevent water accumulation. This experiment included three treatments, namely, AM fungi, patch heterogeneity, and substrate heterogeneity. Specifically, the AM fungi treatment was inoculated with (M^+) or without (M^-) AM fungus *Glomus etunicatum* Becker & Gerd. The patch heterogeneity treatment was conducted through three patch sizes in a microcosm, e.g., homogeneous patch (Homo), heterogeneity-large patch (Hetl), and heterogeneity-small patch (Hets) (Figure 1). The substrate heterogeneity treatment was conducted through three different substrates: gravel-free substrate of 100% soil (GF) for Hetl and Hets patches, the gravel-low substrate mixed with 80% soil and 20% gravel (GL) for Homo patch, and the gravel-high substrate mixed with 60% soil and 40% gravel (GH) for Hetl and Hets patches. Although there were three different substrates, the total amount of soil and gravel components was the same in each treatment microcosm (An equal amount of soil and gravel was placed in each square microcosm used a measuring cylinder), except for the patch size. Three seeds of *B. pilosa* were sown into each grid patch in each microcosm. A 50 g inoculum of G. etunicatum was added into the M^+ treatment, whereas, to the M^- treatment was added 50 g of autoclaved G. etunicatum inoculum; this treatment also received an additional 10 mL filtrate by weighing 50 g of inoculum with double-layer filter paper, in order to ensure that M^+ and M^- treatments had a consistent microflora, except for the target AM fungus. One plant each cell was retained after germination, and there was a total of 16 seedlings in each microcosm.

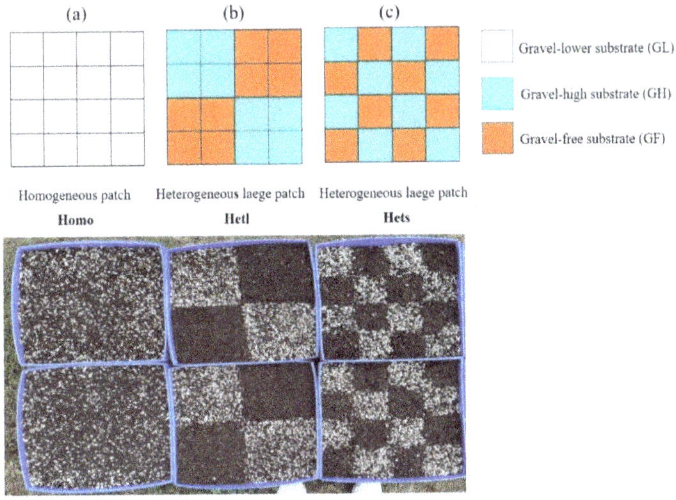

Figure 1. Schematic representation of experimental design grid patch in the square microcosm. The (**a**–**c**) represent homogeneous patch, heterogeneous-large patch and heterogeneous-small patch, respectively. The experiment was conducted using a square device divided into 16 patches. Homo = homogeneous patch. Hetl = heterogeneous-large patch. Hets = heterogeneous-small patch. The GL substrate was filled with a mixed substrate of 80% soil + 20% gravel, and the GF substrate was filled with a substrate of 100% soil; the GH substrate was filled with a mixed substrate of 60% soil + 40% gravel.

The soil and gravel were collected from a typical karst habitat near Guiyang. Nutrient concentrations were calculated per kilogram of soil sample using the Bao [52] method, as 0.622 g of total nitrogen, 0.315 g of available nitrogen (AN), 1.274 g of total phosphorus (TP), 0.163 g of available phosphorus (AP), 37.794 g of total potassium (TK), and 0.532 g of available potassium (AK). For one hour, the collected soil and gravel were air dried and sterilized at 0.14 Mpa at 126 °C. In particular, experimental gravels diameter were approximately 2~4 mm when sieved alternatively and developed from the weathering of carbonate rock. The seed of *B. pilosa* was collected from a typical karst habitat as the pioneer species in the primary succession stage in Huaxi District, Guiyang, Guizhou, China. Additionally, the *G. etunicatum* inoculum purchased from the Institute of Nutrition Resources, Beijing Academy of Agricultural and Forestry Sciences, BGA0046, was propagated with Trifolium repens through a sterilized limestone soil substrate for four months, which included approximately 150 spores per 10 g of soil, hyphal pieces, and colonized root segments. The eight replicates of treatments included a total of 48 microcosms containing 468 plants in this experiment. Treatments were incubated in a greenhouse for 12 weeks at the western campus of Guizhou University (106°22′ E, 29°49′ N, 1120 m above the sea level), and then were harvested for measurement.

2.2. Measurements of Mycorrhizal Colonization Rate, Dry Weight, and Traits of Root

We adopted the method described by He et al. [38] to determine the mycorrhizal colonization rate. Specifically, fresh plant roots were selected and stained by fuchsin after cleaning the roots in 10% KOH. After staining, the percentage of AM colonization was determined with the gridline intersection method. The root dry weight of *B. pilosa* was determined by weighing the root material after drying at 80 °C for constant weight. Plant nitrogen concentration was determined by the Kjeldahl method, and phosphorus was determined by Molybdenum-Antimony colorimetry [52]. Plant root morphological traits were measured using a root scanning analysis system (STD 1600 Epsom USA; Winrhigo -Version 410 B) to obtain root length, average diameter, surface area, volume, branching points, and tips of roots. The specific root length, the root area, and the root volume were calculated by the root length, area, and volume divided by the root dry weight, respectively [53].

2.3. Statistical Analysis

Three-way ANOVA was applied for the effects of mycorrhizal fungus (M^+ vs. M^-), patch heterogeneity (Homo vs. Hetl vs. Hets), and substrate heterogeneity (GL vs. GF vs. GH), and their interactions on root traits and nutrients of nitrogen and phosphorus. Statistical analyses were performed using the SPPS 18.0 software. All graphs were drawn using the Origin 8.0 software, and data were tested for normality and homogeneity of variance before analysis.

3. Results

3.1. Mycorrhizal Colonization Rate for *B. pilosa* Seedling Roots in Different Heterogeneous Patches

The root mycorrhizal colonization rates of *B. pilosa* under M^+ treatment were different, but the root mycorrhizal colonization rates were zero under all M^- treatments (Table 1), for which the AM mycelium and spores were not observed. Under the substrate heterogeneity treatments, the root mycorrhizal colonization rates of the gravel-free substrate of 100% soil (GF) in Hetl and Hets treatments were significantly lower than that of the gravel-high substrate mixed with 60% soil and 40% gravel (GH); however, the colonization of gravel-low substrate mixed with 80% soil and 20% gravel (GL) was significantly greater than GF of Hetl, but not for Hets. The mycorrhizal colonization of Hets was significantly greater than that of Hetl, for both GF and GH treatments, indicating the heterogeneity-small patch could facilitate mycorrhizal colonization of *B. pilosa* in our study. In addition, the interaction of patch heterogeneity and substrate heterogeneity significantly affected root mycorrhizal colonization rates of *B. pilosa*. This study showed that the patch heterogeneity in spatial size and substrate heterogeneity with gravel content increased root mycorrhizal colonization of plant seedlings.

Table 1. The root mycorrhizal colonization rates (%) of *Bidens pilosa*.

Treatments	Homo	Hetl		Hets	
	GL	GF	GF	GF	GH
M^+	58.75 ± 0.66xbcβγ	48.13 ± 0.97xdΔ	48.13 ± 0.97xdΔ	58.21 ± 0.67xcγ	63.79 ± 0.49xaα
M^-	0 yaα	0 yaα	0 yaα	0 yaα	0 yaα

M^+ = *B. pilosa* was inoculated with a mycorrhizal fungus. M^- = *B. pilosa* was not inoculated with a mycorrhizal fungus. Homo = homogeneous patch. Hetl = heterogeneous-large patch. Hets = heterogeneous-small patch. GL = gravel-low substrate. GF = gravel-free substrate. GH = gravel-high substrate. The different lowercase letters (x, y) indicate significant differences between M^+ and M^- treatments of *B. pilosa* ($p \leq 0.05$). The different lowercase letters (a–d) indicate significant differences between Homo, Hetl, and Hets patches ($p \leq 0.05$); The different lowercase letters (α–Δ) indicate significant differences between GL, GF, and GH substrates ($p \leq 0.05$). The values are "mean ± SD".

3.2. The Dry Weight of *B. pilosa* Seedling Roots in Different Heterogeneous Patches

The mycorrhizal fungus treatments significantly affected the root dry weight of *B. pilosa* (Table 2). The root dry weight in the M^+ treatment was significantly greater than in M^- treatment (Figure 2). The patch heterogeneity treatments significantly affected the root dry weight (Table 2). Under the M^+ treatment, the root dry weight in Homo was significantly greater than that in Hets and Hetl of GF; under the M^- treatment, there was no significant difference among treatments of different patches (Figure 2). The substrate heterogeneity treatments significantly affected the dry weight of roots (Table 2). Under the M^+ treatment, the dry weight of roots in GL and GH was found to be higher than in GF, for both Hetl and Hets; under the M^- treatment, GH in Hetl and Hets was significantly greater than GF in Hets (Figure 2). The interaction of patch heterogeneity and substrate heterogeneity significantly affected the dry weight of *B. pilosa* roots (Table 2). Overall, AM fungi significantly improved the dry weight of *B. pilosa* roots. The dry weight of roots in the heterogeneity-small patch was greater than that in the homogeneous and heterogeneity-large patches under the gravel-low and gravel-high substrates regulated by AM fungi; the dry weight of roots in the gravel-low and gravel-high substrates was higher than that of the gravel-free substrate regulated by AM fungi.

Table 2. Three-way ANOVA for the effect of mycorrhizal fungus (M^+ vs. M^-), patch heterogeneity (Homo vs. Hetl vs. Hets), and soil substrate heterogeneity (GL vs. GF vs. GH) on the dry weight of roots.

Treatments	Df	Root Dry Weight	
		F	p
Mycorrhizal fungi (M)	1	760.150	<0.001 ***
Patch heterogeneity (P)	2	4.469	0.037 *
Substrate heterogeneity (S)	2	45.832	<0.001 ***
M × P	2	4.658	0.034 *
M × S	2	44.903	<0.001 ***
P × S	4	1.206	0.275
M × P × S	4	1.269	0.263

Abbreviations: M = mycorrhizal fungus treatments; P = patch heterogeneity treatments; S = substrate heterogeneity treatments; * or *** indicates a significant difference at $p \leq 0.05$ or $p \leq 0.001$.

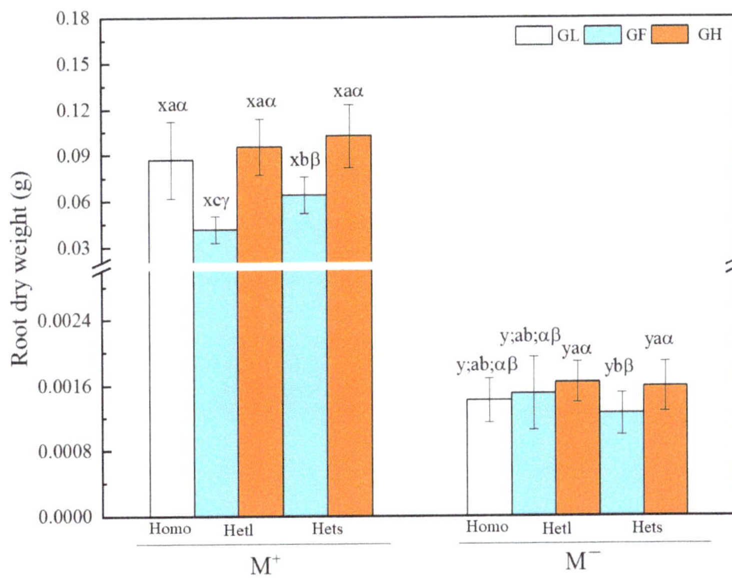

Figure 2. The dry weight of *B. pilosa* roots. Abbreviations: M^+ = *B. pilosa* was inoculated with a mycorrhizal fungus. M^- = *B. pilosa* was not inoculated with a mycorrhizal fungus. Homo = homogeneous patch. Hetl = heterogeneous-large patch. Hets = heterogeneous-small patch. GL = gravel-low substrate. GF = gravel-free substrate. GH = gravel-high substrate. The different lowercase letters (x, y) indicate significant differences between M^+ and M^- treatments of *B. pilosa* ($p \leq 0.05$). The different lowercase letters (a–c) indicate significant differences between Homo, Hetl, and Hets patches ($p \leq 0.05$); The different lowercase letters (α–γ) indicate significant differences between GL, GF, and GH substrates ($p \leq 0.05$). The error bars represent the standard deviation (SD).

3.3. *B. pilosa* Roots Length, Surface Area, Average Diameter, and Volume in Different Heterogeneous Patches

The mycorrhizal fungus treatments significantly affected the morphological traits of *B. pilosa* roots (Table 3). The length, surface area, volume, and average diameter in roots of the M^+ treatment were significantly more than those of M^- under all heterogeneous treatments (Figure 3a–d). The patch heterogeneity treatments significantly affected roots' length and surface area (Table 3). The M^+ treatment showed that the root length and surface area of Homo were significantly higher than those of Hetl in GF and GH, and those of Hets in GF. The average root diameter of Hetl and Hets in GH was significantly more prominent than that of Homo, and the root volume of Hets in GH was significantly more than that of Hets in GH. The M^- treatment showed that the root length and surface area of Hetl in GH were higher than those of Hets in GH and GF, and the root average diameter and volume of Hets in GH were significantly more than those of Homo (Figure 3a–d). The substrate heterogeneity treatments significantly affected the surface area, volume, and average diameter of roots (Table 3). Under M^+ treatment, the root length and surface area of GL were significantly greater than those of GF and GH under Hetl, and of GF under Hets; the root volume of GH under Hets was significantly larger than that of GF under Hetl and Hets; and the root average diameter of GH was significantly more than that of GL and GF, for both Hetl and Hets (Figure 3a-d). Under M^- treatment, the root surface area of GH under Hetl was significantly greater than that of GL, and the root volume and average diameter of GH under Hets were significantly greater than those of GF under Hets and GL (Figure 3b–d). The interaction of M × P significantly affected the root length and surface area; and the interaction of M × S significantly affected the surface area, average diameter, and volume of roots. However, the interactions of S × P and M × S × P did not significantly affect the root's phenotypic traits (Table 3). Overall, AM fungi significantly increased the length, surface area, average diameter, and volume of *B. pilosa* roots. These root morphological indexes were higher in the homogeneous and heterogeneity-small patches than in the heterogeneity-large patch under the gravel-low and gravel-high substrates, and were more prominent in gravel-low and gravel-high substrates than in gravel-free substrate.

Table 3. Three-way ANOVA for the effect of mycorrhizal fungus (M^+ vs. M^-), patch heterogeneity (Homo vs. Hetl vs. Hets), and soil substrate heterogeneity (GL vs. GF vs. GH) on root length, surface area, and volume, and average diameter.

Treatments	df	Root Length (cm)		Root Surface Area (cm^2)		Root Volume (cm^3)		Average Diameter (mm)	
		F	*p*	*F*	*p*	*F*	*p*	*F*	*p*
Mycorrhizal fungi (M)	1	545.125	<0.001 ***	518.377	<0.001 ***	501.696	<0.001 ***	1346.734	<0.001 ***
Patch heterogeneity (P)	2	9.943	0.002 **	6.612	0.012 *	0.703	0.404	0.243	0.623
Substrate heterogeneity (S)	2	3.805	0.054	7.061	0.009 **	7.126	<0.001 ***	28.376	<0.001 ***
M × P	2	10.970	<0.001 **	6.955	0.010 **	0.714	0.401	1.755	0.189
M × S	2	3.747	0.056	6.683	0.011 *	6.981	0.010 **	15.542	<0.001 ***
P × S	4	0.550	0.461	0.067	0.796	1.721	0.193	0.470	0.495
M × P × S	4	0.872	0.353	0.072	0.789	1.701	0.196	2.992	0.087

Abbreviations: M = mycorrhizal fungus treatments; P = patch heterogeneity treatments; S = substrate heterogeneity treatments; * or ** or *** indicates a significant difference at $p \leq 0.05$ or $p \leq 0.01$ or $p \leq 0.001$.

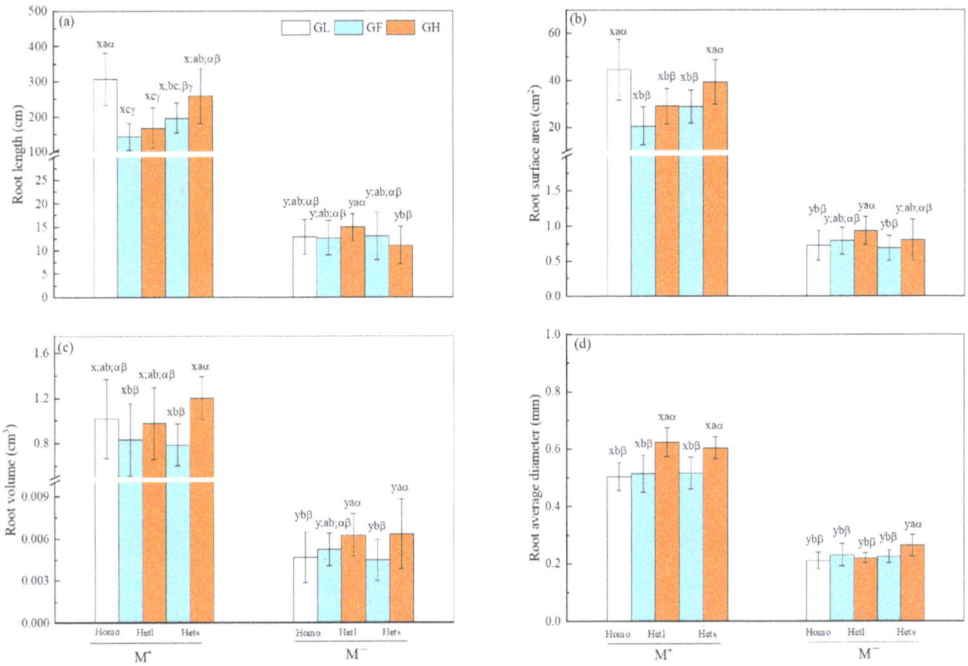

Figure 3. Phenotypic traits of *B. pilosa* roots. (**a–d**) represent the root length, surface area, volume and average diameter, respectively. See Figure 2 for an explanation of M^+, M^-, Homo, Hetl, Hets, GL, GF, and GH, lowercase letters (x, y) and (a, b, c) and Greek alphabet (α, β, γ).

3.4. Root Tips and Branching Points of B. pilosa

The mycorrhizal fungals treatments significantly affected the numbers of root tips and branching points of *B. pilosa* (Table 4). The results showed that AM fungus significantly improved the numbers of root tips and branching points in all heterogeneous treatments (Figure 4a–b). The patch heterogeneity treatments significantly affected the numbers of root tips and branching points (Table 4). Under the M^+ treatment, the numbers of root tips and branching points of Homo and Hets in GH were significantly higher than those of Hetl in GF and GH; under the M^- treatment, there were no significant differences among patches of different sizes (Figure 4a–b). For the substrate heterogeneity treatments, under M^+, the numbers of root tips and branching points of GL and GH under Hets were greater than those of GF and GH under Hetl, and of GF under Hets treatment; under M^-, there were no significant differences in root tips and branching points among different soil substrates (Figure 4a–b). In addition, the interaction of M × P significantly affected the numbers of root tips and branching points (Table 4). Altogether, AM fungi significantly increased the number of root tips and root branching points. The root tips and branching points were more remarkable in the homogeneous and heterogeneity-small patches than in the heterogeneity-large patches under the gravel-low and gravel-high substrates. They were higher in gravel-low and gravel-high substrates under homogeneous and heterogeneity-small patches than in gravel-free substrates.

Table 4. Three-way ANOVA for the effect of mycorrhizal fungus (M^+ vs. M^-), patch heterogeneity (Homo vs. Hetl vs. Hets), and soil substrate heterogeneity (GL vs. GF vs. GH) on root tips and branching points.

Treatments	df	Root Tips		Root Branching Points	
		F	*p*	*F*	*p*
Mycorrhizal fungi (M)	1	800.275	<0.001 ***	459.121	<0.001 ***
Patch heterogeneity (P)	2	7.012	0.010 **	16.386	<0.001 ***
Substrate heterogeneity (S)	2	1.980	0.163	2.308	0.132
M × P	2	7.661	0.007 **	17.632	<0.001 ***
M × S	2	1.604	0.209	2.209	0.141
P × S	4	1.748	0.190	1.437	0.234
M × P × S	4	1.525	0.220	1.602	0.209

Abbreviations: M = mycorrhizal fungus treatments; P = patch heterogeneity treatments; S = substrate heterogeneity treatments; ** or *** indicates a significant difference at $p \leq 0.01$ or $p \leq 0.001$.

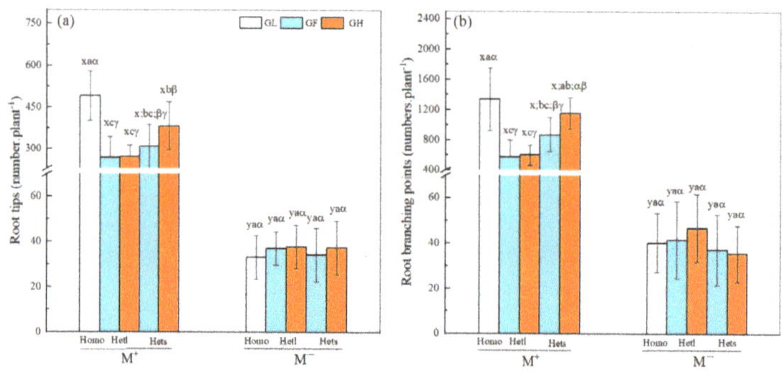

Figure 4. Numbers of root tips and branching points of *B. pilosa*. (**a**,**b**) represent the root tips and branching points, respectively. See Figure 2 for an explanation of M^+, M^-, Homo, Hetl, Hets, GL, GF and GH, lowercase letters (x, y) and (a, b, c) and Greek alphabet (α, β, γ).

3.5. *B. pilosa* Specific Root Traits in Different Heterogeneous Treatments

The mycorrhizal fungals treatments significantly affected the specific root traits of *B. pilosa* (Table 5). AM fungus significantly improved the specific root volume while decreasing the specific root length and surface area in all heterogeneous treatments (Figure 5). For patch heterogeneity treatments, under the M^+ treatment, the specific root length and surface area in Homo were significantly greater those in Hets and Hetl under GH treatment (Figure 5a,b), and the specific root volume in Hetl under GH treatment was significantly greater than those in Homo under GL treatment and in Hets under GH treatment (Figure 5c). Under the M^- treatment, the specific root length in Homo was significantly greater than that in Hets under the GH treatment (Figure 5a). The specific root surface area and volume had no significant difference among these patches (Figure 5b,c). The soil substrate heterogeneity treatments significantly affected the specific root traits (Table 5). Unambiguously, under the M^+ treatment, the specific root length and surface area in GL under Homo and GF under Hetl were significantly higher than those in GH under Hetl and GH under Hetl (Figure 5c). Under the M^- treatment, the specific root length in GL under Homo and GF under Hets were significantly greater than that in GH under Hets (Figure 5a), whereas the specific root surface area and volume had no significant difference among these substrate treatments (Figure 5b,c). In addition, the interactions of M × S and S × P significantly affected specific root volume, and the interaction of M × S × P significantly affected specific root length and volume (Table 5). The results showed that the AM fungi significantly improved the specific root volume while decreasing the specific root length and area. The specific root length, specific root area, and specific root volume in the heterogeneity-large patch were prominently higher than in the heterogeneity-small patch under the gravel-free substrate, and in the heterogeneity-small patch were higher than in the heterogeneity-large patch under gravel-low and gravel-high substrates; the specific root length and specific root area in a gravel-low substrate for the homogeneous patch were greater than those in gravel-free substrate for the heterogeneity-large patch and heterogeneity-small patch.

Table 5. Three-way ANOVA for the effect of mycorrhizal fungus (M^+ vs. M^-), patch heterogeneity (Homo vs. Hetl vs. Hets), and soil substrate heterogeneity (GL vs. GF vs. GH) on specific root traits.

Treatments	df	Specific Root Length (cm/g)		Specific Root Area (cm^2/g)		Specific Root Volume (cm^3/g)	
		F	*p*	*F*	*p*	*F*	*p*
Mycorrhizal fungi (M)	1	231.635	<0.001 ***	11.610	<0.001 ***	163.225	<0.001 ***
Patch heterogeneity (P)	2	0.017	0.898	0.002	0.968	3.425	0.068
Substrate heterogeneity (S)	2	7.143	0.009 **	4.341	0.040 *	7.568	0.007 **
M × P	2	0.180	0.673	0.613	0.436	3.646	0.060
M × S	2	0.126	0.724	2.788	0.099	9.035	0.003 **
P × S	4	2.440	0.122	0.041	0.840	8.273	0.005 **
M × P × S	4	7.533	0.007 **	2.319	0.131	7.897	0.006 **

Abbreviations: M = mycorrhizal fungus treatments; P = patch heterogeneity treatments; S = substrate heterogeneity treatments; * or ** or *** indicates a significant difference at $p \leq 0.05$ or $p \leq 0.01$ or $p \leq 0.001$.

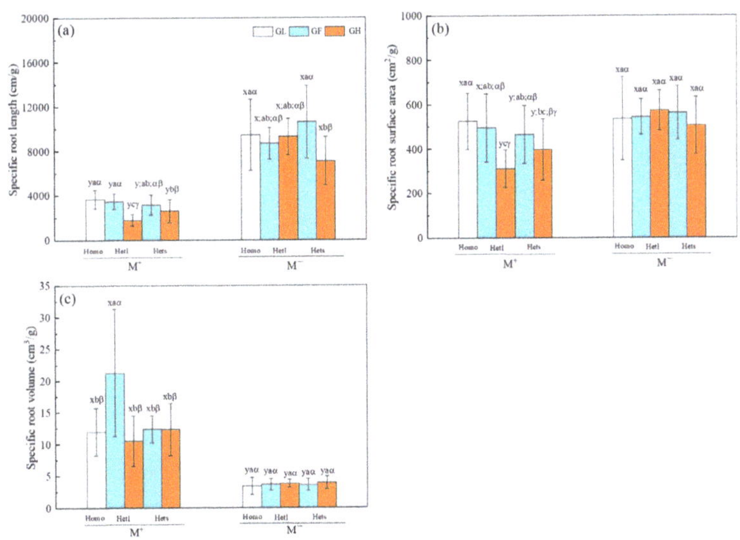

Figure 5. Specific traits of *B. pilosa* roots. (**a–c**) represent specific root length, surface area and volume, respectively. See Figure 2 for an explanation of M^+, M^-, Homo, Hetl, Hets, GL, GF, and GH, lowercase letters (x, y) and (a, b, c) and Greek alphabet (α, β, γ).

3.6. N, P and K Acquisitions of *B. pilosa* Roots

The mycorrhizal fungus treatments affected N, P, and K acquisitions through *B. pilosa* roots (Table 6). The AM fungus significantly enhanced the root's ability to acquire the N, P, and K acquisitions in all heterogeneous treatments (Figure 6). The patch heterogeneity treatments significantly affected the root N, P, and K acquisitions of *B. pilosa* (Table 6). Under the M^+ treatment, the root N acquisition in Homo was significantly greater than that in Hetl under GF treatment, whereas that in Homo under GL treatment and in Hetl under GH treatment was significantly lower than that in Hets under GH treatment (Figure 6a). The root P acquisition in Homo was significantly greater than that in Hets and in Hetl under GF treatments (Figure 6b). The root K acquisition in Homo was significantly greater than that in Hets and Hetl under GF treatments, and that in Homo under GL treatment and in Hets under GH treatment was significantly greater than that in Hetl under GF treatment (Figure 6c). Under the M^{-} treatment, the root N acquisition in Hets under GH treatment was significantly greater than that in Homo under GL treatment and in Hetl under GH treatment (Figure 6a). The root P acquisition in Hetl under GH treatment was significantly greater than that in Homo (Figure 6b). The soil substrate heterogeneity treatments significantly affected the root N, P, and K acquisitions of *B. pilosa* (Table 6). Specifically, under the M^{+} treatment, the root N, P, and K acquisitions in GF were significantly lower than those in GH and in GL under Hetl, Hets, and Homo (Figure 6). Under the M^{-} treatment, the root N acquisition in GL was significantly greater than that in GH under Hetl, and that in GH under Hets was significantly greater than that in GL (Figure 6a); the root P acquisition in GH under Hetl was significantly higher than that in GL (Figure 6b); the root K acquisition in GF was significantly more than that in GH under Hets (Figure 6c). In addition, the interactions of M × S and M × P significantly affected the root N, P, and K acquisitions of *B. pilosa* (Table 6). In conclusion, AM fungi significantly improved plant root N, P, and K acquisitions and promoted plant root accumulation of nutrients in heterogeneity-small patches compared to in heterogeneity-large patches, and in gravel-low and gravel-high substrates compared to in gravel-free substrate.

Table 6. Three-way ANOVA for the effect of mycorrhizal fungus (M^{+} vs. M^{-}), patch heterogeneity (Homo vs. Hetl vs. Hets), and soil substrate heterogeneity (GL vs. GF vs. GH) on the N and P acquisitions.

Treatments	df	Root N Acquisition		Root P Acquisition		Root K Acquisition	
		F	*p*	*F*	*p*	*F*	*p*
Mycorrhizal fungi (M)	1	549.296	<0.001 ***	594.266	<0.001 ***	594.687	<0.001 ***
Patch heterogeneity (P)	2	9.226	0.003 **	7.540	0.007 **	13.492	<0.001 ***
Substrate heterogeneity (S)	2	30.304	<0.001 ***	33.139	<0.001 ***	31.153	<0.001 ***
M × P	2	8.905	0.004 **	8.270	0.005 **	13.390	<0.001 ***
M × S	2	30.042	<0.001 ***	31.502	<0.001 ***	31.481	<0.001 ***
P × S	4	0.279	0.599	0.028	0.868	0.050	0.824
M × P × S	4	0.160	0.690	0.032	0.858	0.051	0.821

Abbreviations: M = mycorrhizal fungus treatments; P = patch heterogeneity treatments; S = substrate heterogeneity treatments; ** or *** indicates a significant difference at $p \leq 0.01$ or $p \leq 0.001$.

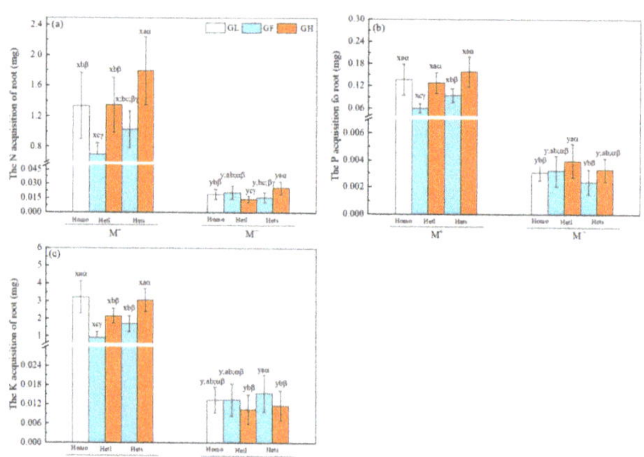

Figure 6. Acquisitions of N, P, and K through *B. pilosa* roots. (**a**–**c**) represent the nitrogen, phosphorus and potassium acquisition of plant roots, respectively. See Figure 2 for an explanation of M^{+}, M^{-}, Homo, Hetl, Hets, GL, GF, and GH, lowercase letters (x, y) and (a, b, c) and Greek alphabet (α, β, γ).

4. Discussion

4.1. Effect of AM Fungi on Root Dry Weight, Morphology, and Nutrients Uptakes of B. pilosa in Heterogeneous Soil

In our experiment, AM fungi significantly affected the dry weight, morphology, and nutrients of *B. pilosa* roots (Tables 2–5). It has been well demonstrated that AM fungal mycelia can accompany plant roots to obtain soil resources in a broader range to improve plant nutrients [50,54]. For example, AM fungi can enhance the N and P uptake of host plants by transporting soil nutrients outside the rhizosphere [55,56] and can improve the K uptake of plant roots [57]. He et al. [58] also showed that AM fungi significantly promoted the N and P content of karst plant roots. These studies suggest that AM fungi play a crucial role in facilitating nutrient uptake by host plants, and the results of our experiment verified that AM fungi significantly increased the N, P, and K acquisitions of *B. pilosa* roots under all heterogeneous treatments. Many studies have documented that AM fungi can promote plant growth by improving plant nutrient acquisition [55,56]. In the present study, AM fungi significantly promoted the growth and development of *B. pilosa* roots, manifested as an increase in dry weight of the M^+ treatment compared with the M^- treatment (Figure 2). This is consistent with the results of Hussain et al. [59], who found that the root biomass of tea plant was significantly increased via AM fungi. Studies have shown that AM fungi can change root morphology due to the strong plasticity of plant roots [50]. In the current experiment, AM fungi significantly increased the length, surface area, average diameter, volume, tips, and branching points of *B. pilosa* roots (Figures 3 and 4); this result verified our first hypothesis that AM fungi can promote root growth and nutrient uptakes of plants in heterogeneous karst soil. Our findings are similar to those of He et al. [38], in that AM fungi significantly changed the root morphology by increasing the length, surface area, and volume of host plant roots in karst soil. The specific root length can reflect the absorption capacity of the root systems [50], and the larger the specific root length, the stronger the ability of plants to absorb water and nutrients; however, it is less dependent on mycorrhiza [60]. AM fungi significantly reduced the specific root length of *B. pilosa* in our study (Figure 5a), consistent with the results of He et al. [61], who found that inoculation with AM fungi decreased the specific root length of host plants. This may be due to the belowground mycorrhizal network formed by AM fungi replacing part of the functions of plant roots [62].

4.2. Effects of Heterogeneous Patch Size on Plant Root Traits and Nutrients Uptake Regulated by AM Fungi

Roots are critical functional organs of plants, and the root system has strong plasticity, which can be affected by the patch scale of environmental spatial heterogeneity [62,63]. Studies have shown that root traits of plants can have different responses along with the separation of heterogeneous niche axis, despite these responses depending on the matching of root systems and nutrient patches [62,63]. Our study showed that the dry weight, length, surface area, volume, tips, and branch points of *B. pilosa* roots in Hetl of GH were higher than those in Hets and Homo under the M^- treatment (Figures 2–4). Day et al. [28] and Hutchings and Wijesinghe [64] also verified that plants growing in heterogeneous soil patches have greater root biomass, length, and surface area in large patches rather than small patches. This could be because the smaller soil patches lead to the appropriate morphological response of plant roots to local conditions not being completed before plants grow beyond a given patch [21]. Therefore, the opportunities for plants to obtain resources in small patches may not be maximized; conversely, plants growing in large patches may obtain more resources, resulting in more remarkable growth in root systems. However, in this experiment, the N, P, and K acquisitions of *B. pilosa* roots treated with M^- showed no significant difference among different patch treatments (Figure 6a–c). This result is inconsistent with our second hypothesis. Cui and Caldwell [65] and Hodge et al. [66] also found that soil patch size did not affect plants' N and P uptakes, although it differentially affected the root morphology. This may be because the smaller patch differences in heterogeneous treatments cannot affect plant nutrient acquisition [66]. Previous studies have shown that the physiological and morphological plastic response of plant roots to a heterogeneous soil patch is affected by the action of AM fungi [46,50,67]. In this study, AM fungi reversed the root response of *B. pilosa* to soil patches of different sizes compared with the M^- treatment. Specifically, the dry weight, and N, P, and K acquisitions of *B. pilosa* roots of Homo and Hets in GH were greater than those of Hetl under the M^+ treatment (Figures 1–4). The epitaxial root hyphae of AM fungi can extend to the soil area outside the root systems to obtain more resources and promote plant growth [55,56]. Furthermore, in small-scale patches with a higher degree of heterogeneity, AM hyphae can contact more soil patches, which stimulates the proliferation and extension of hyphae [47,68]. These may effectively explain the greater dry weight and nutrient acquisitions of *B. pilosa* roots in homogeneous and heterogeneous small patches with the soil–gravel mixed substrate in this experiment. In addition, the study showed that AM fungi will inevitably cause changes in plant phenotype while promoting plant nutrients [69]. In this experiment, the length, surface area, volume, tips, and branching points of *B. pilosa* roots of Homo and Hets in GH were greater than those in Hetl under the M^+ treatment (Figures 3 and 4). These results were similar to those of Xia et al. [50], suggesting that the epitaxial root mycelia absorbed more nutrients to promote the growth and development of host plant root phenotypes in karst soil. Root length and root surface area can represent the capacity of plants to absorb soil resources, and the greater the length and surface area in roots, the stronger the absorption capacity of plants [70,71]. The increase in the number of root tips can improve the ability of plants to utilize soil resources in situ, and the growth in lateral roots can increase the total root length and expand the utilization area of soil resources by plants [72,73]. Research suggests plants in heterogeneous patches with higher fragmentation need higher root growth to absorb soil resources from adjacent patches, especially in the patches mixed with soil and gravel [62,74]. Mycorrhizal fungi have been widely recognized to promote root development in heterogeneous soils [62,68]. These may effectively explain the greater morphological growth and development of *B. pilosa* roots in homogeneous and heterogeneity-small patches with the soil–gravel mixed substrate through AM fungi.

4.3. *Effects of Substrate Heterogeneity on Plant Root Morphology and Nutrients Uptake Regulated by AM Fungi*

In this study, the substrate heterogeneity significantly affected the root dry weight, root morphology, and root N, P, and K acquisitions of *B. pilosa* (Tables 2–5). This may be because AM fungi may regulate the effect of gravel content on the root dry weight and nutrient acquisition. Previous research showed that the gravel content significantly affects the root biomass and nutrients uptake [24,75,76] by changing the hydrological process (e.g., water use efficiency) and nutrient mobility in soil [30,31]. For example, the presence of gravel increases the soil porosity and affects root growth [31,32]; in addition, the high gravel content in the soil causes the deficiency of soil water and nutrients [32]. Therefore, excessive gravel content can inhibit plant root growth and nutrient acquisition [77,78]. This is consistent with our results showing that the root dry weight, root N, P, and K acquisitions of *B. pilosa* in the gravel-free substrate under Hetl were more than those in GL substrate under Homo when not inoculated with AM fungi; whereas those in GL substrate under Homo were higher than those in GF substrate under Hetl when inoculated with AM fungi (Figures 2 and 6a–c). One possible reason for this is that AM fungi can promote plant root growth and accumulate nutrients in the nutrient-poor area with higher gravel content. Many studies showed that the benefits of mycorrhizal fungi are presumably meager in the nutrient-rich area while being more active in enhancing plant root growth and nutrient acquisition in the nutrient-poor area [79,80]. Another possible reason is that it is difficult for the plant roots to access nutrients in soil with higher gravel content; in contrast, the mycelium of AM fungi is thinner than the root system, and therefore more able to access nutrients than the root system, thus promoting the growth of plant roots [62,81].

In addition, root morphology can reveal the ability to absorb nutrients for plants [62,82], and AM fungi may also regulate the effect of gravel content on the root morphology. In this study, the root length, root surface area, root tips, and root branching points in GH were greater than those in GF under the M^+ treatment (Figure 3a,b and Figure 4a,b), which indicates that, under high gravel content conditions, the root system has a greater nutrient acquisition capacity via combination with AM fungi than in the gravel-free substrate condition. One possible reason for this is that plant roots have a foraging behavior [83]. Therefore, under high gravel content conditions within heterogeneous patches, plant roots adapt to the heterogeneous resource environment via high root length, root surface area, root tips, and root branching points to access nutrients in neighboring or even more distant soil substrate patches [74]. Moreover, the specific root traits can indicate the trade-off between plant benefits and costs [84]. In the present study, under the Hetl condition, the specific root length and specific root surface area in GH were higher than those in GF under the M^- treatment, whereas in GH they were lower than those in GF under the M^+ treatment (Figure 5a,b). High specific root length can enhance nutrient acquisition by permitting the exploration of higher soil volume per unit carbon investment in root length [85]; thus, the plants have higher specific root length and area to access more resources under high gravel content conditions than under soil substrate conditions. However, after inoculation with the AM fungi, the mycelium partially replaced the role of the root system [54]; thus, plants may be more dependent on mycelium to uptake nutrients under high gravel content conditions with lower specific root length and area than under soil substrate conditions. Therefore, this study result verified our third hypothesis that the root systems of plants have greater nutrient uptake, and phenotypic growth in the gravel-low and gravel-high substrates was higher than that in gravel-free substrate through AM fungi.

5. Conclusions

In our study, AM fungi enhanced the root development and nutrient uptake of *B. pilosa* by significantly increasing its dry weight, length, surface area, average diameter, volume, tips, branching points, and N, P, and K acquisitions. The heterogeneity-small patch with soil and gravel increased the dry weight, length, surface area, tips, branching points, and N, P, K acquisitions of *B. pilosa* roots compared with heterogeneity-large patch regulated by AM fungi; and the gravel-low substrate and gravel-high substrate treatments also increased these root indexes of *B. pilosa* compared with gravel-free substrate regulated by AM fungi. Overall, the *B. pilosa* roots have higher growth and nutritional acquisition benefits in heterogeneity-small patches and soil with gravels regulated by AM fungi in karst soil. In conclusion, soil patch heterogeneity with gravels promotes root morphological development and nutrient utilization to karst plants associated with arbuscular mycorrhizal fungi. These results will contribute to studies on ecosystem and vegetation restoration in degraded karst areas.

Author Contributions: Conceptualization, Q.L. and Y.H.; Methodology, X.X.; Software, T.X.; Validation, K.S. and M.U.; Formal Analysis, Y.G. and B.W.; Investigation, Q.L. and X.L.; Resources, W.R.; Data curation, Y.S.; Writing—Original Draft Preparation, Q.L.; Writing—Review and Editing, Q.L.; Visualization, X.H.; Supervision, Y.H.; Project Administration, Y.H.; Funding Acquisition, Y.H. All authors have read and agreed to the published version of the manuscript.

Funding: This research was funded by the National Natural Science Foundation of China (NSFC: 31660156; 31360106), the First-class Disciplines Program on Ecology of Guizhou Province (GNYL [2017] 007), the Guizhou Hundred-level Innovative Talents Project (Qian-ke-he platform talents [2020] 6004), the Science and Technology Project of Guizhou Province ([2021] General-455; [2016] Supporting-2805), the Talent-platform Program of Guizhou Province ([2017] 5788; [2018] 5781).

Data Availability Statement: Not applicable.

Acknowledgments: We thank the Institute of Nutrition Resources, Beijing Academy of Agricultural and Forestry Sciences (NO. BGA0046) for providing *Glomus etunicatum* for use in our experiments.

Conflicts of Interest: The authors declare no conflict of interest.

References

1. Huang, Q.H.; Cai, Y.L. Spatial pattern of Karst rock desertification in the Middle of Guizhou Province, Southwestern China. *Environ. Geol.* **2006**, *52*, 1325–1330. [CrossRef]
2. Zhou, L.; Wang, X.; Wang, Z.; Zhang, X.; Chen, C.; Liu, H. The challenge of soil loss control and vegetation restoration in the karst area of southwestern China. *Int. Soil. Water. Conserv. Res.* **2020**, *8*, 26–34. [CrossRef]
3. Zhang, W.; Zhao, J.; Pan, F.; Li, D.; Chen, H.; Wang, K. Changes in nitrogen and phosphorus limitation during secondary succession in a karst region in southwest China. *Plant Soil* **2015**, *391*, 77–91. [CrossRef]
4. Zhang, X.Y.; Yue, Y.S.; Zhang, X.D.; Kai, M.; Herbert, S.J. Spatial variability of nutrient properties in black soil of northeast China. *Pedosphere* **2007**, *17*, 19–29. [CrossRef]
5. Chen, S.; Zhou, Z.; Yan, L.; Li, B. Quantitative Evaluation of Ecosystem Health in a Karst Area of South China. *Sustainability* **2016**, *8*, 975. [CrossRef]
6. Zhu, L.; He, S.; Li, J. Weathering-pedogenesis of Carbonate Rocks and Its Environmental Effects in Subtropical Region. *Acta Geol. Sin.-Engl.* **2008**, *82*, 982–993. [CrossRef]
7. Loreau, M.; Naeem, S.; Inchausti, P.; Bengtsson, J.; Grime, J.P.; Hector, A.; Hooper, D.U.; Huston, M.A.; Raffaelli, D.; Schmid, B.; et al. Biodiversity and ecosystem functioning: Current knowledge and future challenges. *Science* **2001**, *284*, 804–808. [CrossRef]
8. Questad, E.J.; Foster, B.L. Coexistence through spatio-temporal heterogeneity and species sorting in grassland plant communities. *Ecol. Lett.* **2008**, *11*, 717–726. [CrossRef]
9. Price, J.N.; Gazol, A.; Tamme, R.; Hiiesalu, I.; Pärtel, M.; Brody, A. The functional assembly of experimental grasslands in relation to fertility and resource heterogeneity. *Funct. Ecol.* **2014**, *28*, 509–519. [CrossRef]
10. Tilman, D.; Pacala, S. *The Maintenance of Species Richness in Plant Communities*; University of Chicago Press: Chicago, IL, USA, 1993; pp. 13–25.
11. Fitter, A.; Hodge, A.; Robinson, D. Plant response to patchy soils. In *Ecological Consequences of Environmental Heterogeneity*; Blackwell Science: Oxford, UK, 2000; pp. 71–90.
12. Eilts, J.A.; Mittelbach, G.G.; Reynolds, H.L.; Gross, K.L. Resource Heterogeneity, Soil Fertility, and Species Diversity: Effects of Clonal Species on Plant Communities. *Am. Nat.* **2011**, *177*, 574–588. [CrossRef]
13. Xue, W.; Huang, L.; Yu, F.-H. Spatial heterogeneity in soil particle size: Does it affect the yield of plant communities with different species richness? *J. Plant Ecol.* **2016**, *9*, 608–615. [CrossRef]
14. Xue, W.; Huang, L.; Yu, F.; Bezemer, T.M. Intraspecific aggregation and soil heterogeneity: Competitive interactions of two clonal plants with contrasting spatial architecture. *Plant Soil* **2018**, *425*, 231–240. [CrossRef]
15. Liu, Y.; Qi, W.; He, D.; Xiang, Y.; Liu, J.; Huang, H.; Chen, M.; Tao, J. Soil resource availability is much more important than soil resource heterogeneity in determining the species diversity and abundance of karst plant communities. *Ecol. Evol.* **2021**, *11*, 16680–16692. [CrossRef]
16. Désilets, P.; Houle, G. Effects of resource availability and heterogeneity on the slope of the species-area curve along a floodplain-upland gradient. *J. Veg. Sci.* **2005**, *16*, 487–496. [CrossRef]
17. Fransen, B.; De, K.H. Long-term disadvantages of selective root placement: Root proliferation and shoot biomass of two perennial grass species in a 2-year experiment. *J. Ecol.* **2001**, *89*, 711–722. [CrossRef]
18. Facelli, E.; Facelli, J.M. Soil phosphorus heterogeneity and mycorrhizal symbiosis regulate plant intra-specific competition and size distribution. *Oecologia* **2002**, *133*, 54–61. [CrossRef]
19. Day, K.; Hutchings, M.; John, E. The effects of spatially heterogeneous nutrient supply on yield, intensity of competition and root placement patterns in *Briza media* and *Festuca ovina*. *Funct. Ecol.* **2003**, *17*, 454–463. [CrossRef]
20. Tamme, R.; Gazol, A.; Price, J.N.; Hiiesalu, I.; Pärtel, M.; Roxburgh, S. Co-occurring grassland species vary in their responses to fine-scale soil heterogeneity. *J. Veg. Sci.* **2016**, *27*, 1012–1022. [CrossRef]
21. Hutchings, M.J.; John, E.A. The effects of environmental heterogeneity on root growth and root/shoot partitioning. *Ann. Bot.* **2004**, *94*, 1–8. [CrossRef]
22. Qian, Y.; Luo, D.; Gong, G.; Han, L.; Ju, G.; Sun, Z. Effects of Spatial Scale of Soil Heterogeneity on the Growth of a Clonal Plant Producing Both Spreading and Clumping Ramets. *J. Plant. Growth. Regul.* **2013**, *33*, 214–221. [CrossRef]
23. Mi, M.; Shao, M.; Liu, B. Effect of rock fragments content on water consumption, biomass and water-use efficiency of plants under different water conditions. *Ecol. Eng.* **2016**, *94*, 574–582. [CrossRef]
24. Masoni, A.; Ercoli, L.; Mariotti, M.; Pampana, S. Nitrogen and phosphorus accumulation and remobilization of durum wheat as affected by soil gravel content. *Cereal. Res. Commun.* **2008**, *36*, 157–166. [CrossRef]
25. David, R.; Angela, H.; Griffths, B.S.; Fitter, A.H. Plant root proliferation in nitrogen-rich patches confers competitive advantage. *P. Roy. Soc. B-Biol. Sci.* **1999**, *266*, 431–435. [CrossRef]
26. Pan, F.; Liang, Y.; Wang, K.; Zhang, W. Responses of Fine Root Functional Traits to Soil Nutrient Limitations in a Karst Ecosystem of Southwest China. *Forests* **2018**, *9*, 743. [CrossRef]

27. Wijesinghe, D.K.; Hutchings, M.J. The Effects of Spatial Scale of Environmental Heterogeneity on the Growth of a Clonal Plant: An Experimental Study with Glechoma Hederacea. *J. Ecol.* **1997**, *85*, 17–28. [CrossRef]
28. Day, K.; John, E.; Hutchings, M. The effects of spatial pattern of nutrient supply on the early stages of growth in plant populations. *J. Ecol.* **2003**, *91*, 305–315. [CrossRef]
29. Alagna, A.; Fernandez, T.V.; Anna, G.D.; Magliola, C.; Mazzola, S.; Badalamenti, F. Assessing Posidonia oceanica seedling substrate preference: An experimental determination of seedling anchorage success in rocky vs. sandy substrates. *PLoS ONE* **2015**, *10*, e0125321. [CrossRef]
30. Zhang, Y.; Zhang, M.; Niu, J.; Li, H.; Xiao, R.; Zheng, H.; Bech, J. Rock fragments and soil hydrological processes: Significance and progress. *Catena* **2016**, *147*, 153–166. [CrossRef]
31. Shi, Z.; Xu, L.; Wang, Y.; Yang, X.; Jia, Z.; Guo, H.; Xiong, W.; Yu, P. Contribution of rock fragments on formation of forest soil macropores in the stoney mountains of the Loess Plateau, China. *Afric. J. Biotechnol.* **2012**, *11*, 9350–9361.
32. Alameda, D.; Villar, R. Linking root traits to plant physiology and growth in Fraxinus angustifolia Vahl. seedlings under soil compaction conditions. *Environ. Exp. Bot.* **2012**, *79*, 49–57. [CrossRef]
33. Bengough, A. Root growth and function in relation to soil structure, composition, and strength. In *Root Ecology*; Springer: Berlin/Heidelberg, Germany, 2003; pp. 151–171.
34. Laliberté, E. Below-ground frontiers in trait-based plant ecology. *New Phytol.* **2016**, *213*, 1597–1603. [CrossRef]
35. Van Der Heijden, M.G.A.; Bardgett, R.D.; Van Straalen, N.M. The unseen majority: Soil microbes as drivers of plant diversity and productivity in terrestrial ecosystems. *Ecol. Lett.* **2008**, *11*, 296–310. [CrossRef]
36. Manuel, D.B.; Maestre, F.T.; Reich, P.B.; Jeffries, T.C.; Gaitan, J.J.; Encinar, D.; Berdugo, M.; Campbell, C.D.; Singh, B.K. Microbial diversity drives multifunctionality in terrestrial ecosystems. *Nat. Commun.* **2016**, *7*, 10541. [CrossRef]
37. Manuel, D.B.; Reich, P.B.; Trivedi, C.; Eldridge, D.J.; Abades, S.; Alfaro, F.D.; Bastida, F.; Berhe, A.A.; Cutler, N.A.; Gallardo, A.; et al. Multiple elements of soil biodiversity drive ecosystem functions across biomes. *Nat. Ecol. Evol.* **2020**, *4*, 210–220. [CrossRef]
38. He, Y.; Zhong, Z. Effects of Water Stress and AM Inoculation on Root Morphological Characteristics in Cinnamomum camphora Seedlings. *J. Southwest. Univ.* **2012**, *34*, 33–39. [CrossRef]
39. Ryan, M.H.; Tibbett, M.; Edmonds-Tibbett, T.; Suriyagoda, L.D.B.; Lambers, H.; Cawthray, G.R.; Pang, J. Carbon trading for phosphorus gain: The balance between rhizosphere carboxylates and arbuscular mycorrhizal symbiosis in plant phosphorus acquisition. *Plant Cell Environ.* **2012**, *35*, 2170–2180. [CrossRef]
40. Mei, L.; Yang, X.; Zhang, S.; Zhang, T.; Guo, J. Arbuscular mycorrhizal fungi alleviate phosphorus limitation by reducing plant N:P ratios under warming and nitrogen addition in a temperate meadow ecosystem. *Sci. Total Environ.* **2019**, *686*, 1129–1139. [CrossRef]
41. Shen, K.; Cornelissen, J.H.C.; Wang, Y.; Wu, C.; He, Y.; Ou, J.; Tan, Q.; Xia, T.; Kang, L.; Guo, Y.; et al. AM Fungi Alleviate Phosphorus Limitation and Enhance Nutrient Competitiveness of Invasive Plants via Mycorrhizal Networks in Karst Areas. *Front. Ecol. Evol.* **2020**, *8*, 125. [CrossRef]
42. Chen, W.; Koide, R.T.; Eissenstat, D.M. Nutrient foraging by mycorrhizas: From species functional traits to ecosystem processes. *Funct. Ecol.* **2018**, *32*, 858–869. [CrossRef]
43. Smith, S.E.; Read, D.J. *Mycorrhizal Symbiosis*, 3rd ed.; Academic Press: New York, NY, USA, 2008.
44. Shi, Z.; Wang, F.; Zhang, C.; Yang, Z. Exploitation of phosphorus patches with different phosphorus enrichment by three arbuscular mycorrhizal fungi. *J. Plant Nutr.* **2011**, *34*, 1096–1106. [CrossRef]
45. Yang, Z.; Midmore, D.J. Modelling plant resource allocation and growth partitioning in response to environmental heterogeneity. *Ecol. Model.* **2005**, *181*, 59–77. [CrossRef]
46. Croft, S.A.; Hodge, A.; Pitchford, J.W. Optimal root proliferation strategies: The roles of nutrient heterogeneity, competition and mycorrhizal networks. *Plant Soil* **2012**, *351*, 191–206. [CrossRef]
47. Simard, S.W.; Beiler, K.J.; Bingham, M.A.; Deslippe, J.R.; Philip, L.J.; Teste, F.P. Mycorrhizal networks: Mechanisms, ecology and modelling. *Fungal. Biol. Rev.* **2012**, *26*, 39–60. [CrossRef]
48. Liang, Y.; He, X.; Chen, C.; Feng, S.; Liu, L.; Chen, X.; Zhao, Z.; Su, Y. Influence of plant communities and soil properties during natural vegetation restoration on arbuscular mycorrhizal fungal communities in a karst region. *Ecol. Eng.* **2015**, *82*, 57–65. [CrossRef]
49. He, Y.; Cornelissen, J.H.C.; Wang, P.; Dong, M.; Ou, J. Nitrogen transfer from one plant to another depends on plant biomass production between conspecific and heterospecific species via a common arbuscular mycorrhizal network. *Environ. Sci. Pollut. Res. Int.* **2019**, *26*, 8828–8837. [CrossRef]
50. Xia, T.; Wang, Y.; He, Y.; Wu, C.; Shen, K.; Tan, Q.; Kang, L.; Guo, Y.; Wu, B.; Han, X. An invasive plant experiences greater benefits of root morphology from enhancing nutrient competition associated with arbuscular mycorrhizae in karst soil than a native plant. *PLoS ONE* **2020**, *15*, e0234410. [CrossRef]
51. Zhang, Z.; Zhang, J.; Xu, G.; Zhou, L.; Li, Y. Arbuscular mycorrhizal fungi improve the growth and drought tolerance of *Zenia insignis* seedlings under drought stress. *New For.* **2019**, *50*, 593–604. [CrossRef]
52. Bao, S.D. *Soil and Agricultural Chemistry Analysis*; China Agriculture Press: Beijing, China, 2000.
53. Wang, Y.; Zhong, Q.; Bin, X.; Zhang, Z.; Cheng, D. Effect of adding a combination of nitrogen and phosphorus on fine root morphology and soil microbes of *Machilus pauhoi* seedling. *Acta Ecol. Sin.* **2018**, *38*, 8–2271. [CrossRef]

54. Nottingham, A.T.; Turner, B.L.; Winter, K.; Chamberlain, P.M.; Stott, A.; Tanner, E.V. Root and arbuscular mycorrhizal mycelial interactions with soil microorganisms in lowland tropical forest. *FEMS Microbiol. Ecol.* **2013**, *85*, 37–50. [CrossRef]
55. Lin, J.; Wang, Y.; Sun, S.; Mu, C.; Yan, X. Effects of arbuscular mycorrhizal fungi on the growth, photosynthesis and photosynthetic pigments of Leymus chinensis seedlings under salt-alkali stress and nitrogen deposition. *Sci. Total Environ.* **2017**, *576*, 234–241. [CrossRef]
56. Wang, Y.; Chen, D.; Yan, R.; Yu, F.; Van Kleunen, M. Invasive alien clonal plants are competitively superior over co-occurring native clonal plants. *Perspect. Plant. Ecol.* **2019**, *40*, 125484. [CrossRef]
57. Zhang, H.; Wei, S.; Hu, W.; Xiao, L.; Tang, M. Arbuscular mycorrhizal fungus *Rhizophagus irregularis* increased potassium content and expression of genes encoding potassium channels in *Lycium barbarum*. *Front. Plant Sci.* **2017**, *8*, 440. [CrossRef]
58. He, Y.J.; Zhong, Z.C.; Liu, J.M.; Liu, J.C.; Jin, J.; Song, H.X. Response of n and p absorption on *Broussonetia papyrifera* seedlings to inoculate vesicular-arbuscular mycorrhizal fungus. *Acta Ecol. Sin.* **2007**, *27*, 4840–4847. [CrossRef]
59. Hafiz, A.; Zhang, Q.; Saddam, H.; Li, H.; Ahmed, W.; Zhang, L. Effects of Arbuscular Mycorrhizal Fungi on Maize Growth, Root Colonization, and Root Exudates Varied with Inoculum and Application Method. *J. Soil. Sci. Plant Nut.* **2021**, *21*, 1577–1590. [CrossRef]
60. Ostonen, I.; Rosenvald, K.; Helmisaari, H.S.; Godbold, D.; Parts, K.; Uri, V.; Lõhmus, K. Morphological plasticity of ectomycorrhizal short roots in *Betula sp* and *Picea abies* forests across climate and forest succession gradients: Its role in changing environments. *Front. Plant Sci.* **2013**, *4*, 335. [CrossRef]
61. He, Y.M.; Fan, X.M.; Zhang, G.Q.; Li, B.; Li, T.G.; Zu, Y.Q.; Zhan, F. Effects of arbuscular mycorrhizal fungi and dark septate endophytes on maize performance and root traits under a high cadmium stress. *S. Afr. J. Bot.* **2020**, *134*, 415–423. [CrossRef]
62. Hodge, A. The plastic plant: Root responses to heterogeneous supplies of nutrients. *New Phytol.* **2004**, *162*, 9–24. [CrossRef]
63. Farley, R.; Fitter, A. The responses of seven co-occurring woodland herbaceous perennials to localized nutrient-rich patches. *J. Ecol.* **1999**, *87*, 849–859. [CrossRef]
64. Hutchings, M.J.; Wijesinghe, D.K. Performance of a clonal species in patchy environments: Effects of environmental context on yield at local and whole-plant scales. *Evol. Ecol.* **2007**, *22*, 313–324. [CrossRef]
65. Cui, M.; Caldwell, M.M. Nitrate and phosphate uptake by *Agropyron desertorum* and *Artemisia tridentata* from soil patches with balanced and unbalanced nitrate and phosphate supply. *New Phytol.* **1998**, *139*, 267–272. [CrossRef]
66. Hodge, A.; Robinson, D.; Griffiths, B.; Fitter, A. Nitrogen capture by plants grown in N-rich organic patches of contrasting size and strength. *J. Exp. Bot.* **1999**, *50*, 1243–1252. [CrossRef]
67. Zhang, Z.; Zhang, J.; Huang, Y. Effects of arbuscular mycorrhizal fungi on the drought tolerance of *Cyclobalanopsis glauca* seedlings under greenhouse conditions. *New For.* **2014**, *45*, 545–556. [CrossRef]
68. Hodge, A.; Berta, G.; Doussan, C.; Merchan, F.; Crespi, M. Plant root growth, architecture and function. *Plant Soil* **2009**, *321*, 153–187. [CrossRef]
69. Lin, S.; Sun, X.; Wang, X.; Dou, C.; Li, Y.; Luo, Q.; Sun, L.; Jin, L. Mycorrhizal studies and their application prospects in China. *Acta Prata. Sin.* **2013**, *22*, 310.
70. Mommer, L.; Visser, E.J.W.; Van Ruijven, J.; De Caluwe, H.; Pierik, R.; De Kroon, H. Contrasting root behaviour in two grass species: A test of functionality in dynamic heterogeneous conditions. *Plant Soil* **2011**, *344*, 347–360. [CrossRef]
71. Shao, Y.; Zhang, D.; Hu, X.; Wu, Q.; Jiang, C.; Xia, T.; Gao, X.B.; Kuča, K. Mycorrhiza-induced changes in root growth and nutrient absorption of tea plants. *Plant Soil. Environ.* **2018**, *64*, 283–289. [CrossRef]
72. Campbell, B.; Grime, J.; Mackey, J.A. trade-off between scale and precision in resource foraging. *Oecologia* **1991**, *87*, 532–538. [CrossRef]
73. Hodge, A. Plastic plants and patchy soils. *J. Exp. Bot.* **2006**, *57*, 401–411. [CrossRef]
74. Rajaniemi, T.K.; Reynolds, H.L. Root foraging for patchy resources in eight herbaceous plant species. *Oecologia* **2004**, *141*, 519–525. [CrossRef]
75. Rytter, R. Stone and gravel contents of arable soils influence estimates of C and N stocks. *Catena* **2012**, *95*, 153–159. [CrossRef]
76. Suo, G.; Xie, Y.; Zhang, Y.; Luo, H. Long-term effects of different surface mulching techniques on soil water and fruit yield in an apple orchard on the Loess Plateau of China. *Sci. Hortic.* **2019**, *246*, 643–651. [CrossRef]
77. Hanson, C.T.; Blevins, R. Soil water in coarse fragments. *Soil Sci. Soc. Am. J.* **1979**, *43*, 819–820. [CrossRef]
78. Clark, L.; Whalley, W.; Barraclough, P. How do roots penetrate strong soil. In *Roots: The Dynamic Interface between Plants and the Earth*; Springer: Berlin/Heidelberg, Germany, 2003; pp. 93–104.
79. Singh, S.; Kapoor, K. Inoculation with phosphate-solubilizing microorganisms and a vesicular-arbuscular mycorrhizal fungus improves dry matter yield and nutrient uptake by wheat grown in a sandy soil. *Biol. Fert. Soils* **1999**, *28*, 139–144. [CrossRef]
80. Lambers, H.; Teste, F.P. Interactions between arbuscular mycorrhizal and non-mycorrhizal plants: Do non-mycorrhizal species at both extremes of nutrient availability play the same game. *Plant Cell Environ.* **2013**, *36*, 1911–1915. [CrossRef] [PubMed]
81. Fitter, A. Costs and benefits of mycorrhizas: Implications for functioning under natural conditions. *Experientia* **1991**, *47*, 350–355. [CrossRef]
82. Barber, S.; Silberbush, M. *Plant Root Morphology and Nutrient Uptake*; American Society of Agronomy: Madison, WI, USA, 1984; Volume 49, pp. 65–87. [CrossRef]
83. Hutchings, M.; De Kroon, H. *Heterogeneous Soil-Resource Distribution and Plant Responses from Individual-Plant Growth to Ecosystem Functioning*; Springer: Berlin/Heidelberg, Germany, 1994; pp. 451–476.

84. Luo, Y.; Zhou, J.; Li, Y.; Li, Q.; Li, D.; Xiao, Z.; Zhang, F.; Hu, Z. Effects of tephra gravel content on growth of flue-cured tobacco at seedling stage. *J. South. Agric.* **2014**, *45*, 570–574.
85. Laliberté, E.; Lambers, H.; Burgess, T.I.; Wright, S.J. Phosphorus limitation, soil-borne pathogens and the coexistence of plant species in hyperdiverse forests and shrublands. *New Phytol.* **2015**, *206*, 507–521. [CrossRef]

agronomy

MDPI

Article

Regulation of Cell Wall Degradation and Energy Metabolism for Maintaining Shelf Quality of Blueberry by Short-Term 1-Methylcyclopropene Treatment

Han Yan [1], Rui Wang [1,*], Ning Ji [1], Jiangkuo Li [2], Chao Ma [1], Jiqing Lei [1], Liangjie Ba [1], Guangzhong Wen [3] and Xiaobo Long [3]

[1] College of Food and Pharmaceutical Engineering, Guiyang University, Guiyang 550003, China
[2] Tianjin Key Laboratory of Postharvest Physiology and Storage of Agricultural Products, National Engineering and Technology Research Center for Preservation of Agricultural Produce, Tianjin 301699, China
[3] Agriculture and Rural Bureau of Majiang County, Qiandongnan Miao and Dong Autonomous Prefecture 557600, China
* Correspondence: gyuruiwang@163.com

Abstract: In order to study a short-term and efficient technology by 1-methylcyclopropene (1-MCP) in blueberry, the fruit was treated with 0, 0.5, 1 and 3 μL/L 1-MCP for 2 h then stored at 25 ± 1 °C with 40–50% relative humidity (RH) for 9 d. The weight loss, decay incidence, respiration rate, firmness, soluble solid content (SSC), titratable acid (TA), Brix-acid ratio (BAR), sensory evaluation, content of cell wall polysaccharide, activities of cell wall composition-related enzymes and energy metabolism in blueberry were determined during shelf life. The results showed that the weight loss, decay incidence and respiration rate were reduced by 3 μL/L 1-MCP treatment. Compared to other groups, the firmness, the content of TA and anthocyanins were maintained in 3 μL/L 1-MCP-treated blueberry. In contrast, the SSC and BAR were lower compared to those untreated. However, the sensory evaluation of "taste" and "aroma" value showed no differences in all fruits. The content of protopectin, cellulose and hemicellulose was higher in 1-MCP-treated blueberry, accompanied by a decrease in polygalacturonase (PG) and pectin methyl esterase (PME) activity. The content of water-soluble pectin (WSP) was lower in 1-MCP-treated blueberry than untreated ones. The activity of phenylalanine ammonia lyase (PAL), peroxidase (POD), cinnamyl alcohol dehydrogenase (CAD) and 4-coumarate-CoA ligase (4CL) was higher in 1-MCP-treated blueberry than the untreated, which induced more serious lignification. The results of energy metabolism also showed that the 1-MCP treatment could ensure sufficient intracellular energy supply. The 3 μL/L 1-MCP treatment could maintain the shelf quality and retard decomposition of cell wall polysaccharide by ensuring sufficient intracellular energy supply and inhibiting cell wall-degrading enzymes activity. Taken together, this study highlighted an efficient and short-term 1-MCP treatment technique.

Keywords: blueberry; 1-MCP; cell wall polysaccharide; shelf life; karst fruit

Citation: Yan, H.; Wang, R.; Ji, N.; Li, J.; Ma, C.; Lei, J.; Ba, L.; Wen, G.; Long, X. Regulation of Cell Wall Degradation and Energy Metabolism for Maintaining Shelf Quality of Blueberry by Short-Term 1-Methylcyclopropene Treatment. *Agronomy* **2023**, *13*, 46. https://doi.org/10.3390/agronomy13010046

Academic Editor: Sophie Parks

Received: 1 December 2022
Revised: 19 December 2022
Accepted: 20 December 2022
Published: 23 December 2022

1. Introduction

Guizhou province, which is one of the world's three major karst landscape areas, is located in southwest China, the center of east Asia. Karst plateau mountains typically have thin soil layers, sinkholes, steep slopes and low fertility in their soil [1]. However, Shan et al. (2020) reported that long-term cultivation of fruit plantations decreased mineralization and nitrification rates in calcareous soil in the karst region [2], which is conducive to improving the local karst ecosystem. In 2021, the scale of blueberry plantation had reached 1.5×10^4 hm^2 in Guizhou province, which produced 69,000 tons of fresh fruit worth over 1.5 billion yuan (the monetary unit of China). It is very important for Guizhou Province to alleviate poverty and revitalize the countryside.

Blueberry (*Vaccinium* spp.) is rich in phenolic acids and anthocyanins, which are natural antioxidant [3,4] and widely cultivated in Guizhou. However, blueberry is susceptible to mechanical damage and perishables because of its thin skin [5]. On the other hand, postharvest blueberry has strong respiration, which accelerates the softening, weight loss and decay [6]. Since fruit firmness is an important indicator in the commercial quality of blueberry, excess softening can result in notable economic loss. Thus, it will be of great value to maintain the firmness of postharvest blueberry, for this can improve fruit quality and storability as well as provide more time for transportation and sales. It is well known that the decrease in firmness is caused by water loss and the change in structure and composition of cell wall [6]. Among them, the polysaccharide (pectin, cellulose, hemicellulose, lignin) metabolism of cell wall is demonstrated to be the primary factor causing fruit softening [7]. In recent years, several technologies have been used to maintain fruit quality and delay softening of blueberry, such as 1-MCP, sodium nitroprusside, acidic electrolyzed oxidizing water dipping and modified atmosphere storage [3,7–9].

An ethylene receptor blocker, 1-MCP can inhibit endogenous ethylene production and the physiological and biochemical reactions of fruit by irreversibly binding with the ethylene receptor [10]. Previous studies showed that 1-MCP could delay softening and inhibit the polysaccharide decomposition of the cell wall in blueberry, pears, apple and plum [11–14]. Ortiz et al. (2018) reported that 1 μL/L 1-MCP treatment for 12 h could alleviate cell wall degradation in blueberry, resulting in a firmness of 16% higher than the control [14]. Moreover, Grozeff et al. (2017) reported that a combination of nitric oxide and 1-MCP (1 μL/L, 12 h) could maintain the firmness and extend storage-life accompanied with ASA and glutathione content in blueberry [15]. Many similar studies had been previously reported [8,16]. Indeed, the application concentration of 1-MCP in previous studies was usually between 0.3 and 1.0 μL/L, and the treatment time was between 12 and 24 h, which made them similar. In practical production, shortening processing time would not only be beneficial for reducing production costs but also provide more time for transportation and sales.

Cell wall polysaccharides include a variety of components, such as pectin, cellulose, hemicellulose and lignin [6]. In postharvest fruit, the pectin and cellulose disintegrated gradually during storage. Furthermore, various abiotic stresses, such as low temperature, high carbon dioxide and physiological damage could also disturb the cellular homeostasis [17]. For blueberry, the softening of fruits is triggered by the accumulation of ethylene during storage and the disassembly of cell wall polysaccharides, and softening has been delayed by 1-MCP treatment, which inhibited activities of cell wall-degrading enzymes, such as PG and PME [12,14]. The deposition of lignin is the main reason for lignification. Lignin is produced by PAL, POD, CAD and 4CL when plants are subjected to biotic/abiotic stress [18,19]. Suo et al. (2018) reported that the risk of lignification increased in 1-MCP-treated 'Hongyang' kiwifruit [20]. However, to the best of our knowledge, we have not found an available report concerning the lignification of blueberry fruit after 1-MCP treatment.

In order to develop a short-term and efficient treatment technique for 1-MCP in blueberry, the present study investigated the effects of short-term treatment with a high concentration of 1-MCP on maintaining shelf quality and delaying softening in blueberry after harvest. The preservation mechanisms of action were explored in relationship to enzyme activities and cell wall metabolism. Moreover, the energy metabolism in postharvest blueberry was also identified by a chromatographic technique.

2. Materials and Methods

2.1. Blueberry Samples and Treatment

'Britewell' blueberry fruits were picked from a seven-year old orchard in Majiang (107°59′15″ N, 26°49′23″ E), China. Uniform fruits without injuries were harvested at the same maturity stage, in terms of color and size. Fruits were put into trays (125 ± 2 g/tray) and transported to the laboratory within 2 h after harvest. Fruits were randomly divided into four groups of 3750 g each (30 trays/group). The four groups of blueberries were placed in polypropylene plastic containers (1 m × 1 m × 1 m) (LEYI Inc., Shanghai, China) and fumigated for 2 h at controlled temperature (25 ± 1 °C), as follows: (1) untreated, (2) 0.5 μL/L 1-MCP (SmartFresh^{SM} Inc., Wilmington, Delaware, USA), (3) 1 μL/L 1-MCP and (4) 3 μL/L 1-MCP. After treatment, a shelf experiment was carried out for 9 days at 25 ± 1 °C with 40–50% RH. A total of 360 fruits from each group were placed separately for the determination of weight loss, decay incidence and respiration rate. The detection was performed at 0, 3, 5, 7 and 9 d, in turn. Samples were frozen in liquid nitrogen immediately and then stored at −80 °C for the subsequent measurements.

2.2. Determination of Weight Loss, Decay Incidence and Respiration Rate

Weight loss (%) (n = 60) was calculated as follows: $\frac{(m_0-m)}{m_0} \times 100$, where m_0 is the initial weight and m is the final weight of each sample of fruit. Decay percentage was measured by counting the number of fruits decayed in each group. The decayed fruits were considered rotten if they had visible fungal growth or bacterial lesions on their surface.

The respiration rate was calculated for a constant sixty fruits per replication. Sixty fruits were placed in a 1.0 L sealed plastic box at 25 ± 1 °C for 4 h. The CO_2 concentration was determined by a gas analyzer (Checkpoint 3 Premium, Mocon Inc., Minneapolis, MN, USA).

2.3. Determination of Firmness, SSC, TA and Anthocyanin Content

Fruit firmness (n = 18) was measured with a texture analyzer (SMS Inc., London, England). The texture analyzer was fitted with P/2N probe. Fruits were deformed for a distance of 6 mm at a speed of 2 mm s^{-1}, and a 0.049 N trigger force was used.

To determine SSC, blueberry juice was extracted from the blueberries (n = 18) after grinding and centrifugation (10,000× g, 10 min, 4 °C), and its soluble solid content (SSC) was measured using a digital hand-held refractometer (PAL-1, Atago Inc., Tokyo, Japan). Total titratable acid (TA) was determined by a previously described method (Hui et al., 2018) [17]. Then, BAR was calculated by using following formula: BAR = SSC/TA.

For anthocyanin measurement, 10 g of frozen blueberry fruit was ground, and 0.1 g of powder was added to 1 mL methanol (containing 1% HCl) (Aladdin Inc., Shanghai, China). Homogenized samples were centrifuged (10,000× g, 10 min, 20 °C) for 10 min. The absorbance was measured at 250 nm and 700 nm. The anthocyanin content (mg/g) was calculated using the method of Ge et al. [3].

2.4. Determination of Cell Wall Composition

2.4.1. Cellulose and Hemicelluloses

Cellulose and hemicellulose contents (mg/g) were measured with an assay kit (Norminkoda Biotechnology Co., Ltd., Wuhan, China). The reagents were added in accordance with the manufacturer's instructions, and the absorbance at 620 nm was measured.

2.4.2. Lignin

Lignin content was measured according to the method of Ge et al. (2019), with modifications [3]. Frozen tissues (2.0 g) were weighed and added to 3.0 mL of 95% cold ethanol (pre-cooled in an explosion-proof refrigerator at 4 °C) and then centrifuged (12,000 g, 20 min, 4 °C). The sediment was washed three times with 95% ethanol (Aladdin Inc., Shanghai, China) and rinsed three times with the mixture of ethanol and n-hexane (1:2) (Aladdin Inc., Shanghai, China). The precipitate was collected and dried at 80 °C, then dissolved in 1.0 mL acetic acid (Aladdin Inc., Shanghai, China)and incubated at 80 °C for 30 min; the reaction was stopped by the addition of 1.5 mL 2.0 mol/L NaOH (Aladdin Inc., Shanghai, China). The content of lignin was expressed as $\Delta_{OD}280$/g FW.

2.4.3. Pectin Content

The WSP and protopectin content was measured according to the method of Hamauzu et al. (2011) [21]. In principle, galacturonic acid reacted with carbazole to produce an absorption peak at 530 nm, and the absorbance value at 530 nm is positively correlated with WSP and protopectin content.

2.5. Enzyme Activity Assays

2.5.1. PAL, POD, CAD and 4CL Activity

The activity of PAL was assayed following the procedure as described by Assis et al. [9], with some modifications. The reaction mixture, consisting of 1.0 mL of the supernatant, was mixed with 2 mL of borate buffer (50 mmol L^{-1}, pH 8.8) (Aladdin Inc., Shanghai, China), and 1 mL of 1-phenylalanine (20 mmol L^{-1}) (Aladdin Inc., Shanghai, China), was incubated at 40 °C for 1 h. The reaction was terminated by heating the reaction mixture in boiling water for 1 min, and absorbance was read at 290 nm by a UV spectrophotometer (Cary 60, Agilent, Santa Clara, California, USA). The PAL activity was expressed as U/mg.

The enzyme activities of 4CL, CAD and POD were measured according to the methods previously described [22,23]. POD, CAD and 4CL activities were expressed as U/mg.

2.5.2. PG and PME Activity

PG activity was measured according to a previous method (Ji et al., 2021) [24]. A volume of 0.1 mL supernatant (the enzyme solution was inactivated by a boiling water bath as a control) was mixed with 0.4 mL poly-galacturonic acid (1%, *w*/*v*; 37 °C) (Aladdin Inc., Shanghai, China), and the mixture was maintained for 30 min at 37 °C. After cooling, 2.5 mL 3,5-dinitrosaliacylic acid (DNS) (Aladdin Inc., Shanghai, China)reagent was added and the absorbance value was measured at 540 nm. PG activity was expressed as U/mg.

PE activity was determined according to a previous method (Ji et al., 2021) [24]. Frozen samples (0.2 g) were homogenized in 1 mL of a 8.8% (*w*/*v*) NaCl solution (Aladdin Inc., Shanghai, China)in an ice bath. After centrifugation (10,000× *g*, 20 min, 4 °C), the supernatant was maintained at pH 7.5 with 0.1 M NaOH solution (Aladdin Inc., Shanghai, China). The absorbance changes in 1 minute of the reaction mixture containing 1 mL of 0.5% (*w*/*v*) pectin (Sigma-Aldrich CN Inc., Shanghai, China)), 0.1 mL bromothymol blue (0.01%) (Aladdin Inc., Shanghai, China)and 0.2 mL supernatant were measured and recorded at 620 nm. PE activity was expressed as U/mg.

2.6. Determination of ATP, ADP and AMP Content and Energy Charge

ATP, ADP and AMP were measured according to our previous method described by Li et al. (2014). Results were expressed as mg/kg FW. Energy charge (EC) was calculated by using the following formula: EC = (ATP + 1/2ADP)/(ATP + ADP + AMP).

2.7. Sensory Evaluation

Sensory evaluation was conducted using a nine-point hedonic scale and performed in a sensory laboratory at Guiyang University. Ten trained and regular blueberry consumers (there are five men and five women, ranging in age from 20 to 25) were asked to evaluate each sample for the sensory attributes of appearance, color, flavor and taste. For scoring criteria, we referred to Nirmal et al. (2020) [25]. Each attribute was scored as follows: 9 = Like extremely; 8 = Like very much; 7 = Like; 6 = Like slightly; 5 = Neither like nor dislike; 4 = Dislike slightly; 3 = Dislike moderately; 2 = Dislike; 1 = Dislike extremely.

2.8. Statistical Analysis

The whole experiment was designed following a randomized experimental design with triplicate samples collected on each sampling date. Data were presented as arithmetic means with standard errors. Statistical analysis was carried out using SPSS version 21 (IBM, Armonk, New York, USA). All the tests were performed with a level of significance of 0.05.

3. Results

3.1. Weight Loss, Decay Incidence and Respiration Rate

3.1.1. Weight Loss

As revealed in Figure 1a, weight loss was increased in untreated and 1-MCP-treated blueberries. We found that the weight losses of untreated, 1 μL/L 1-MCP-treated and 3 μL/L 1-MCP-treated blueberries were not significantly different between 3 d and 5 d but were lower in the 1 μL/L 1-MCP-treated blueberries than in untreated blueberries between 7 d and 9 d ($p < 0.05$). However, weight loss in the 0.5 μL/L 1-MCP-treated blueberries was significantly higher than in the others.

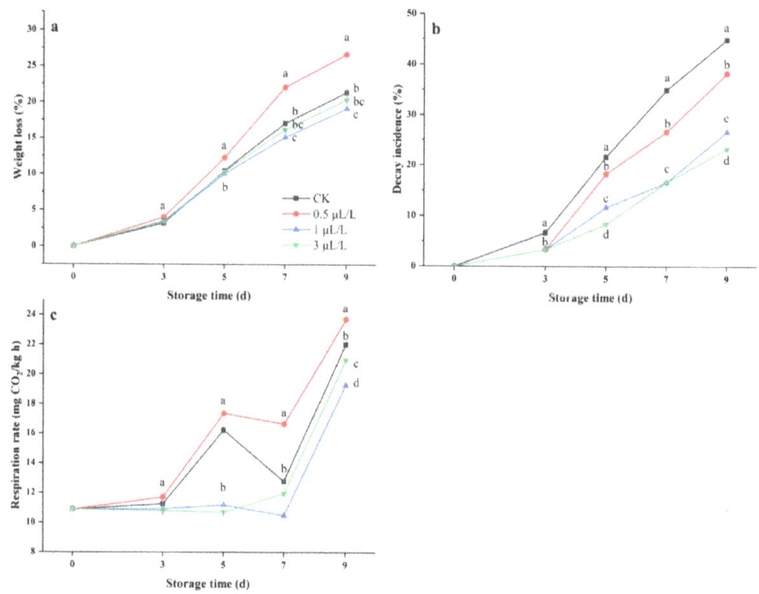

Figure 1. Effect of 1-MCP treatment on weight loss (**a**), decay incidence (**b**) and respiration rate (**c**) in blueberries during shelf life at 25 ± 1 °C. Data are the means of three replicates ± standard deviation. Values with different letters are significantly different according to Duncan's test ($p < 0.05$).

The foregoing results indicated that treatment with 1 μL/L 1-MCP suppressed blueberry weight loss in later storage times.

3.1.2. Decay Incidence

As revealed in Figure 1b, the incidence of decay was increased in all groups. However, the rate of decay was slower for 1 μL/L 1-MCP-treated and 3 μL/L 1-MCP-treated blueberries than for the untreated and 0.5 μL/L 1-MCP-treated groups. The decay incidence of untreated blueberries reached 20% in 5 d, while 1 μL/L 1-MCP-treated and 3 μL/L 1-MCP-treated blueberries reached this level only at the end of storage.

Therefore, we found that the decay incidence was effectively suppressed both by 1 μL/L and by 3 μL/L 1-MCP.

3.1.3. Respiration Rate

As revealed in Figure 1c, the respiration rate remained relatively constant both in untreated blueberries and in 0.5 μL/L 1-MCP-treated blueberries from 0 d to 3 d. Similarly, the respiration rate remained relatively constant both in 1 μL/L 1-MCP-treated and in 3 μL/L 1-MCP-treated blueberries from 0 d to 5 d. However, the rate was significantly lower for 1 μL/L 1-MCP-treated and 3 μL/L 1-MCP-treated blueberries than for untreated blueberries and 0.5 μL/L 1-MCP-treated blueberries at 5 d and 9 d. Hence, according to this observation, the highest respiration rate obviously resulted in weight loss for the 0.5 μL/L 1-MCP-treated group.

3.2. Quality Parameters and Anthocyanin Content

3.2.1. Firmness

As revealed in Figure 2a, blueberry firmness was 1.53 N at 0 d and decreased as the storage time increased. For untreated blueberries, firmness rapidly decreased from 0 d to 3 d, then remained relatively constant until 5 d, and finally decreased until the end of storage. Relative to untreated blueberries, 1-MCP treatment delayed the decrease in firmness within a certain range. The firmness dropped to 0.89 N, 0.94 N, 1.20 N and 1.3 N by the end of storage for the untreated, 0.5 μL/L 1-MCP-treated, 1 μL/L 1-MCP-treated and 3 μL/L 1-MCP-treated blueberries, respectively.

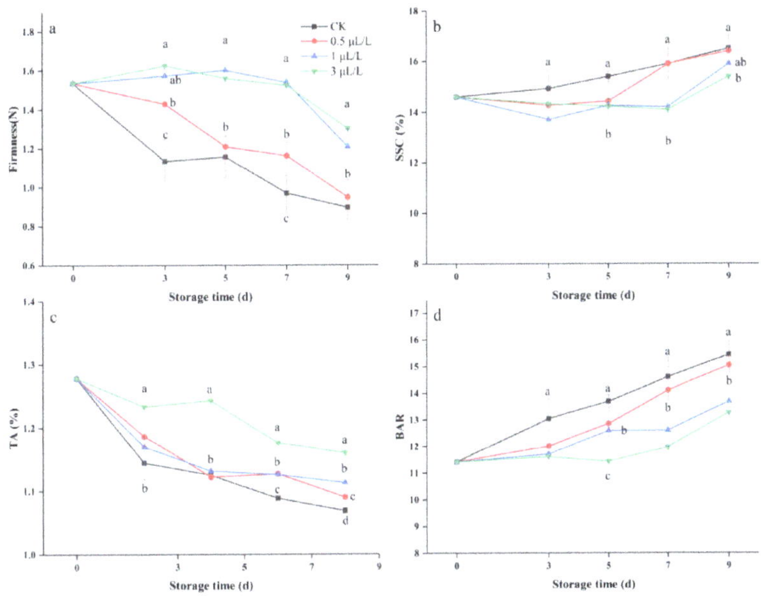

Figure 2. Effect of 1-MCP treatment on firmness (**a**), SSC (**b**), TA (**c**) and BAR (**d**) in blueberries during shelf life at 25 ± 1 °C. Data are the means of three replicates ± standard deviation. Values with different letters are significantly different according to Duncan's test ($p < 0.05$).

The foregoing results indicated that 1μL/L1-MCP treatment and 3 μL/L 1-MCP treatment greatly delayed the decrease in firmness in postharvest blueberries.

3.2.2. SSC, TA and BAR

As revealed in Figure 2b, blueberry SSC was 14.60% at 0 d and slightly increased as the storage time went on. There were no significant differences between all treatment groups at 3 d. However, we found that blueberry SSC was suppressed by 3 μL/L 1-MCP treatment, which was significantly lower than in untreated blueberries. Blueberry TA content was 1.27% at 0 d, then decreased in all treatment groups until the end of storage. No difference in TA content was found among untreated, 0.5 μL/L 1-MCP-treated and 1 μL/L 1-MCP-treated blueberries, other than at 7 d. In contrast, 3 μL/L 1-MCP treatment greatly delayed the decrease in TA content, which was significantly higher than the other treatment groups over the whole storage time. The BAR was 11.42 at 0 d and increased, as a result of increased SSC and decreased TA, over the whole storage time. The maximum values were 15.43, 15.04, 13.68 and 13.27 at 9 d for untreated, 0.5 μL/L 1-MCP-treated, 1 μL/L 1-MCP-treated and 3 μL/L 1-MCP-treated blueberry, respectively. There was no difference between untreated and 0.5 μL/L 1-MCP-treated blueberries. In contrast, the BAR of the 3 μL/L 1-MCP-treated blueberries was significantly lower than that of untreated blueberries from 5 d to 9 d ($p < 0.05$).

The foregoing results indicated that the changes in SSC, TA and BAR were delayed by 1-MCP treatment, but 0.5 μL/L 1-MCP treatment was not effective.

3.2.3. Anthocyanin Content

As revealed in Figure 3, anthocyanin content was 0.83 mg/g at 0 d. We found that anthocyanin content reached the maximum at 5 d for all treatment groups. However, 1-MCP treatment accelerated the increase in anthocyanin content. The maximum values were 1.37 mg/g, 1.473 mg/g, 1.616 mg/g and 1.844 mg/g for untreated, 0.5 μL/L 1-MCP-treated, 1 μL/L 1-MCP-treated and 3 μL/L 1-MCP-treated blueberries, respectively. The content was significantly higher in 1 μL/L 1-MCP-treated and 3 μL/L 1-MCP-treated blueberries than in the untreated group ($p < 0.05$).

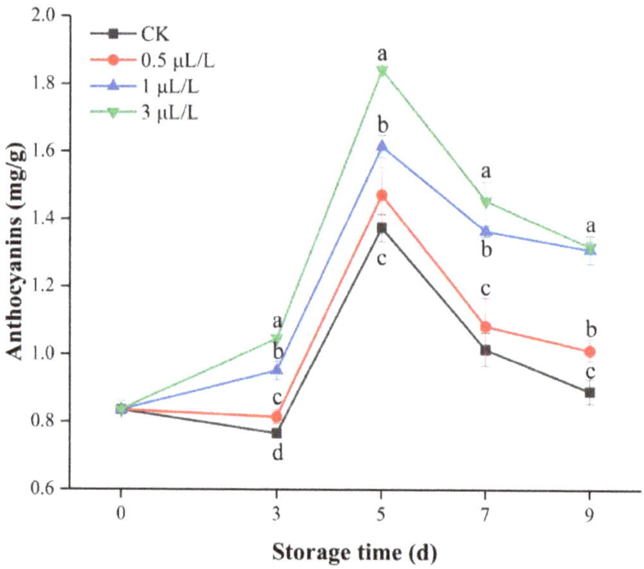

Figure 3. Effect of 1-MCP treatment on anthocyanins in blueberries during storage at 25 ± 1 °C. Data are the means of three replicates ± standard deviation. Values with different letters are significantly different according to Duncan's test ($p < 0.05$).

The foregoing results indicated that 1 μL/L 1-MCP and 3 μL/L 1-MCP treatment significantly upregulated the anthocyanin content, whereas the 0.5 μL/L 1-MCP treatment was not effective.

3.3. *Sensory Evaluation*

In order to directly evaluate the effect of 1-MCP treatment on the shelf quality of blueberries, sensory evaluation is necessary. Taking commodity value and consumer acceptability into account, blueberries with a decay incidence not exceeding 20% were chosen to evaluate. As shown in Table 1, the values for firmness, appearance and aroma decreased as the storage time went on. From 5 d to 7 d, the firmness and appearance value was higher in 3 μL/L 1-MCP-treated blueberries than in untreated fruit ($p < 0.05$), while the aroma value showed no difference between 1-MCP-treated and untreated fruit. In contrast, the taste value of the 1-MCP treated blueberries was slightly increased, while the taste value of untreated blueberries sharply increased at 5 d and was higher than the 1-MCP-treated fruits.

Table 1. Sensory evaluation of blueberry during shelf life at 25 ± 1 °C.

Shelf Time	Treatment	Sensory Attribute			
		Firmness	Taste	Appearance	Aroma
0 d	untreated	6.8 ± 1.09	6.6 ± 0.89	6.4 ± 0.89	6.4 ± 0.54
3 d	untreated	6.8 ± 0.83 [a]	7.0 ± 0.7 [a]	5.8 ± 0.83 [a]	7.0 ± 1.64 [a]
	0.5 μL/L	7.0 ± 0.70 [a]	6.4 ± 0.89 [a]	6.0 ± 0.70 [a]	5.4 ± 0.89 [a]
	1 μL/L	6.8 ± 0.83 [a]	6.2 ± 0.98 [a]	6.6 ± 0.54 [a]	5.8 ± 0.83 [a]
	3 μL/L	6.6 ± 0.74 [a]	5.4 ± 0.78 [a]	6.8 ± 0.83 [a]	6.0 ± 0.83 [a]
5 d	untreated	3.4 ± 0.83 [b]	7.2 ± 0.44 [a]	4.4 ± 0.89 [b]	5.4 ± 0.98 [a]
	0.5 μL/L	3.8 ± 0.83 [b]	6.5 ± 0.44 [ab]	4.6 ± 0.54 [b]	4.8 ± 0.83 [a]
	1 μL/L	6.4 ± 0.70 [a]	6.0 ± 0.54 [ab]	5.6 ± 0.89 [ab]	5.4 ± 0.54 [a]
	3 μL/L	7.0 ± 0.70 [a]	5.6 ± 0.54 [b]	6.6 ± 0.54 [a]	5.2 ± 0.83 [a]
7 d	untreated	-	-	-	-
	0.5 μL/L	-	-	-	-
	1 μL/L	6.0 ± 0.71 [a]	6.2 ± 0.83 [a]	5.2 ± 0.44 [a]	6.0 ± 0.70 [a]
	3 μL/L	6.6 ± 0.54 [a]	5.6 ± 0.89 [a]	6.2 ± 0.83 [a]	5.8 ± 0.44 [a]

Data are the means of nine replicates ± standard deviation. In columns, different superscript letters at the same shelf time indicate a significant difference according to Duncan's test ($p < 0.05$).

3.4. Cell Wall Metabolism and Enzyme Activity

3.4.1. WSP, Protopectin Content and PG and PE Activity

As revealed in Figure 4a, the WSP content in untreated blueberries increased from 0 d to 3 d, then slightly decreased on day 5, and then increased again until the end of storage. A similar trend was observed for the 1-MCP-treated blueberries. The maximum values were 22.31 mg/g, 19.42 mg/g, 14.89 mg/g and 15.06 mg/g for untreated, 0.5 μL/L 1-MCP-treated, 1 μL/L 1-MCP-treated and 3 μL/L 1-MCP-treated blueberry, respectively. The content was significantly higher for 1 μL/L 1-MCP-treated and 3 μL/L 1-MCP-treated blueberries than for untreated and 0.5 μL/L 1-MCP-treated blueberries over the entire storage time ($p < 0.05$). In contrast, protopectin content decreased until the end of storage for all treatments (Figure 4b). For untreated blueberries, it rapidly decreased from 0 d to 5 d then slightly decreased until the end of storage. A similar trend was observed in the 1 μL/L 1-MCP-treated and 3 μL/L1-MCP-treated blueberries. The 3 μL/L 1-MCP treatment significantly suppressed protopectin degradation, which was 2.09 times, 1.62 times and 1.85 times higher than in the untreated blueberries at 5 d, 7 d and 9 d, respectively.

Figure 4c showed that PG activity increased from 0 to 3 d and 5 to 9 d, but decreased at the other storage times regardless of 1-MCP treatment. However, it was significantly lower for the 1 μL/L 1-MCP-treated and 3 μL/L 1-MCP-treated blueberries than for untreated fruit over the whole storage time. The different trend was observed for PE activity, which decreased rapidly at first, then slowly. The PE activity of 1-MCP-treated blueberries was significantly lower than untreated at 3 d and 5 d. In addition, the 3 μL/L1-MCP treatment was most effective in this regard; the PE activity of this group was 45.54% and 30.31% lower than untreated blueberries at 3 d and 5 d, respectively.

The foregoing results indicated that the changes in WSP and protopectin content were maintained by 3 μL/L1-MCP treatment and accompanied by suppressed PG and PE activity.

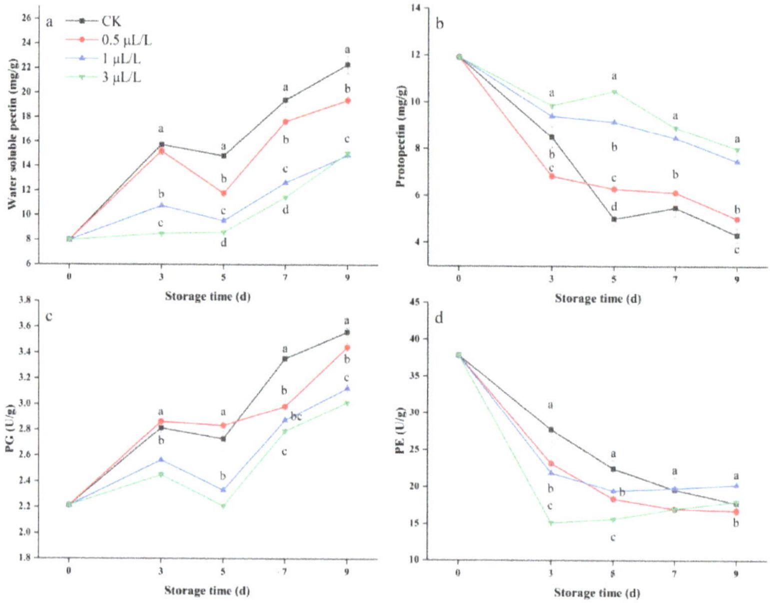

Figure 4. Effect of 1-MCP treatment on WSP (**a**), protopectin (**b**), PG activity (**c**) and PE activity (**d**) in blueberries during shelf life at 25 ± 1 °C. Data are the means of three replicates ± standard deviation. Values with different letters are significantly different according to Duncan's test ($p < 0.05$).

3.4.2. Lignin Content and PAL, POD, CAD and 4CL Enzyme Activity

As revealed in Figure 5, lignin content was 0.27% at 0 d. For untreated blueberries, the lignin content increased from 0 d to 3 d, then decreased until the end of storage; the maximum was 0.98%. The changes in lignin content for 1 μL/L1-MCP-treated and 3 μL/L 1-MCP-treated blueberries were similar to untreated blueberries. The maximum values were 1.34%, 1.51% and 2.08% for 0.5 μL/L 1-MCP-treated, 1 μL/L 1-MCP-treated and 3 μL/L 1-MCP-treated blueberries; these values were 1.36 times, 1.54 times and 2.12 times higher, respectively, than thosef for untreated blueberries.

As revealed in Figure 6a, PAL activity increased with storage time. However, it was 1.24 times, 1.21 times, 1.15 times and 1.21 times higher for the 3 μL/L 1-MCP-treated blueberries than for untreated fruit at 3 d, 5 d, 7 d and 9 d ($p < 0.05$), respectively. CAD and 4CL activity had a similar trend to PAL activity (Figure 4b,d). CAD activity was 2.91 times, 3.67 times, 2.41 times and 1.67 times higher for 3 μL/L 1-MCP-treated blueberries than for untreated blueberries at 3 d, 5 d, 7 d and 9 d ($p < 0.05$), respectively. 4CL activity was 1.52 times, 1.54 times, 1.22 times and 1.16 times higher for 3 μL/L 1-MCP-treated blueberry than for untreated at 3 d, 5 d, 7 d and 9 d ($p < 0.05$), respectively. The POD activity increased with storage time in untreated blueberries. However, there was a decrease from 0 d to 3 d, then an increase until the end of storage. POD activity was significantly higher for untreated blueberries than 1-MCP-treated blueberries at 3 d, but lower at 7 d ($p < 0.05$). The change in 3 μL/L 1-MCP-treated blueberries was most obvious and 1.41 times and 1.28 times higher, respectively, than that of untreated blueberries.

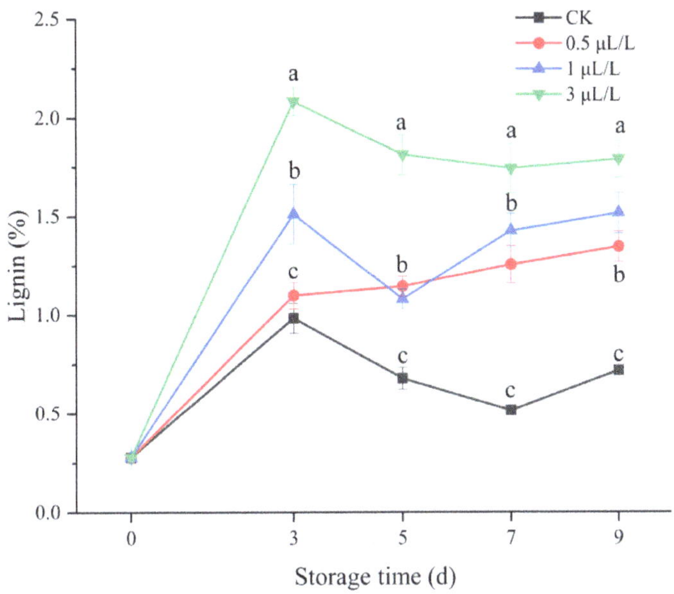

Figure 5. Effect of 1-MCP treatment on the lignin content of blueberries during shelf life at 25 ± 1 °C. Data are the means of three replicates ± standard deviation. Values with different letters are significantly different according to Duncan's test ($p < 0.05$).

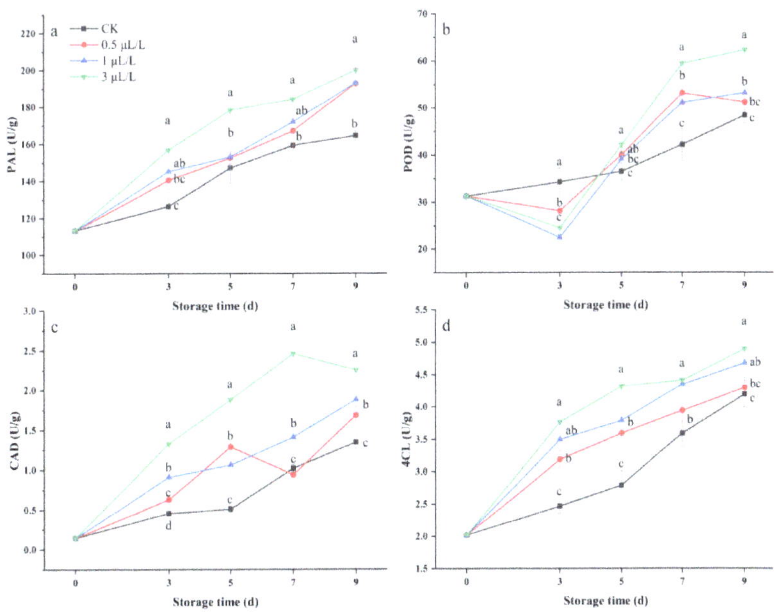

Figure 6. Effect of 1-MCP treatment on the PAL activity (**a**), POD activity (**b**), CAD activity (**c**) and 4CL activity (**d**) in blueberries during shelf life at 25 ± 1 °C. Data are the means of three replicates ± standard deviation. Values with different letters are significantly different according to Duncan's test ($p < 0.05$).

The foregoing results indicated that the lignin content was increased by 1-MAP treatment accompanied by activated PAL, POD, CAD and 4CL activity.

3.4.3. Cellulose and Hemicellulose Content

As revealed in Figure 7a, cellulose content rapidly decreased from 3 d to 5 d regardless of 1-MCP treatment. However, there was a slower decrease for 1-MCP-treated blueberries; cellulose content was 1.43, 1.93 and 1.90 times more than in untreated blueberries at 5 d ($p < 0.05$), respectively. After 5 d, the cellulose content slightly decreased until the end of the storage time both in 0.5 μL/L 1-MCP-treated and untreated blueberries, whereas in 1 μL/L 1-MCP-treated and 3 μL/L 1-MCP-treated blueberries, it increased as the storage time went on.

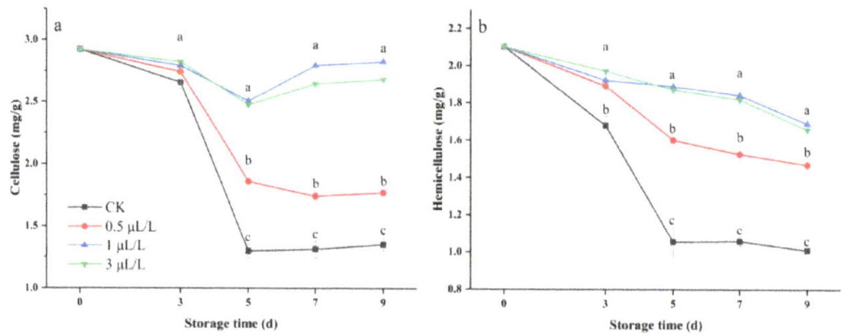

Figure 7. Effect of 1-MCP treatment on cellulose (**a**) and hemicellulose content (**b**) of blueberries during shelf life at 25 ± 1 °C. Data are the means of three replicates ± standard deviation. Values with different letters are significantly different according to Duncan's test ($p < 0.05$).

From Figure 7b, the change in hemicellulose content was similar to that of cellulose, which decreased over the entire storage time, regardless of 1-MCP treatment. However, hemicellulose content was significantly higher in 0.5 μL/L 1-MCP-treated, 1 μL/L 1-MCP-treated and 3 μL/L 1-MCP-treated blueberries than in untreated ones from 3 d to 9 d ($p < 0.05$), and 1.45, 1.67 and 1.64 times higher, respectively, than in untreated blueberries.

3.5. ATP, ADP, and AMP Contents and Energy Charge (EC)

Changes in the levels of ATP, ADP, AMP and EC in untreated and 1-MCP-treated blueberry are shown in Figure 8. In untreated blueberries, the ATP level and EC decreased as the storage time went on, but the level was significantly higher for 1 μL/L 1-MCP-treated and 3 μL/L 1-MCP-treated blueberries than for untreated ones from 3 d to 9 d ($p < 0.05$). In contrast, the AMP content was increased over the whole storage time, regardless of 1-MCP treatment (Figure 8c). However, in 3 μL/L 1-MCP-treated blueberries, AMP content was significantly lower than in untreated fruit by 22.97%, 31.61%, 39.35% and 46.53% at 3 d, 5 d, 7 d and 9 d, respectively. In contrast, the ADP content was increased at first and then decreased in all blueberries (Figure 8b). Obviously, the ADP content was significantly lower for 3 μL/L 1-MCP-treated blueberries than for untreated ones from 5 d to 9 d (3 μL/L 1-MCP-treated/untreated: 6.04/8.11 at 5 d, 5.63/9.09 at 7 d, 5.72/6.67 at 9 d).

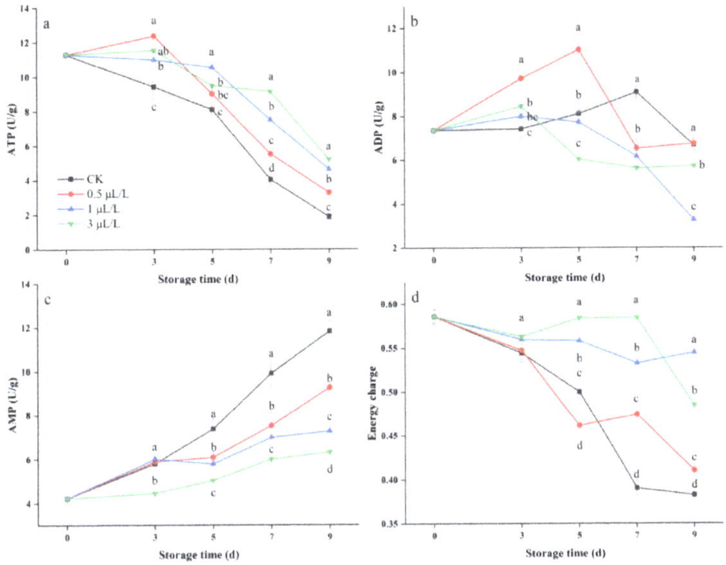

Figure 8. Effect of 1-MCP treatment on the ATP (**a**), ADP (**b**) and AMP (**c**) content and EC (**d**) in blueberries during shelf life at 25 ± 1 °C. Data are the means of three replicates ± standard deviation. Values with different letters are significantly different according to Duncan's test ($p < 0.05$).

4. Discussion

4.1. Effect of 1-MCP Treatment on Blueberry Weight Loss, Decay Incidence and Respiration Rate

Blueberries are quite perishable during storage, and this is accompanied by a deterioration in quality and quantity. As an ethylene receptor inhibitor, 1-MCP is crucial in inhibiting fruit respiration and decay [8,26]. In this work, the weight loss was lower in fruits treated with 1 μL/L and 3 μL/L 1-MCP than in untreated fruits. In contrast, the highest weight loss occurred in the 0.5 μL/L1-MCP-treated blueberries (Figure 1a). Similarly, Grozeff et al. found that 1 μL/L 1-MCP treatment was not effective in reducing weight loss for 'Misty' and 'Blue Cuinex' blueberries, but was effective for 'Blue Chip' blueberries [15]. We speculated that this phenomenon was due to different cultivars having different sensitivity to 1-MCP; which is to say, different concentrations of 1-MCP might have opposite effects on weight loss.

During the postharvest stage, the weight loss and decay incidence increased simultaneously. Generally, a higher respiration rate is particularly important for fruit weight loss and decay. Ji et al. reported that ethanol fumigation treatment effectively delayed the increase in weight loss rate and decay incidence of blueberries simultaneously. As the weight loss rate increases, the decay becomes more severe [24]. The effect of 1 μL/L and 3 μL/L 1-MCP treatment on the respiration rate and decay incidence was obvious, but the effect of 0.5 μL/L 1-MCP treatment was not. Similar results were reported in previous studies on French prunes treated with 1.2 μL/L 1-MCP and apples treated with 1 μL/L 1-MCP [27,28]. In addition, for 0.5 μL/L 1-MCP-treated blueberries, we found that the change in respiration rate was similar to the weight loss. Therefore, the highest weight loss and higher decay incidence in 0.5 μL/L 1-MCP-treated blueberries were ascribed to a higher respiration rate.

4.2. Effect of 1-MCP Treatment on Blueberry Firmness, SSC, TA, BAR, Anthocyanins and Sensory Evaluation

Treatment with 1-MCP has delayed the decrease in firmness [15], inhibited SSC and maintained the TA content [8] of the blueberry fruit. In the present study, the firmness of postharvest blueberry fruit decreased gradually with shelf life (25 ± 1 °C) time. However, 3 μL/L 1-MCP treatment slowed down the decline in firmness (Figure 2a), accompanied by significant effects on other quality traits, such as SSC, TA, BAR and anthocyanins. Similar effects were also observed in previous studies on blueberry [16], kiwifruit [29] and apple [30]. In contrast, Nesmith et al. (2011) found that 1-MCP accelerated the rate of firmness loss in three different cultivars [31]. It seems contradictory, but that is not the case. Different cultivars of blueberry had very different metabolic mechanisms, e.g., the metabolic mechanism of 'Britewell' blueberry has a special structure.

On the other hand, the results of the sensory evaluation supported the reliability of the quality parameters to a certain extent (Table 1). After 1-MCP treatment, although the values of "taste" and "aroma" decreased, there was no difference between treated and untreated ones. Particularly, the "firmness" and "appearance" values were higher for 3 μL/L 1-MCP-treated than untreated ones at 5 d. In fact, the softening of postharvest blueberries was effectively delayed with 3 μL/L 1-MCP treatment (2 h) in this study, and it did not significantly affect the aroma and taste. Compared to the previous studies that applied 1 μL/L 1-MCP for 12 h and 0.3 μL/L 1-MCP for 24 h [15,26], the treatment of the present study was more efficient and thus provided more time for transportation and sales.

4.3. *Effect of 1-MCP Treatment on Cell Wall Metabolism and Enzyme Activity*

Cell wall polysaccharides, mainly including pectin, cellulose and hemicellulose, are important components of cell wall structure and play key roles in cell wall metabolism. In addition, modifications in the composition of cell wall polysaccharides have been associated with changes in fruit firmness [6]. The reduction in fruit firmness is mainly due to the decomposition of protopectin and cellulose [32]. With time, the protopectin and cellulose degrade to WSP and hemicellulose, respectively. In the present study, the content of protopectin decreased and WSP increased over the whole storage time. Particularly, 3 μL/L 1-MCP treatment inhibited the decomposition of protopectin, such that the content of protopectin and WSP was significant higher and lower, respectively, than in untreated ones (Figure 5a,b). Accordingly, treatment with 3 μL/L 1-MCP also effectively inhibited the activity of PG and PME (Figure 4c,d). Similarly, the content of cellulose and hemicellulose was higher in 3 μL/L 1-MCP-treated blueberries than in untreated ones over the whole storage time (Figure 7). Moreover, blueberry firmness was significantly higher than the untreated group (Figure 2a). These results were also in accordance with previous reports on blueberry, plum and apple [6,12,13].

To the best of our knowledge, there is little available literature concerning the lignification of blueberries. However, the application of 1-MCP has been reported to increase the risk of lignification in kiwifruit [20]. In contrast, other studies have demonstrated that lignification in bamboo shoots and loquat has been reduced by the application of 1-MCP. [33,34]. Hence, the effect of 1-MCP on lignification is different in different varieties. In the present study, 3 μL/L 1-MCP treatment increased the content of lignin and the activities of PAL, POD, CAD and 4CL (Figures 5 and 6); thus, lignification was more serious than in untreated blueberries. However, no negative effects of lignification on food quality were found in the sensory evaluation results (Table 1). In fact, the content of lignin was correlated with fruit firmness [35]. The results of this study indicated that higher lignin content could be responsible for maintaining higher blueberry firmness in the 1-MCP treatment groups.

4.4. *Effect of 1-MCP Treatment on ATP, ADP and AMP Content and Energy Charge (EC)*

Adequate cellular energy supply is an important factor in the quality and storability of fruit [36]. In ETC respiration, ATP is decomposed to ADP by ATPases and produces free phosphate ions to coincide with energy release [37]. In the present study, 1-MCP-treated blueberries showed higher ATP and EC levels than untreated ones (Figure 8a,d). In contrast, the ADP and AMP levels were inhibited by 1-MCP treatment (Figure 8b,c); this effect was particularly obvious with 3 μL/L 1-MCP treatment. Thus, we suggested that the higher quality of the 3 μL/L 1-MCP-treated blueberries is due to a sufficient supply of intracellular energy. In previous studies, Cheng et al. (2015) reported that 1-MCP treatment alleviated chilling injury and maintained postharvest quality by regulating energy metabolism in 'Nanguo' pears [26]. Similar results had also been reported in kiwifruit and tomato [38,39]: fruits with better quality were the result of adequate mitochondrial energy supply.

5. Conclusions

This study explores a short duration and efficient treatment technique for 1-MCP in blueberry. These results shed light on the valuable efficiency of short-term 1-MCP treatment (3 μL/L, 2 h) for reducing the softening of blueberry fruit during shelf life. The 3 μL/L 1-MCP treatment maintained better shelf quality and delayed softening, which was related to higher firmness and sufficient energy supply, as well as inhibition of the activity of cell-wall degradation, including PG and PME. In addition, the lignin content was increased by 3 μL/L 1-MCP treatment because the activities of PAL, POD, CAD and 4CL were induced. The results suggested that short-term 1-MCP treatment was an effective technique for delaying the postharvest softening process and extending the shelf life of blueberry fruit. As a result, there has been an improvement in production efficiency.

Author Contributions: Conceptualization and methodology: H.Y. and R.W.; investigation: H.Y., R.W., N.J., L.B., G.W. and X.L.; formal analysis: H.Y., J.L. (Jiangkuo Li) and J.L. (Jiqing Lei); data curation: H.Y. and C.M.; writing—original draft preparation: H.Y.; writing—review and editing: H.Y. and R.W. All authors have read and agreed to the published version of the manuscript.

Funding: This research was funded by the Guizhou Province Key Technology Research and Development and Application of Innovation Base for Agricultural Products Primary Processing (Qi keZhong Yin Di (2020) 4018) and the Discipline and Master's Site Construction Project of Guiyang University by Guiyang City Financial Support Guiyang University (SY-2020).

Data Availability Statement: Not applicable.

Conflicts of Interest: The authors declare no conflict of interest.

References

1. Yue, Y.M.; Qi, X.K.; Wang, K.L.; Liao, C.J.; Tong, X.W.; Brandt, M.; Liu, B. Large scale rocky desertification reversal in South China karst. *Prog. Phys. Geogr. Earth Environ.* **2022**, *46*, 661–675. [CrossRef]
2. Shan, Z.; Yin, Z.; Yang, H.; Zuo, C.; Zhu, T. Long-Term Cultivation of Fruit Plantations Decreases Mineralization and Nitrification Rates in Calcareous Soil in the Karst Region in Southwestern China. *Forests* **2020**, *11*, 1282. [CrossRef]
3. Ge, Y.H.; Li, X.; Li, C.Y.; Tang, Q.; Duan, B.; Cheng, Y.; Hou, J.B.; Li, J.R. Effect of sodium nitroprusside on antioxidative enzymes and the phenylpropanoid pathway in blueberry fruit. *Food Chem.* **2019**, *295*, 607–612. [CrossRef]
4. Doyle, J.W.; Nambeesan, S.U.; Malladi, A. Physiology of Nitrogen and Calcium Nutrition in Blueberry (*Vaccinium* sp.). *Agronomy* **2021**, *11*, 765. [CrossRef]
5. Zhou, Q.; Ma, C.; Cheng, S.; Wei, B.; Liu, X.; Ji, S. Changes in antioxidative metabolism accompanying pitting development in stored blueberry fruit. *Postharvest Biol. Technol.* **2014**, *88*, 88–95. [CrossRef]
6. Olmedo, P.; Zepeda, B.; Rojas, B.; Silva-Sanzana, C.; Delgado-Rioseco, J.; Fernandez, K.; Balic, I.; Arriagada, C.; Moreno, A.A.; Defilippi, B.G.; et al. Cell Wall Calcium and Hemicellulose Have a Role in the Fruit Firmness during Storage of Blueberry (*Vaccinium* spp.). *Plants* **2021**, *10*, 553. [CrossRef]
7. Chen, Y.; Hung, Y.-C.; Chen, M.; Lin, H. Effects of acidic electrolyzed oxidizing water on retarding cell wall degradation and delaying softening of blueberries during postharvest storage. *LWT* **2017**, *84*, 650–657. [CrossRef]
8. Xu, F.X.; Liu, Y.F.; Dong, S.Z.; Wang, S.H. Effect of 1-methylcyclopropene (1-MCP) on ripening and volatile compounds of blueberry fruit. *J. Food Process. Preserv.* **2020**, *44*, e14840. [CrossRef]
9. Assis, J.S.; Maldonado, R.; Munoz, T.; Escribano, M.I.; Merodio, C. Effect of high carbon dioxide concentration on PAL activity and phenolic contents in ripening cherimoya fruit. *Postharvest Biol. Technol.* **2001**, *23*, 33–39. [CrossRef]
10. Gong, H.J.; Fullerton, C.; Billing, D.; Burdon, J.J.S.H. Retardation of 'Hayward' kiwifruit tissue zone softening during storage by 1-methylcyclopropene. *Sci. Hortic.* **2020**, *259*, 108791. [CrossRef]
11. Chen, Y.H.; Sun, J.Z.; Lin, H.T.; Hung, Y.C.; Zhang, S.; Lin, Y.F.; Lin, T. Paper-based 1-MCP treatment suppresses cell wall metabolism and delays softening of Huanghua pears during storage. *J. Sci. Food Agric.* **2017**, *97*, 2547–2552. [CrossRef]
12. Lin, Y.; Lin, Y.; Lin, H.; Lin, M.; Li, H.; Yuan, F.; Chen, Y.; Xiao, J. Effects of paper containing 1-MCP postharvest treatment on the disassembly of cell wall polysaccharides and softening in Younai plum fruit during storage. *Food Chem.* **2018**, *264*, 1–8. [CrossRef]

13. Win, N.M.; Yoo, J.; Naing, A.H.; Kwon, J.-G.; Kang, I.-K. 1-Methylcyclopropene (1-MCP) treatment delays modification of cell wall pectin and fruit softening in "Hwangok" and "Picnic" apples during cold storage. *Postharvest Biol. Technol.* **2021**, *180*, 111599. [CrossRef]
14. Ortiz, C.M.; Franceschinis, F.; Gergoff Grozeff, G.E.; Chan, H.; Labavitch, J.M.; Crisosto, C.; Vicente, A.R. Pre-treatment with 1-methylcyclopropene alleviates methyl bromide-induced internal breakdown, softening and wall degradation in blueberry. *Postharvest Biol. Technol.* **2018**, *146*, 90–98. [CrossRef]
15. Gergoff Grozeff, G.E.; Alegre, M.L.; Senn, M.E.; Chaves, A.R.; Simontacchi, M.; Bartoli, C.G. Combination of nitric oxide and 1-MCP on postharvest life of the blueberry (*Vaccinium* spp.) fruit. *Postharvest Biol. Technol.* **2017**, *133*, 72–80. [CrossRef]
16. Lisanti, M.T.; Mataffo, A.; Scognamiglio, P.; Teobaldelli, M.; Iovane, M.; Piombino, P.; Rouphael, Y.; Kyriacou, M.C.; Corrado, G.; Basile, B. 1-Methylcyclopropene Improves Postharvest Performances and Sensorial Attributes of Annurca-Type Apples Exposed to the Traditional Reddening in Open-Field Melaio. *Agronomy* **2021**, *11*, 1056. [CrossRef]
17. Chu, W.; Gao, H.; Chen, H.; Fang, X.; Zheng, Y. Effects of cuticular wax on the postharvest quality of blueberry fruit. *Food Chem.* **2018**, *239*, 68–74. [CrossRef]
18. Yang, B.; Fang, X.; Han, Y.; Liu, R.; Chen, H.; Gao, H. Analysis of lignin metabolism in water bamboo shoots during storage. *Postharvest Biol. Technol.* **2022**, *192*, 111989. [CrossRef]
19. Tang, Y.H.; Liu, F.; Xing, H.C.; Mao, K.Q.; Chen, G.; Guo, Q.Q.; Chen, J.R. Correlation Analysis of Lignin Accumulation and Expression of Key Genes Involved in Lignin Biosynthesis of Ramie (*Boehmeria nivea*). *Genes* **2019**, *10*, 389. [CrossRef]
20. Suo, J.T.; Li, H.; Ban, Q.Y.; Han, Y.; Meng, K.; Jin, M.J.; Zhang, Z.K.; Rao, J.P. Characteristics of chilling injury-induced lignification in kiwifruit with different sensitivities to low temperatures. *Postharvest Biol. Technol.* **2018**, *135*, 8–18. [CrossRef]
21. Hamauzu, Y.; Mizuno, Y. Non-extractable Procyanidins and Lignin are Important Factors in the Bile Acid Binding and Radical Scavenging Properties of Cell Wall Material in some Fruits. *Plant Foods Hum. Nutr.* **2011**, *66*, 70–77. [CrossRef] [PubMed]
22. Moerschbacher, B.M.; Noll, U.M.; Flott, B.E.; Reisener, H.J. Lignin biosynthetic-enzymes in stem rust infected, resistant and susceptible near-isogenic wheat lines. *Physiol. Mol. Plant Pathol.* **1988**, *33*, 33–46. [CrossRef]
23. Wang, C.; Hu, D.; Liu, X.B.; She, H.Z.; Ruan, R.W.; Yang, H.; Yi, Z.L.; Wu, D.Q. Effects of uniconazole on the lignin metabolism and lodging resistance of culm in common buckwheat (*Fagopyrum esculentum* M.). *Field Crops Res.* **2015**, *180*, 46–53. [CrossRef]
24. Ji, Y.; Hu, W.; Liao, J.; Xiu, Z.; Jiang, A.; Guan, Y.; Yang, X.; Feng, K. Ethanol vapor delays softening of postharvest blueberry by retarding cell wall degradation during cold storage and shelf life. *Postharvest Biol. Technol.* **2021**, *177*, 111538. [CrossRef]
25. Nirmal, N.P.; Mereddy, R.; Webber, D.; Sultanbawa, Y. Biochemical, antioxidant and sensory evaluation of *Davidsonia pruriens* and *Davidsoina jerseyana* fruit infusion. *Food Chem.* **2020**, *342*, 128349. [CrossRef] [PubMed]
26. Chiabrando, V.; Giacalone, G. Shelf-life extension of highbush blueberry using 1-methylcyclopropene stored under air and controlled atmosphere. *Food Chem.* **2011**, *126*, 1812–1816. [CrossRef]
27. Xiong, Z.F.; Li, H.; Liu, Z.Y.; Li, X.H.; Gui, D.L. Effect of 1-MCP on postharvest quality of French prune during storage at low temperature. *J. Food Process. Preserv.* **2019**, *43*, e14011. [CrossRef]
28. DeEll, J.R.; Lum, G.B.; Mostofi, Y.; Lesage, S.K. Timing of Ethylene Inhibition Affects Internal Browning and Quality of 'Gala' Apples in Long-Term Low Oxygen Storage. *Front. Plant Sci.* **2022**, *13*, 914441. [CrossRef]
29. Zhang, Y.; Wang, K.; Xiao, X.; Cao, S.; Chen, W.; Yang, Z.; Shi, L. Effect of 1-MCP on the regulation processes involved in ascorbate metabolism in kiwifruit. *Postharvest Biol. Technol.* **2021**, *179*, 111563. [CrossRef]
30. Al Shoffe, Y.; Nock, J.F.; Zhang, Y.; Watkins, C.B. Physiological disorder development of 'Honeycrisp' apples after pre- and post-harvest 1-methycyclopropene (1-MCP) treatments. *Postharvest Biol. Technol.* **2021**, *182*, 111703. [CrossRef]
31. MacLean, D.D.; NeSmith, D.S. Rabbiteye Blueberry Postharvest Fruit Quality and Stimulation of Ethylene Production by 1-Methylcyclopropene. *Hortscience* **2011**, *46*, 1278–1281. [CrossRef]
32. Wang, L.; Sang, W.; Xu, R.; Cao, J. Alteration of flesh color and enhancement of bioactive substances via the stimulation of anthocyanin biosynthesis in 'Friar' plum fruit by low temperature and the removal. *Food Chem.* **2020**, *310*, 125862. [CrossRef] [PubMed]
33. Song, L.L.; Gao, H.Y.; Chen, W.X.; Chen, H.J.; Mao, J.L.; Zhou, Y.J.; Duan, X.W.; Joyce, D.C. The role of 1-methylcyclopropene in lignification and expansin gene expression in peeled water bamboo shoot (*Zizania caduciflora* L.). *J. Sci. Food Agric.* **2011**, *91*, 2679–2683. [CrossRef] [PubMed]
34. Cao, S.F.; Zheng, Y.H.; Wang, K.T.; Rui, H.J.; Shang, H.T.; Tang, S.S.J.J.o.P.; Science, H. The effects of 1-methylcyclopropene on chilling and cell wall metabolism in loquat fruit. *J. Agric. Food Chem.* **2010**, *85*, 147–153. [CrossRef]
35. Cai, C.; Xu, C.J.; Li, X.; Ferguson, I.; Chen, K.S. Accumulation of lignin in relation to change in activities of lignification enzymes in loquat fruit flesh after harvest. *Postharvest Biol. Technol.* **2006**, *40*, 163–169. [CrossRef]
36. Pinhero, R.G.; Paliyath, G.; Yada, R.Y.; Murr, D.P. Modulation of phospholipase D and lipoxygenase activities during chilling. Relation to chilling tolerance of maize seedlings. *Plant Physiol. Biochem.* **1998**, *36*, 213–224. [CrossRef]
37. Cheng, S.; Wei, B.; Zhou, Q.; Tan, D.; Ji, S. 1-Methylcyclopropene alleviates chilling injury by regulating energy metabolism and fatty acid content in 'Nanguo' pears. *Postharvest Biol. Technol.* **2015**, *109*, 130–136. [CrossRef]

38. Ali, M.; Raza, M.A.; Li, S.G.; Huan, C.; Zheng, X.L. 1-Methylcyclopropene treatment controls ethanol accumulation associated with regulation of mitochondrial energy metabolism in kiwifruit (*Actinidia deliciosa*) cv. "Bruno" during storage at room temperature. *J. Food Biochem.* **2020**, *44*, e13273. [CrossRef]
39. Zhang, W.T.; Li, Z.; Du, M.L.; Zhang, X.L.; Tian, Y.Q.; Wang, J.E. 1-Methylcyclopropene (1-MCP) retards the senescence of Pteridium aquilinum var. latiusculum by regulating the cellular energy status and membrane lipid metabolism. *Food Sci. Nutr.* **2021**, *9*, 4349–4363. [CrossRef]

agronomy

MDPI

Article

Study on the Regulation Mechanism of 1-MCP Combined with SO_2 Treatment on Postharvest Senescence of Bamboo Shoots (*Chimonobambusa quadrangularis*) in Karst Mountain Area

Jinyang Xu [1], Ning Ji [1,*], Rui Wang [1], Chao Ma [1], Jiqing Lei [1], Ni Zhang [1], Renchan Liu [1] and Yunbing Deng [2]

1 College of Food and Pharmaceutical Engineering, Guiyang University, Guiyang 550003, China; jinyangxu1998@163.com (J.X.)
2 Suiyang Taiping Gaoshan Ecological Traditional Chinese Medicine Planting Co., Ltd., Zunyi 563300, China
* Correspondence: jining552100@163.com

Abstract: Fresh bamboo shoots (*Chimonobambusa quadrangularis*) are subjected to senescence (e.g., lignification and browning) during postharvest storage. This study investigated the effects of 1-MCP and SO_2 treatment on bamboo shoot senescence and its regulation mechanism in order to extend bamboo shoot storage time. 1-MCP and SO_2 treatments significantly inhibited the browning and lignification of fresh bamboo shoots during storage, according to the results. Its lower browning index and lignin content are directly related to its lower lignin content compared to the CK control group. The browning index and lignin content of the 1-MCP + SO_2 treatment during the late storage period were 90.55% and 81.50% of the CK treatment, respectively. The result of the in-depth analysis suggested that 1-MCP and SO_2 treatments reduced nutrient loss and maintained the nutritional value of bamboo shoots by inhibiting respiration and physiological metabolism. The PPO activity was inhibited to inhibit the browning process. Moreover, the scavenging ability of ROS was enhanced, the accumulation of MDA was inhibited, and the senescence of bamboo shoots was delayed after higher contents of total flavonoids and ascorbic acid were maintained and the activities of ascorbic acid peroxidase and superoxide dismutase were stimulated. Furthermore, lignin biosynthesis was hindered, and the lignification of bamboo shoots was delayed after the activities of POD and PAL were inhibited. In brief, 1-MCP + SO_2 treatment is capable of inhibiting the physiological metabolism, browning, and lignification of bamboo shoots, maintaining good quality during storage, and delaying the senescence of bamboo shoots. Clarifying the senescence mechanism of bamboo shoots is of great significance for expanding the bamboo shoot industry and slowing down rocky desertification in karst mountainous areas.

Keywords: lignification; browning; physiological metabolism; senescence mechanism; karst mountainous area

Citation: Xu, J.; Ji, N.; Wang, R.; Ma, C.; Lei, J.; Zhang, N.; Liu, R.; Deng, Y. Study on the Regulation Mechanism of 1-MCP Combined with SO_2 Treatment on Postharvest Senescence of Bamboo Shoots (*Chimonobambusa quadrangularis*) in Karst Mountain Area. *Agronomy* **2023**, *13*, 1122. https://doi.org/10.3390/agronomy13041122

Academic Editor: Fabio Fiorani

Received: 15 March 2023
Revised: 10 April 2023
Accepted: 12 April 2023
Published: 14 April 2023

1. Introduction

Nearly 15 percent of the world's land is covered by karst landforms. To be specific, they are extensively distributed in Southeast Asia, the Mediterranean region of Europe, Central America, southeast North America, and southwest China [1,2]. Guizhou Province is the center of karst landforms in southwest China. The ecological environment of the karst area in Guizhou has progressively degraded under the destruction of human beings over the past few years, thus causing rocky desertification [3,4]. Relevant research has suggested that planting bamboo forests is capable of effectively delaying rocky desertification, taking on critical significance for soil and water conservation, nutrient balance, and so forth, while leading to the production of edible fresh bamboo shoots, such that higher ecological and social benefits can be achieved [5].

Chimonobambusa quadrangularis refers to a vital bamboo shoot species that is primarily produced in the karst mountainous areas of southwest China. The fresh bamboo shoots of

Chimonobambusa quadrangularis are the buds of bamboo. They have gained wide popularity for their rich dietary fiber and unique flavor, and they are considered a type of healthy vegetable [6,7]. However, fresh bamboo shoots are extremely intolerant to postharvest storage, and the texture and appearance will deteriorate in a short time after harvest. Harvest damage and severe physiological metabolism have been confirmed as the direct causes of the rapid senescence of bamboo shoots after harvest. Lignification and browning serve as crucial criteria for measuring the senescence of bamboo shoots after harvest [8,9]. The preservation technology of fresh bamboo shoots was studied to prolong the storage period, such that the fresh bamboo shoots after storage still had ideal edible value and brought higher benefits to growers.

A lignification process increases the firmness of postharvest fruits and vegetables, and this process is directly caused by lignin biosynthesis [10]. Phenylpropanoid metabolism takes on great significance in the postharvest lignification of fruits and vegetables. Phenylalanine ammonia lyase (PAL) and peroxidase (POD) are essential enzymes in phenylpropanoid metabolism [11]. PAL refers to the rate-limiting enzyme for lignin formation, providing precursors for lignin formation [12]. POD is an essential enzyme in catalyzing lignin biosynthesis in the cell wall, and it can polymerize lignin monomers into lignin polymers [13]. Browning is a vital feature of postharvest senescence in fruits and vegetables. The main reason is that the scavenging ability of reactive oxygen species (ROS) is reduced, and the accumulation of malondialdehyde (MDA), a product of membrane lipid peroxidation, reduces the integrity of cell membranes, such that the phenolic substances in fruits and vegetables come into contact with polyphenol oxidase (PPO) to produce browning [14]. Existing research has suggested that a wide variety of antioxidant enzymes in postharvest fruits and vegetables (e.g., ascorbate peroxidase (APX) and superoxide dismutase (SOD)) can scavenge ROS in fruits and vegetables and slow down senescence during storage [15]. Accordingly, activating the active oxygen scavenging ability and inhibiting lignin biosynthesis in bamboo shoots can delay the rapid senescence of bamboo shoots during storage.

It has been shown that 1-MCP competes with ethylene for binding sites on ethylene receptors in plant tissues; this inhibits ethylene's physiological role and slows postharvest senescence of fruits and vegetables. Extensive studies have been conducted on its application to a wide variety of fruits and vegetables [16]. Bamboo shoots are not climacteric fruits and vegetables, but relevant research has suggested that ethylene may induce senescence metabolism in bamboo shoots, and using 1-methylcyclopropene (1-MCP) to treat bamboo shoots can improve their quality after harvest [17]. However, due to the complexity of antioxidants (e.g., phenols), during storage, 1-MCP treatment has rarely been reported to improve the antioxidant capacity of bamboo shoots [18,19]. In recent studies, 1-MCP treatment has been shown to delay total phenol and anthocyanin accumulation in peach fruits [20]. As a result of 1-MCP treatment, phenolic compounds in strawberry and loquat were significantly increased, suggesting that the response of 1-MCP to the antioxidant metabolism of different fruits and vegetables was different [21,22]. Thus, the effect of 1-MCP on improving the antioxidant capacity and activating the active oxygen scavenging ability of fresh bamboo shoots after harvest should be investigated in depth.

Existing research has suggested that harvest damage is a vital factor causing postharvest decay of fruits and vegetables. Pathogenic microorganisms infect fruits and vegetables through traumatic sites, induce stress resistance, aggravate growth and metabolism, and accelerate senescence [23]. Postharvest damage to bamboo shoots is inevitable due to its unique harvesting method. A fungicide, sulfur dioxide (SO_2), can effectively inhibit microorganism growth and maintain fruit quality [24]. Two methods of SO_2 treatment for long-term fruit and vegetable storage are described below: with sulfur dioxide slow-release paper and by repeated application of the gas on the storage room's fumigation [25,26]. A fumigated storage room can affect fruit and vegetable flavors, damage fruits and vegetables, and leave excessive sulfite residues [27]. Sodium pyrosulfite ($Na_2S_2O_5$) contained in SO_2 slow-release paper is wrapped in a special paper, and the appropriate dose of SO_2 is released slowly to inhibit the growth of pathogenic bacteria. High doses of SO_2 can cause bleaching, damage, and taste residue in fruits and vegetables. Accordingly, EU countries have proposed a limit on SO_2 dosage (26.2 mg/m^3) to maintain the optimal quality [28].

Therefore, 1-MCP or SO_2 treatment alone can maintain fruit and vegetable quality more effectively during storage, though the mechanism might be different. An effective combination of the two may provide a novel solution for storing fruits and vegetables. Furthermore, little research has been conducted on how 1-MCP and SO_2 combined treatments affect bamboo shoot postharvest quality. In this study, a combination of 1-MCP, SO_2, and 1-MCP plus SO_2 was used to treat bamboo shoots after harvest. Fresh bamboo shoots were examined for changes in physiology and enzyme activity during storage, as well as gene expression, to explore the regulatory mechanism of 1-MCP combined with SO_2 treatment in maintaining the postharvest quality and delaying the senescence of bamboo shoots. Expanding the planting area of local bamboo shoots provides a new solution for delaying rocky desertification in mountainous areas of southwest China.

2. Materials and Methods

2.1. Plant Material

The fresh bamboo shoots (*Chimonobambusa quadrangularis*) were collected from Taibai Town, Suiyang County, Zunyi City, Guizhou Province (28°40′50.40″ N, 107°09′26.00″ E). Moreover, the bamboo shoots with no pests and diseases, moderate size, and strong growth were selected. The amounts of 1-MCP (SmartFreshSM Inc., Wilmington, DE, USA) and UVASYS (Grapete Inc., Cape Town, South Africa) applied and the PE spontaneously modified atmosphere bag were all determined in accordance with the previous experimental results. The specific experimental results are presented in supplementary materials.

The fresh bamboo shoots were randomly assigned to two groups (60 kg in each group), which were placed in a closed space with a 1 m square plastic tent. One group was fumigated with 1.0 $\mu\cdot L^{-1}$ 1-MCP for 2 h, and the other group was not treated. The respective group of bamboo shoots was weighed and loaded into a 0.02 mm PE modified atmosphere bag (24 bags in the respective group), and then pre-cooled in a fresh-keeping warehouse at a temperature of (1.0 $\pm$ 0.3) °C and a humidity of (90 $\pm$ 5)% for 24 h. After precooling, the 1-MCP group fell into two parts: one part was put into 1/4 UVASYS and then recorded as 1-MCP + UVASYS treatment, and the other part was not treated and then recorded as 1-MCP treatment. Moreover, the non-treatment group was assigned to two parts: some of them were placed into 1/4 UVASYS as UVASYS treatment, and some of them were not treated as CK treatment (Uvasys®, Cape Town, South Africa, with a 37.55% share of sodium metabisulfite, over 98% pure, and a 356 $\times$ 260 mm measurement. The release was divided into two parts, 40% was first released in the fast phase of 24–48 h, and then 60% was released slowly. Leaflet was slowly released to maintain a concentration of 10.48 mg/m^3 SO_2 release [29]). Subsequently, the modified atmosphere bag was bagged and then stored in the fresh-keeping warehouse for 60 days, and the physiological and biochemical indexes were measured every 15 days. In addition, the frozen samples were stored at −80 °C until further biochemical analysis could be carried out.

2.2. Respiratory Rate Measurement

The respiratory rate was examined using the static method [30]. Randomly select 2.0 kg of fresh bamboo shoots, weigh them, and place them in a 30 L sealed plastic container. Let them stand for 2 h at 25 °C. Use a residual oxygen meter ((Checkpoint 3 Premium, Mocon Inc., Minneapolis, MN, USA) to measure the CO_2 concentration in the container and measure the respiration rate as the amount of CO_2 increased per kilogram of fresh bamboo shoots per hour.

2.3. Weight Loss Rate Measurement

Weight loss was used to determine the weight loss rate [31], and the sample was weighed on day 0 of the respective treatment and at the end of each storage interval.

2.4. Color Feature

Bamboo shoot color change was measured by total color difference (ΔE) and browning index (BI). In the test, a color difference meter (Konica Minolta Inc., Tokyo, Japan) was used to measure the color difference using the international standard CIE-L* a* b* color system, and the parameter ΔE was adopted. BI refers to a crucial color change parameter that represents the number of enzymatic and non-enzymatic browning reactions after treatment and during storage [32].

$$BI = \frac{[100 \times (X - 0.31)]}{0.172} \qquad X = \frac{a^* + 1.75 \times L^*}{5.645 \times L^* + a^* - 3.012 \times b^*} \tag{1}$$

$$\Delta E = \sqrt{(L^* - L0^*)^2 + (a^* - a0^*)^2 + (b^* - b0^*)^2} \tag{2}$$

2.5. Firmness

The firmness test reference LI method was optimized [33]. The analysis was conducted every 15 days using a texture analyzer ((SMS Inc., London, UK). The P-2N probe was selected to test the middle of the shelled bamboo shoots. 15 bamboo shoots were randomly selected for the respective treatment, the test depth was 8 mm, the speed before the test was 5 $mm \cdot s^{-1}$, the speed during the test was 5 $mm \cdot s^{-1}$, the speed after the test was 5 $mm \cdot s^{-1}$, and the result is expressed in grams (g).

2.6. Lignin Content

Using acetyl bromide, the amount of lignin in the sample was determined [34]. The 100.0 mg sample was dried to a constant weight. A total of 1000 μL of acetyl bromide reagent and 40 μL of perchloric acid were added (Macklin Inc., Shanghai, China). In a water bath at 80 °C for 40 min, 500 μL of a 2 $mol \cdot L^{-1}$ sodium hydroxide solution and 500 μL of glacial acetic acid solution were added to end the reaction(Macklin Inc., Shanghai, China). They were thoroughly mixed and centrifuged for 5 min at 10,000 r/min^{-1} (DragonLabS Inc., Beijing, China). To measure the absorbance of the solution, glacial acetic acid was used as a blank.

2.7. MDA Content

MDA was determined using the thiobarbituric acid method and improved [35]. In order to obtain the sample extract, a 3.0 g (FW) sample was accurately weighed and ground into a homogenate in 10 mL (Vt) of 10.0% trichloroacetic acid (TCA) (Macklin Inc., Shanghai, China). A reaction solution (Va) was prepared by adding 5 mL of a 0.6% TBA solution to 5 mL of the extraction solution (Vs) (Macklin Inc., Shanghai, China). A water bath at 95 °C for 15 min was used to conduct the reaction after the solution was shaken and mixed. Next, the reaction solution was placed in an ice bath. The solution cooled quickly, and it was centrifuged at 5000× *g* for 10 min. 5 mL of 10% TCA and 5 mL of 0.6% TBA solution were mixed in a centrifuge tube, cooled, and centrifuged together with the sample as a blank control. Using the supernatant, absorbances at 532 nm, 600 nm, and 450 nm were measured, and the MDA content was calculated using Equation (3):

$$\text{MDA content } (\mu mol \cdot kg^{-1}) = \frac{[6.452 \times (A532 - A600) - 0.559 \times A450] \times Vt \times Va}{Vs \times FW} \tag{3}$$

2.8. Soluble Protein Content and Free Amino Acid Content

The content of soluble protein in bamboo shoots was determined using the Coomassie Brilliant Blue G-250 method [36]. The sample was homogenized with 5 mL of distilled water containing 1.0 g of the sample. Next, it was centrifuged at 4 °C for 20 min at 10,000 $r \cdot min^{-1}$. A total of 1.0 mL of the sample was taken to extract the supernatant. After mixing the sample with 5.0 mL of Coomassie brilliant blue G-250 solution (Macklin Inc., Shanghai,

China), the mixture was placed for 2 min. Afterward, the content of soluble proteins was determined through spectrophotometry at a wavelength of 595 nm.

Using the ninhydrin color method with minor modifications, free amino acids were determined [37]. Approximately 3.0 g of sample were weighed, ground into a homogenate under 10 mL of a 10% acetic acid solution (Macklin Inc., Shanghai, China), diluted with 100 mL of distilled water , shaken, and then filtered. The filtrate was mixed with ninhydrin and ascorbic acid reagents and heated in a boiling water bath for 15 min(Macklin Inc., Shanghai, China). A 95% ethanol solution was added after cooling. Then, the mixed solution was shaken up and diluted with 60% ethanol to 20 mL (Macklin Inc., Shanghai, China). The solution without any filtrate was taken as a blank control, and 570 nm was used to measure absorbance.

2.9. Determination of Ascorbic Acid Content and Total Flavone Content

2,6-dichloro-indophenol was titrated and then improved [38]. Five grams of sample were accurately weighed, and a small amount of a 20 g·L^{-1} oxalic acid solution was added (Macklin Inc., Shanghai, China). A homogenate of the sample was ground in an ice bath and filtered, and the filtrate exhibited a constant volume of 100 mL. A calibrated solution of 2,6-dichloro-indophenol was titrated with 20 mL of the filtrate in a 100 mL conical flask (Macklin Inc., Shanghai, China). Next, the sample showed a slight red color without fading for 15 s. A total of 20 g·L^{-1} of oxalic acid solution was employed as the blank control.

The content of total flavonoids varied slightly in accordance with the method of DU et al. [39]. A 3.0 g sample was accurately weighed and ground into a homogenate in 70% ethanol. The solution was diluted to 25 mL. After extraction for 1 h in the dark, the extract was transferred to a centrifuge tube and centrifuged at 10 °C for 10 min at 10,000 r·min^{-1}. 2 mL of 0.1 mol·L^{-1} $AlCl_3$ solution and 3 mL of 1 mol·L^{-1} CH_3COOK solution were added to 1 mL of 1 mL of extract solution (Macklin Inc., Shanghai, China), diluted to a volume with 75% ethanol solution, shaken up, and then through spectrophotometer colorimetry (Macklin Inc., Shanghai, China), the samples were placed at ambient temperature for 30 min and then measured at a 420 nm wavelength.

2.10. Total Phenol Content

In accordance with the method of BAE et al. [40]. 3.0 g (m) of sample was accurately weighed, and 10 mL of 70% absolute ethanol was quickly added. In an ice bath, the sample was ground into a homogenate (Macklin Inc., Shanghai, China). 70% absolute ethanol was adopted to achieve a constant volume of 25 mL. A centrifuge tube containing 50 mL of the mixture was placed in the dark for two hours after standing at 25 °C. In a dark place for 30 s, 2.0 mL of supernatant and 3.0 mL of Folin phenol reagent were added to the centrifuged material and centrifuged for 10 min at 10 °C. After adding 6.0 mL of 12% Na_2O_3 solution, 25 mL of distilled water was added (Macklin Inc., Shanghai, China). Moreover, the absorbance (A) at 760 nm was measured after the solution stood at 25 °C for 1 h. The absorbance of the gradient concentration of gallic acid was obtained under the same conditions, and the standard curve was prepared. The results are expressed in (mg·g^{-1}), and the standard curve equation and total phenol calculation equation are expressed as follows:

$$Y = 0.365x + 0.0165 \qquad R^2 = 0.9972 \tag{4}$$

$$M = \frac{(A - 0.1688) \times 25 \times 25}{0.2112 \times m \times 1000} \tag{5}$$

2.11. Enzyme Activity Measurement

The PPO activity was determined using the method outlined by Yang et al. [41]. A 0.05 mmol·L^{-1} sodium phosphate buffer solution was added to 10 mg of sample to form a homogenate (Macklin Inc., Shanghai, China), which was then dissolved in a 0.1 mmol sodium phosphate buffer solution for 10 min and centrifuged at 4 °C for 30 min. After

10 min in a 37 °C water bath, three milliliters of supernatant, 3.9 milliliters of sodium phosphate buffer, and one milliliter of catechol solution were added, and two milliliters of 20% TCA solution was added after that (Macklin Inc., Shanghai, China). The absorbance value was measured at 420 nm, and the enzyme solution was replaced with 0.05 mmol·L^{-1} sodium phosphate buffer as a blank (Macklin Inc., Shanghai, China).

APX activity was determined and modified using the method of Milena et al. [42]. A 3.0 g of sample was ground into a homogenate with 10 mL of extraction buffer (100 mmol·L^{-1} potassium phosphate buffer, 0.1 mmol·L^{-1} EDTA, 1 mmol·L^{-1} ascorbic acid, 20% PVPP) in an ice bath, and centrifuged at 4 °C for 30 min at 10,000 r·min^{-1} (Macklin Inc., Shanghai, China). A total of 0.6 mL of supernatant and 5.2 mL of reaction buffer (50 mmol·L^{-1} potassium phosphate buffer, 0.1 mmol·L^{-1} EDTA, and 1 mmol·L^{-1} ascorbic acid) (Macklin Inc., Shanghai, China) were taken and fully mixed. A solution of 0.3 mL of 20 mmol·L^{-1} H_2O_2 was added (Macklin Inc., Shanghai, China), and the oxidation rate of ascorbic acid at 290 nm was measured to determine the enzyme activity. The activity unit was U·g^{-1}·min^{-1}.

POD activity was determined using LIU et al.'s method with minor modification [43]. 3.00 g of sample were accurately weighed, ground into a homogenate, and dissolved in 10 mL of acetic acid sodium acetate extraction buffer (each liter contained 40 g PVPP, 1% Triton X-100, and 1 mmol PEG) (Macklin Inc., Shanghai, China). It was centrifuged for 30 min at 4 °C at 10,000 r·min^{-1}. 1 mL of supernatant, 6 mL of 25 mmol·L^{-1} guaiacol solution, and 400 μL 0.5 mol·L^{-1} H_2O_2 were taken, and the absorbance value at 470 nm was quickly measured after mixing (Macklin Inc., Shanghai, China). The active unit U·g^{-1}·min^{-1} was expressed.

PAL activity was slightly modified using the method of Chen et al. [44]. A sample of 3.00 g in 10 mL of 100 mmol·L^{-1} boric acid-borax extraction buffer (each liter contains 40 g PVPP, 35 μL β-Mercaptoethanol, and 2 mmol EDTA) (Macklin Inc., Shanghai, China) was accurately weighed, and the homogenate was ground in an ice bath and centrifuged at 4 °C for 30 min at 10,000 r·min^{-1}. 1 ml of supernatant, 6 ml of 50 mmol·L^{-1} boric acid borax buffer solution, and 1.0 mL of 20 mmol·L^{-1} L-phenylalanine were taken (Macklin Inc., Shanghai, China). After mixing, the absorbance value at 290 nm was quickly measured. The absorbance value at 290 nm was measured after 1 h of reaction in a water bath at 37 °C. The activity is expressed in U·g^{-1}·min^{-1}.

The SOD activity was determined using the method of QI et al. [45]. A sample of 3.00 g was accurately weighed, homogenized in 10 mL of distilled water in an ice bath, then filtered at 4 °C under low temperatures and kept away from light. 2 mL of filtrate and 2.35 mL of 0.1 mol·L^{-1} Tris HCI solution (containing 2.0 mmol·L^{-1} EDTA) (Macklin Inc., Shanghai, China) were taken, and 0.15 mL of 4.5 mol·L^{-1} pyrogallol was quickly mixed. Afterwards, distilled water was used as the blank control in order to measure the absorbance value at 325 nm.

2.12. RT-PCR

The treated bamboo shoots were operated using the method of WU et al. [46]. The *Phyllostachys edulis* genomic database was searched and specific gene primers were designed within the conserved domains (Table 1). The specificity of the melting peak and dissociation curve was analyzed (cFX fluorescence quantitative PCR instrument, Bio-rad company, Inc., CA, USA); (sW-CJ-1FD ultra-clean workbench, Inc., China Sujing Antai Company, Suzhou, China); (nanoDrop2000 ultramicro spectrophotometer, Thermo, Inc., Guangzhou, China). Three biological replicates were used for all qRT-PCR analyses. Using $2^{-\Delta\Delta Ct}$, Ct was used to calculate relative expression levels, and Table 1 provides a list of relevant genes.

2.13. Statistical Analysis

The experiments were conducted in a completely randomized design. The results indicated the mean ± standard deviation (SD) of at least three replicates. IBM SPSS version 21 (IBM, Armonk, New York, NY, USA) and *t*-tests were performed for a one-way ANOVA for statistical analysis. $p < 0.05$ indicated a difference that achieved statistical significance.

3. Results

3.1. Changes of Respiration Rate and Weight Loss Rate

As depicted in Figure 1A, there was an increase in bamboo shoot respiration rate after treatment with CK. In general, CK treatment resulted in a greater respiration rate than other treatment methods; as a result, the respiration rate of the CK treatment significantly exceeded that of the 1-MCP + UVASYS treatment throughout the storage period. The respiration rate of 1-MCP + UVASYS treatment was 242.78 mg $CO_2 \cdot kg^{-1} \cdot h^{-1}$ at 60 d, significantly lower than that of CK, 1-MCP, and UVASYS treatments.

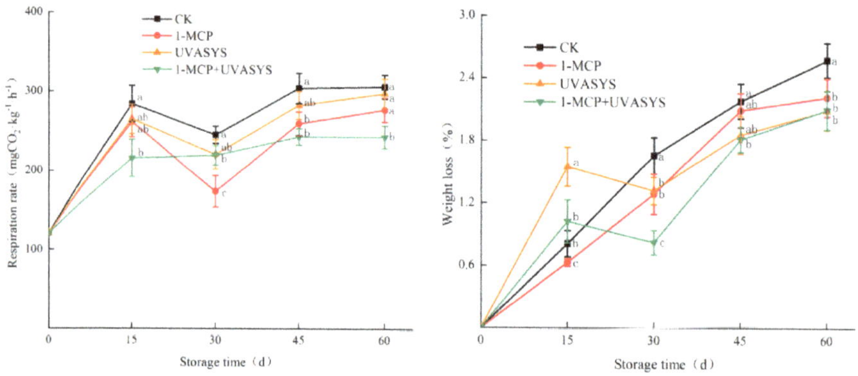

(A) Respiration **(B)** Weight loss

Figure 1. Effects of 1-MCP and UVASYS on respiration rate (**A**) and weight loss rate (**B**) of bamboo shoots during storage. Each data point represents an average ± standard error (n = 3). Different letters indicate significant differences among treatments for each sampling time at $p < 0.05$.

The weight loss rate of bamboo shoots increases with the increase in storage time, as shown in Figure 1B, weight loss rates from CK treatment are higher than those from other treatments in the middle and late stages of storage. 1-MCP + UVASYS treatment was at the lowest level at 30 d, significantly lower than other treatments, and significantly lower than CK treatment from 30 d to 60 d, whereas no significant difference was identified between 1-MCP and UVASYS treatments during the later storage period.

Table 1. The sequences of specific primers used for real-time PCR analysis and conditions.

Species	Gene	Primer Sequences	Size (bp)	Annealing (°C)
Phyllostachys edulis	Actin	F:TGCCCTTGATTATGAGCAGG R:AACCTTTCTGCTCCGATGGT	108	60
Phyllostachys edulis	SOD	F:CTTTCCACTCGCTCCTCCTC R:TGATACGGGCGTTCACTGTT	107	60
Phyllostachys edulis	PPO	F:GATGATTGCCAGTGCCAAGA R:TCGGTGAAGTCGGTGTTGCT	270	60
Phyllostachys edulis	APX	F:CACCAACCGATGAGAAGAA R:GAGTAATTGGCAGCAACGA	103	60
Phyllostachys edulis	PAL	F:GAACAGCACAACCAAGATG R:TCTTTCTAGCCACCGTCGTC	190	60
Phyllostachys edulis	POD	F:CTTCGTCTTTCTCCTCGCATT R:TCTCAAGGTTTGGGCAGATG	95	60

3.2. Changes of Free Amino Acid and Soluble Protein Contents

Bamboo shoots first decreased and then increased in free amino acid content, as shown in Figure 2A. At 45 days, CK treatment had a higher content of free amino acids than other treatments, and it was at the highest level from 45 d to 60 d. At a later stage of storage,

1-MCP, uvasys, and 1-MCP + uvasys treatments slowed down the rise of free amino acids. At 45 days, the 1-MCP + uvasys treatment had a significantly lower level of free amino acids than other treatments.

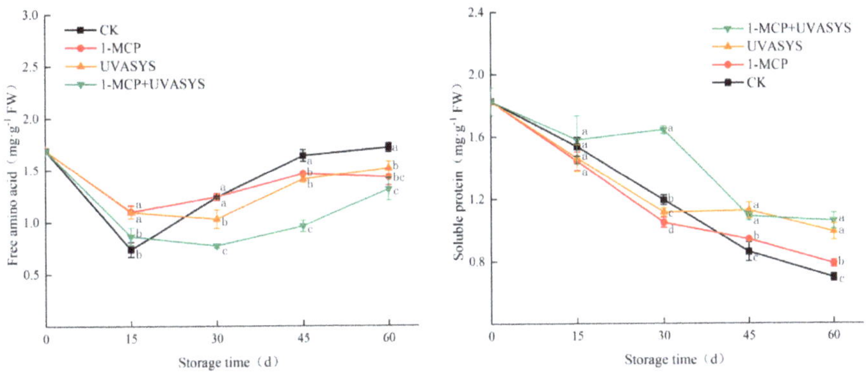

Figure 2. Effect of 1-MCP and UVASYS treatment on free amino acid (**A**) and soluble protein (**B**) content of bamboo shoots during storage. Each data point represents an average ± standard error (n = 3). Different letters indicate significant differences among treatments for each sampling time at $p < 0.05$.

According to Figure 2B, the amount of soluble protein in bamboo shoots decreased continuously as storage time increased. The CK treatment significantly reduced soluble protein content from 30 to 45 days, and the CK treatment was at its lowest level from 45 to 60 days and was significantly lower than the other treatments at 60 d. In general, 1-MCP + UVASYS treatment slowed down the decrease of soluble protein content, and the soluble protein content was 1.64 mg·g^{-1} at 30 d, which notably exceeded other treatments.

3.3. Changes of the Total Color Difference (ΔE) and Browning Index (BI)

Based on Figure 3B, the ΔE of bamboo shoots showed an upward trend with storage time, indicating that the appearance and color of square bamboo shoots changed all the time during storage. Among them, fruits treated with CK rose most rapidly and reached their highest level after 30 days. Despite an increase in ΔE, the ΔE of 1-MCP, UVASYS, and 1-MCP + UVASYS treatments was significantly lower than that of CK treatment after 45 days.

As the storage time increases, the browning degree of bamboo shoots increases (Figure 3A,C). Compared with other treatments, browning indexes were highest for CK treatment at 45 d and 60 d, and CK treatment differed significantly from other treatments. On the 60th day, the browning indices of 1-MCP, UVASYS, and 1-MC + UVASYS treatments were only 94.24%, 86.32%, and 90.56% of CK treatments, respectively.

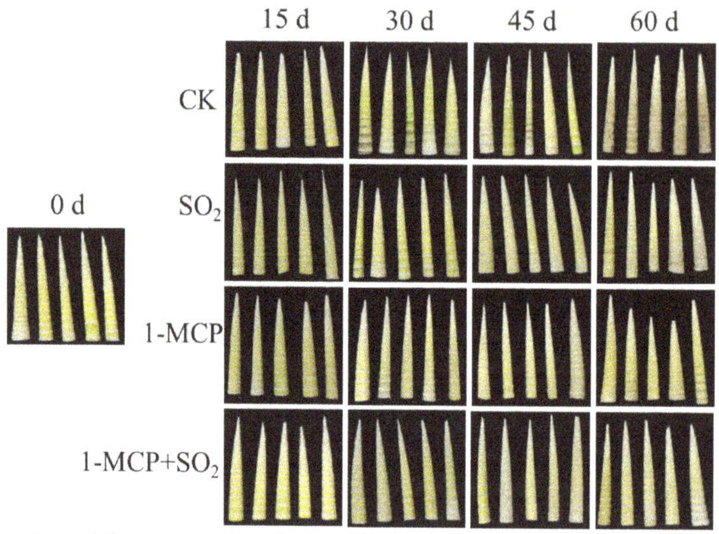

(**A**) Comparison of changers bamboo shoots with different treatments and different storage periods

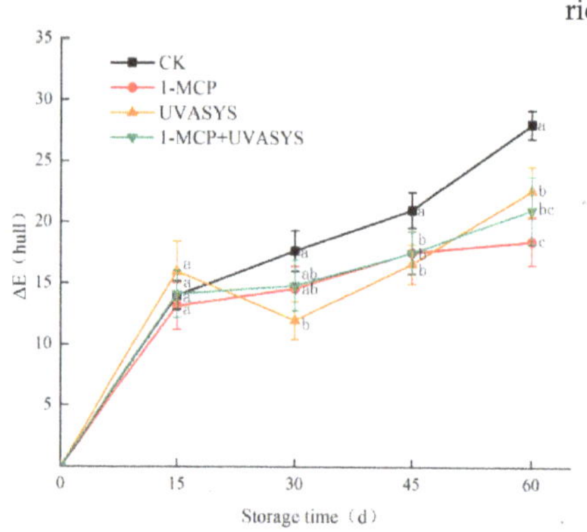

(**B**) ΔE

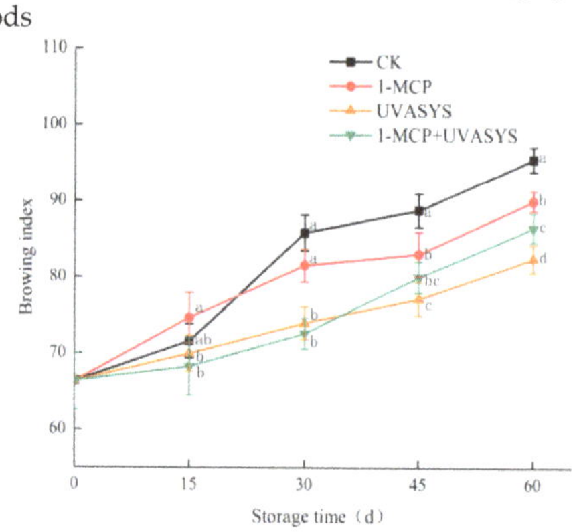

(**C**) BI

Figure 3. Effects of 1-MCP and UVASYS treatment on bamboo shoots during storage: A comparison of changes in shoots with different treatments and storage periods (**A**), ΔE (**B**), and browning Index (**C**). Each data point represents an average ± standard error ($n = 10$). Different letters indicate significant differences among treatments for each sampling time at $p < 0.05$.

3.4. Changes of the Total Phenolic Content and Polyphenol Oxidase (PPO) Activity

As depicted in Figure 4A, CK treatment increased total phenolic content for the first 45 days but then decreased. Overall, the total phenolic content of CK treatment was high throughout the storage period, and the total phenolic content of CK treatment reached its maximum from 45 d to 60 d, and a significant difference was identified with 1-MCP + UVASYS treatment. Moreover, CK treatment had a significantly higher total phenolic content than other treatments at 15 d, suggesting that 1-MCP and UVASYS treatment may delay this increase.

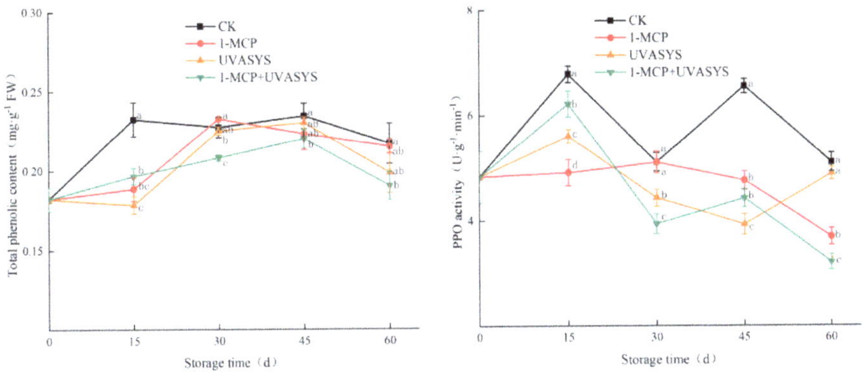

(A) Total phenolic content

(B) PPO activity

Figure 4. Effects of 1-MCP and UVASYS on total phenol (**A**) and PPO (**B**) of bamboo shoots during storage. Each data point represents an average ± standard error (n = 3). Different letters indicate significant differences among treatments for each sampling time at $p < 0.05$.

As depicted in Figure 4B, the overall PPO activity tended to decrease. The CK treatment, however, had a higher level of PPO activity during storage than other treatments, and there were significant differences with other treatments at 15 d and 45 d. In contrast to CK treatment, 1-MCP, UVASYS, and 1-MCP + UVASYS treatments inhibited PPO expression. In comparison with other treatments, the 1-MCP + UVASYS treatment had significantly lower PPO activity at 60 days.

3.5. Changes of the Ascorbic Acid (ASA) and Total Flavonoids Content

Compared to other treatments, the CK treatment had the lowest ASA content during storage, as shown in Figure 5A. At 15 days and 45 days, a significant difference was observed between the CK treatment and the other treatments. The above result indicated that 1-MCP and UVASYS treatments maintained the ASA content to a certain extent, and the ASA content of the 1-MCP + UVASYS treatment notably exceeded that of other treatments at the later stage of storage.

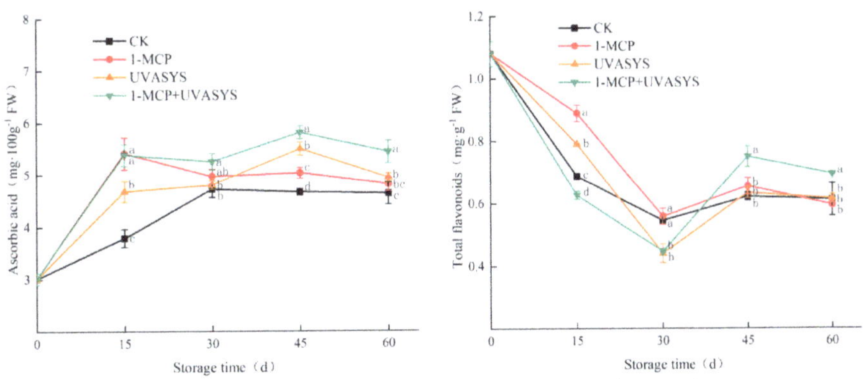

(A) ascorbic acid

(B) total flavonoids content

Figure 5. Effects of 1-MCP and UVASYS treatments on the contents of ascorbic acid (**A**) and total flavonoids (**B**) in bamboo shoots during storage. Each data point represents an average ± standard error (n = 3). Different letters indicate significant differences among treatments for each sampling time at $p < 0.05$.

In general, total flavonoids tended to decrease during storage (Figure 5B).

While the 1-MCP + UVASYS treatment maintained total flavonoid content more effectively during storage than CK, 1-MCP, and UVASYS treatments, between 45 and 60 days, the total flavonoid content significantly exceeded that of the CK, 1-MCP, and UVASYS treatments. After 45 days, 1-MCP + UVASYS treatment had 1.21 times, 1.15 times, and 1.18 times the flavonoid content of CK, 1-MCP, and UVASYS, respectively.

3.6. Changes of the ROS Metabolism

MDA continued to accumulate throughout the storage period, and the content continued to rise (Figure 6A). Compared with other treatments, the increase rate of MDA content in the CK treatment significantly exceeded that of other treatments. In the treatment, the 1-MCP, UVASYS, and 1-MCP + UVASYS treatments inhibited MDA accumulation to some extent, but the 1-MCP + UVASYS treatment had the greatest impact, and the MDA content of CK, 1-MCP, and UVASYS at 45 d only reached 67.35%, 84.96%, and 84.59%, respectively. Throughout the storage period, SOD activity increased and then decreased (Figure 6B). At the early stages of storage, CK treatment had significantly lower SOD activity than other treatments. As a result of 1-MCP, UVASYS, and 1-MCP + UVASYS treatments, SOD activity was stimulated and antioxidant capacity was enhanced during storage. SOD activity of the 1-MCP + UVASYS treatment was noticeably higher at 60 days than that of other treatments, suggesting that the 1-MCP + UVASYS treatment more effectively maintained SOD activity. A decrease in APX activity was observed during storage (Figure 6C), and CK treatment had the lowest APX activity during storage and was significantly lower than other treatments at 60 days. Accordingly, APX activity was better maintained in 1-MCP, UVASYS, and 1-MCP + UVASYS treatments during storage than in the other three treatments.

3.7. Changes of Lignification Related Indexes

During storage, the firmness and lignin content showed a rising trend (Figure 7A,B). Compared with the other treatments, while 1-MCP + UVASYS was at the lowest level during the entire storage period overall, the CK treatment significantly exceeded the other treatments in the late storage period. While being stored, PAL activity first increased and then decreased (Figure 7C). Among them, the PAL activity of the 1-MCP + UVASYS treatment was at the lowest level in the whole storage period, significantly different from other treatments at 45 days, and its PAL activity was only 44.35% and 44.10% of that of the CK, 1-MCP, and UVASYS treatments, respectively. 60.97%. POD activity tended to fluctuate throughout the storage period (Figure 7D). At the early storage stage, POD activity was not significantly different between the respective treatments. POD activity of the CK treatment has significantly exceeded that of the other three treatments since 30 days. At 45 days, 1-MCP, UVASYS, and 1-MCP + UVASYS treatments showed 1.32 times, 1.50 times, and 1.57 times activity, respectively. There were no significant differences among the other three treatments during storage.

3.8. 1-MCP and UVASYS Affect the Relative Expression of Enzyme Genes

The change in SOD gene expression coincided with the change in SOD activity over storage, showing an initial increase and then a decrease (Figures 6B and 8A). SOD gene expression could be induced by 1-MCP and UVASYS treatment compared to CK treatment during late storage, and 1-MCP + UVASYS treatment exceeded CK treatment significantly. The results showed that 1-MCP and UVASYS treatments could increase SOD activity by inducing the expression of the SOD gene. The amount of PPO gene expression tended to increase during storage (Figure 8B), in which CK treatment significantly exceeded other treatments in the late storage period, suggesting that 1-MCP and UVASYS treatment could inhibit PPO gene expression, in which 1-MCP + UVASYS treatment was only 41.33% of CK at 60 days. This is consistent with the significantly lower PPO activity observed in the 1-MCP + UVASYS treatment compared to the CK treatment during the later stage of storage (Figure 4B). The expression of the APX gene is shown in Figure 8C. During storage, the

1-MCP and UVASYS treatments both significantly increased the expression of the APX gene compared to the CK treatment, and late storage was significantly different between the two treatments. This result is consistent with the fact that 1-MCP + UVASYS treatment can maintain APX activity well during the later stage of storage and is significantly higher than that of the CK treatment (Figures 6C and 7C). The change trend of PAL gene expression during storage is consistent with that of PAL activity (Figures 7C and 8D). Specifically, 1-MCP + UVASYS treatment was significantly lower than CK treatment in the late storage period. The POD gene expression amount during storage is shown in Figure 8E. In the late storage period, 1-MCP and UVASYS significantly reduced the relative expression amount of POD compared to the CK control group. This result was also consistent with the inhibition of PAL and POD activities observed in bamboo shoots treated with 1-MCP + UVASYS during the later stage of storage.

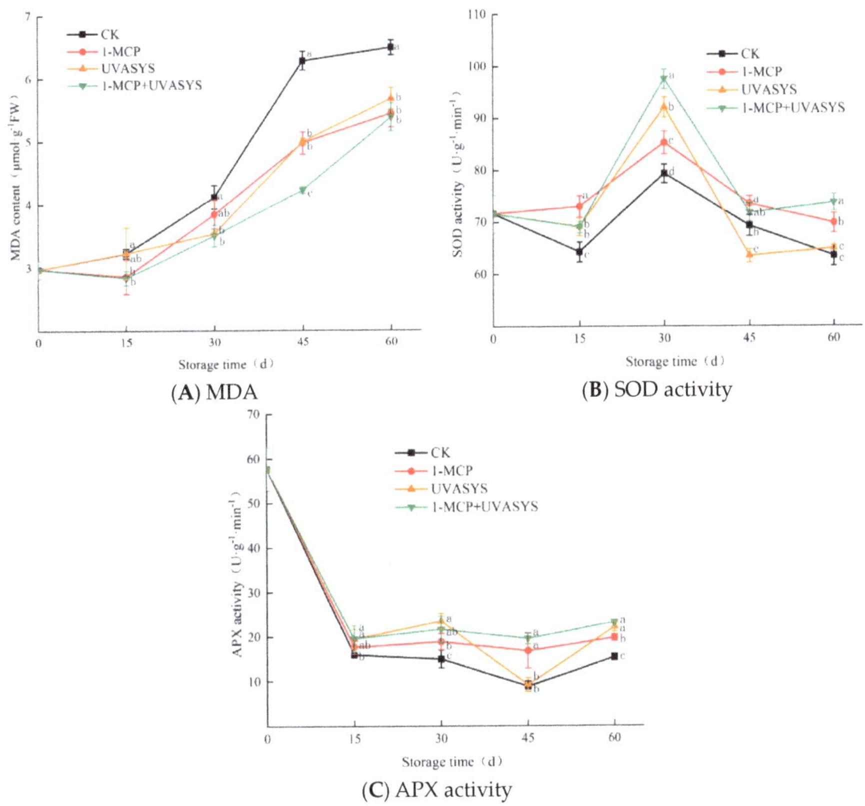

Figure 6. Effects of 1-MCP and UVASYS on MDA (**A**), SOD (**B**) and APX (**C**) of bamboo shoots during storage. Each data point represents an average ± standard error ($n = 3$). Different letters indicate significant differences among treatments for each sampling time at $p < 0.05$.

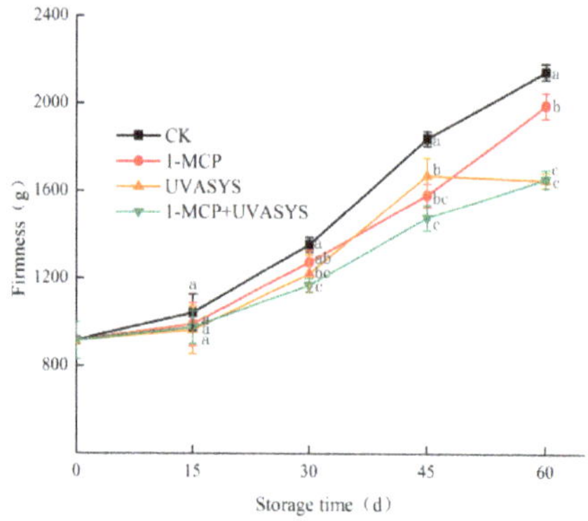

(**A**) Firmness

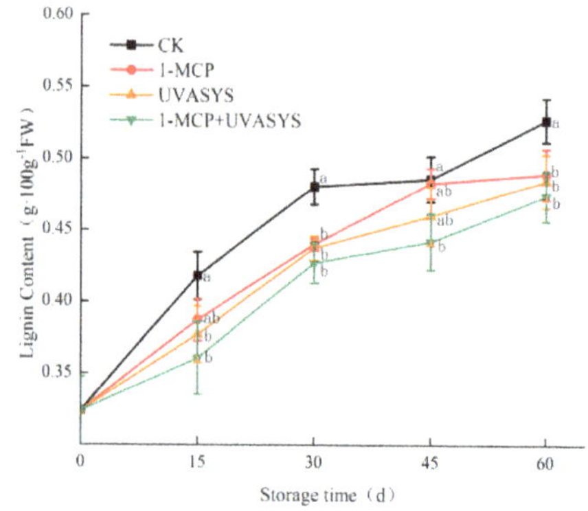

(**B**) Lignin Content

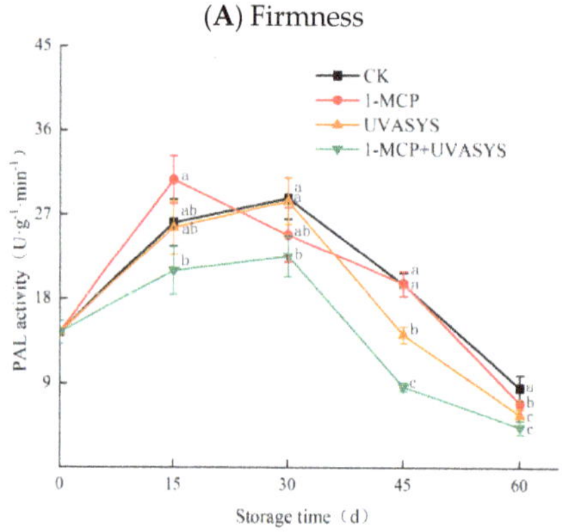

(**C**) PAL activity

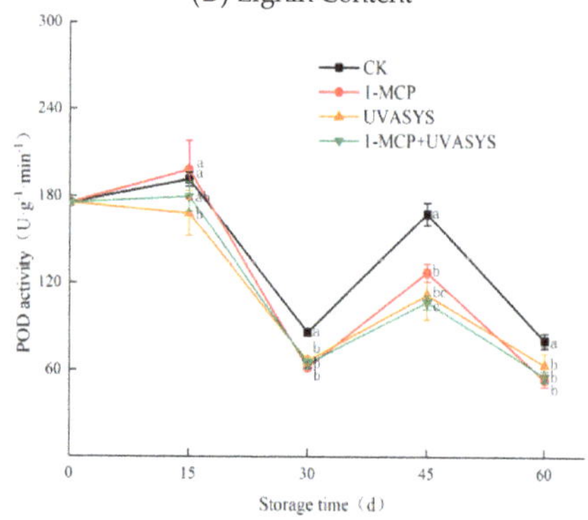

(**D**) POD activity

Figure 7. Effects of 1-MCP and UVASYS on Firmness (**A**), Lignin Content (**B**), PAL activity (**C**) and POD activity (**D**) of bamboo shoots during storage. Each data point represents an average ± standard error (n = 3). Different letters indicate significant differences among treatments for each sampling time at $p < 0.05$.

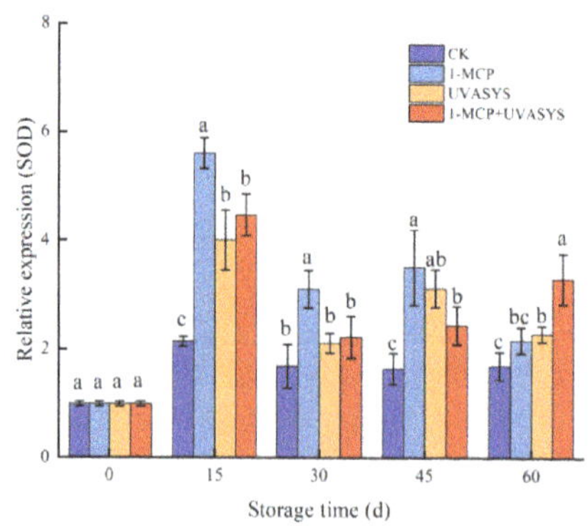

(**A**) SOD Expression

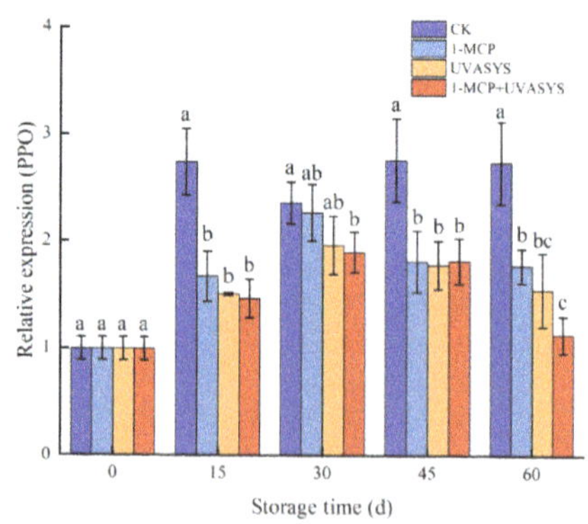

(**B**) PPO Expression

Figure 8. *Cont.*

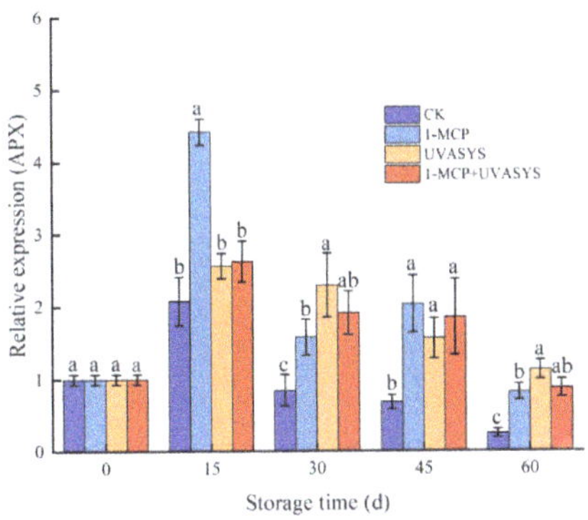

(**C**) APX Expression

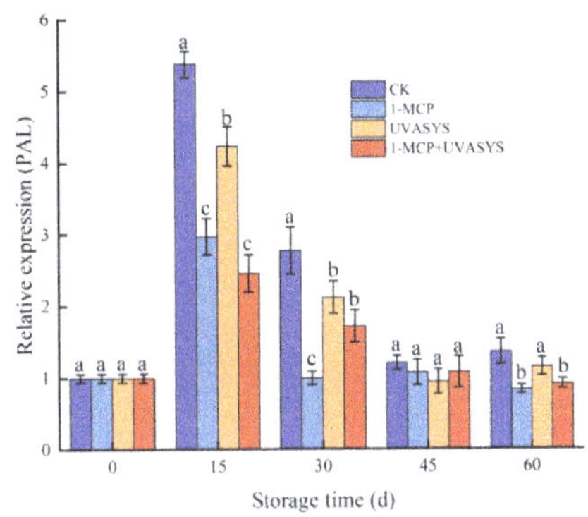

(**D**) PAL Expression

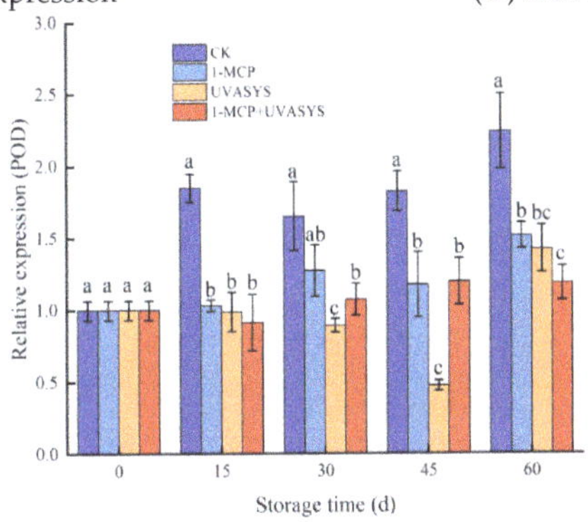

(**E**) POD Expression

Figure 8. Effects of 1-MCP and UVASYS on SOD (**A**), PPO (**B**), APX (**C**), PAL (**D**) and POD (**E**) A graph showing the expression of bamboo shoots over time. Each point represents the average ± standard error (n = 3). Different letters indicate significant differences among treatments for each sampling time at $p < 0.05$.

4. Discussion

4.1. Physiological and Nutritional Quality Changes

The senescence of postharvest fruits and vegetables is considered an irreversible biological process that consumes nutrients until cell death [47]. Fruits and vegetables undergo aerobic respiration as they ripen and age, and it is the main energy source for their cells during this process [48]. Fruits and vegetables will lose quality and senescence quicker after harvest if their respiration rate is excessive. Additionally, it affects the physiological metabolism of fruits and vegetables, determining their storage time [49,50]. Postharvest respiration of bamboo shoots is vigorous, which rapidly consumes the nutrients of the shoots and causes mass loss. Accordingly, inhibiting postharvest respiration rate can prolong the storage time of bamboo shoots [51]. Numerous fruits and vegetables have been found to be inhibited by 1-MCP in previous research [52]. In addition, SO_2 inhibits weight loss during storage of fruits and vegetables [53]. The result of this study suggested that respiration continued during storage; 1-MCP and SO_2 treatments inhibited bamboo shoot respiration during storage compared to CK treatment, inhibited weight loss rate, and maintained soluble protein levels (Figures 1A,B and 2B), consistent with the above description. As a result of the loss of external nutrients and the decomposition of its own protein, the free amino acid content increased at the later stage of storage (Figure 2A). Furthermore, the above-described result is consistent with the study of Li et al. [54]. Thus,

1-MCP and SO_2 can slow down the loss of nutrients by inhibiting respiration to delay senescence and prolong the preservation time of bamboo shoots.

4.2. Appearance Color and Enzymatic Browning

It has been established that browning is a crucial sign of senescence in fruits and vegetables after harvest [55]. Bamboo shoots are prone to browning, especially after peeling and cutting processes. Enzymatic browning is considered a vital cause of bamboo shoot browning [56]. The process of enzyme browning occurs when polyphenol oxidase (PPO) reacts with phenolic compounds, resulting in the browning of fruits and vegetables after harvest [57]. The total color difference (ΔE) and browning index (BI) can indicate the total color change and browning degree of bamboo shoots during storage. Study results indicated that bamboo shoots browned more rapidly as storage time increased, suggesting that bamboo shoots are browning more rapidly (Figure 3A,B). As storage time increases, the ROS scavenging ability of fruits and vegetables decreases, and the MDA content increases, destroying the integrity of the cell membrane in the process. The outflow of phenolic substances made the enzymatic reaction more intense [58]. Based on the increasing amount of total phenol content, these values should be increasing as well. This is consistent with the increase of ΔE and BI values caused by the continuous increase of total phenol content, PPO activity, and gene expression in the late storage period of bamboo shoots in this study (Figures 4A,B and 8B). The results of this study suggest that 1-MCP treatment and SO_2 treatment inhibited PPO activity at a later stage of storage compared with CK treatment, which was consistent with the fluorescence quantitative data of PPO (Figures 4B and 8B). Based on the results, 1-MCP treatment and SO_2 treatment inhibited PPO activity by inhibiting PPO gene expression, thereby inhibiting enzymatic browning and slowing down the increase of ΔE and BI values. Among them, the 1-MCP + SO_2 treatment group exerted the optimal effect, and a significant difference with CK treatment was detected during the later storage period. (Figure 3A,B). This result is consistent with CHEN et al., that is, 1-MCP and SO_2 treatments inhibit postharvest browning of fruits and vegetables [59,60].

4.3. Active Oxygen Metabolism

Cell membrane integrity takes on critical significance to the anti-browning activity of browning-associated enzymes and substrates during storage of fruits and vegetables [61]. The integrity of the cell membrane is significantly correlated with the content of ROS, showing a close association with the content of non-enzymatic antioxidants (e.g., ascorbic acid and total flavonoids) and the activity of APX and SOD [62]. With the senescence of fruits and vegetables, their antioxidant capacity is reduced, and ROS continues to accumulate. Excessive ROS products (H_2O_2, O^{2-}) are capable of causing oxidative stress and destroying cell integrity, leading to oxidative damage of the cell membrane. Membrane lipid peroxidation produces MDA as its final product [63,64]. Ascorbic acid and flavonoids exhibit antioxidant capacity. Relevant research has suggested that ascorbic acid and flavonoids, i.e., the main non-enzymatic antioxidants, are capable of significantly inhibiting the accumulation of MDA in fruits and vegetables, thus preventing ROS accumulation from causing oxidative damage during fruit senescence [65]. To alleviate oxidative damage to cell membranes and slow down MDA accumulation, APX and SOD can remove ROS products (H_2O_2, O^{2-}) [66]. In this study, the ASA content, SOD, APX enzyme activity, and gene expression of bamboo shoots administered with CK were generally at their lowest levels during storage. The bamboo shoots administrated with 1-MCP and SO_2 maintained and induced the content and activity of the corresponding antioxidant substances (Figures 5A, 6B,C and 8A,C), consistent with Wang et al. [67,68]. Bamboo shoots treated with 1-MCP and SO_2 were significantly inhibited compared to bamboo shoots treated with CK in this study (Figure 6A). It has been shown that 1-MCP and SO_2 treatments can delay the senescence of bamboo shoots by increasing ROS scavenging and inhibiting MDA accumulation [69].

4.4. Lignification

Lignification is another major feature of postharvest senescence of bamboo shoots, which is mainly manifested in the deterioration of texture and taste of bamboo shoots caused by the increase in firmness [33]. At present, most model plant studies have found that lignin accumulation is the direct cause of increased firmness and lignification of postharvest fruits and vegetables [70,71]. Phenylpropanoid metabolism is considered the main metabolic pathway of lignin biosynthesis. PAL and POD are the essential enzymes involved in phenylpropanoid metabolism [72]. The key rate-limiting enzyme in the lignin synthesis pathway is PAL, which catalyzes the formation of cinnamic acid derivatives, and POD is responsible for lignin synthesis from lignin monomers [73]. In this study, the firmness of bamboo shoots increased with the increase in storage time. Consistent with the increase in firmness, the lignin content also tended to be increased (Figure 7A,B). This study supports the findings of Yang et al. that bamboo shoots became lignified during storage [74]. Relevant research has suggested that 1-MCP treatment is capable of effectively inhibiting the increase of firmness and lignin during fruit and vegetable storage, which is achieved by inhibiting lignin biosynthesis and its enzyme activity [75]. Research has shown that SO_2 delays the postharvest senescence of fruits and vegetables [76]. At the later stage of storage, 1-MCP treatment and SO_2 treatment significantly inhibited POD and PAL activities (Figure 7A,B), and the relative gene expression changes further verified the above result (Figure 8D,E). As indicated by the results, bamboo shoots were inhibited by 1-MCP and SO_2 treatments due to inhibition of enzyme activity and gene expression in the lignin biosynthesis process. The low level of lignin content and firmness of 1-MCP and SO_2 treatments in this study is consistent with this conclusion (Figure 7A,B). The above results indicate that 1-MCP and SO_2 treatments can effectively delay bamboo shoot senescence and inhibit lignification.

5. Conclusions

The purpose of this study is to extend the storage time of fresh bamboo shoots, increase income for local bamboo growers, promote the development of the bamboo industry, and delay rocky desertification in karst mountainous areas. The results are presented in Figure 9. As indicated by the results, the 1-MCP + SO_2 treatment delayed bamboo shoot senescence and maintained bamboo shoot quality during storage. The treatment inhibited the respiration of bamboo shoots, reduced the physiological metabolism, and slowed down nutrient loss. Additionally, the treatment inhibited the activity of PPO and POD, stimulated the content of ASA and total flavonoids, and stimulated the activity of SOD and APX. As a result, the bamboo shoots that were administered with 1-MCP + SO_2 exhibited higher antioxidant capacity, reduced ROS-induced oxidative damage, prevented MDA accumulation, and significantly reduced browning (Figure 9). Furthermore, the treatment inhibited POD and PAL enzyme activities, lignin production, and firmness increase, ultimately delaying lignification (Figure 9). Therefore, 1-MCP + SO_2 treatment was found to be the most effective method for delaying bamboo shoot senescence during storage, making it a suitable preservation method. Desertification in karst areas has always hindered southwest China's economic development. Planting bamboo forests has been proven to be an effective way to conserve water and improve soil water quality. *Chimonobambusa quadrangularis*, a native plant of southwest China's karst mountainous areas, is favored by consumers for its fresh bamboo shoots. This study explored the preservation method of fresh bamboo shoots, clarified the senescence mechanism, prolonged the storage time, and provided technical support for expanding the planting area of bamboo. This, in turn, stimulates local economic development and delays rocky desertification.

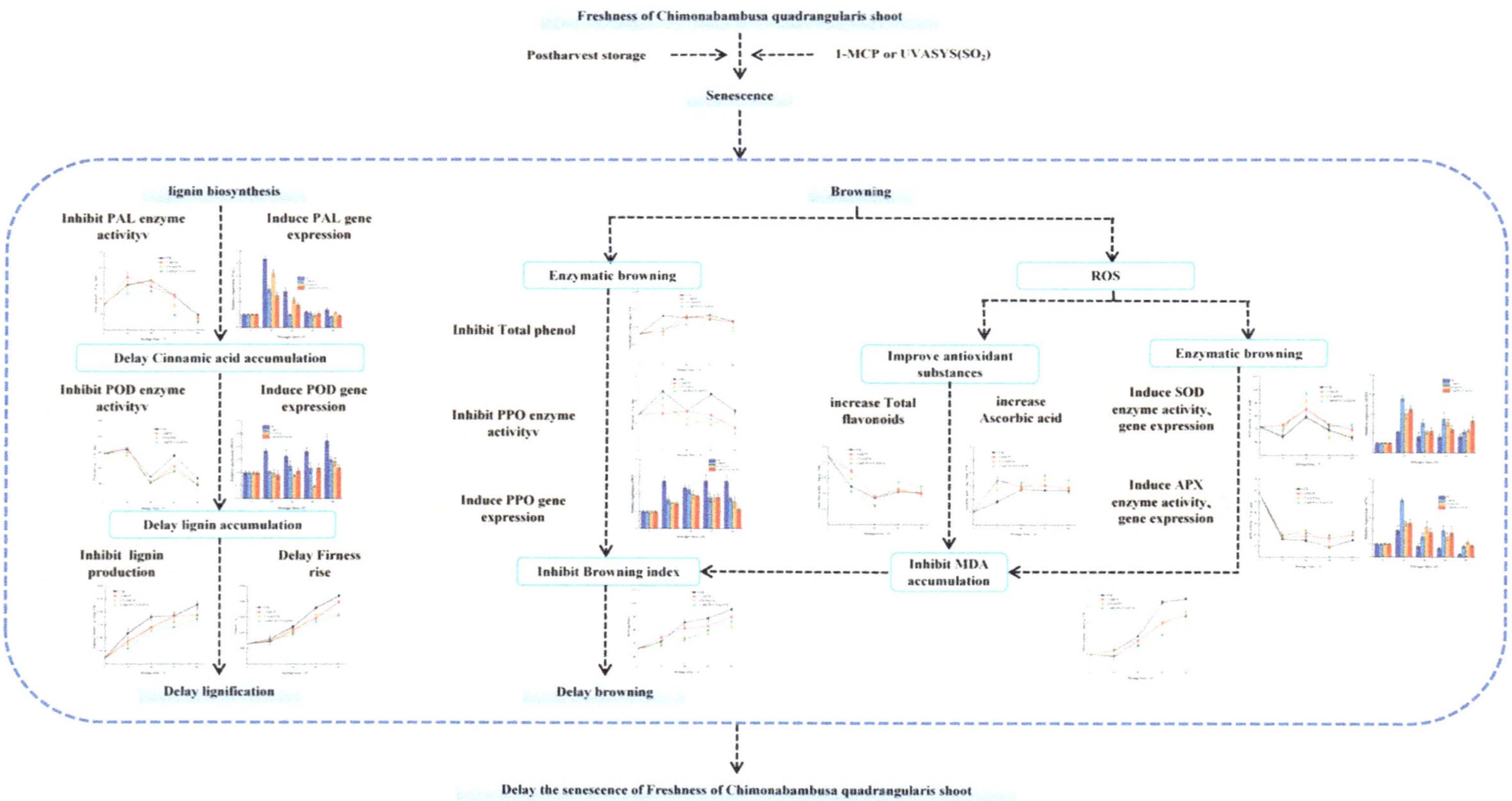

Figure 9. 1-MCP and UVASYS treatment delayed the senescence of Freshness of *Chimonabambusa quadrangularis* shoot after harvest and its molecular regulation mechanism.

Supplementary Materials: The following supporting information can be downloaded at: https://www.mdpi.com/article/10.3390/agronomy13041122/s1, The best spontaneous modified atmosphere bag screening (Figures S1 and S2); 2.The optimal dosage of 1-MCP pre-experiment screening (Figure S3); Pre-experiment screening of optimal UVASYS dosage (Figure S4).

Author Contributions: Conceptualization and methodology: J.X. and N.J.; investigation: J.X., N.J., R.W., Y.D., C.M. and J.L.; formal analysis: J.X., N.Z. and R.L.; data curation: J.X. and N.J.; writing-original draft preparation: J.X.; writing-review and editing: J.X. and N.J. All authors have read and agreed to the published version of the manuscript.

Funding: This research was supported by the Guizhou Provincial Key Technology R&D Program (Qian Ke He Zhi Cheng [2020]1Y139), the young and middle-aged academic backbone project of Guiyang University (GYURC-34[2022.1.1~2024.12.31]), and the Guizhou Province Key Technology Research and Development and Application of Innovation Base for Agricultural Products Primary Processing (Qian Ke Zhong Yin Di [2020] 4018).

Data Availability Statement: Not applicable.

Conflicts of Interest: The authors declare no conflict of interest.

References

1. Christina, L. Climate change. An unsung carbon sink. *Science* **2011**, *334*, 886–887.
2. He, X.Y.; Wang, K.L.; Zhang, W.; Chen, Z.H.; Zhu, Y.G.; Chen, H.S. Positive correlation between soil bacterial metabolic and plant species diversity and bacterial and fungal diversity in a vegetation succession on Karst. *Plant Soil* **2008**, *307*, 123–134. [CrossRef]
3. Wang, S.J.; Liu, Q.M.; Zhang, D.F. Karst rocky desertification in southwestern China: Geomorphology, landuse, impact and rehabilitation. *Land Degrad. Dev.* **2004**, *15*, 115–121. [CrossRef]
4. Bai, X.Y.; Wang, S.J.; Xiong, K.N. Assessing spatial-temporal evolution processes of karst rocky desertification land: Indications for restoration strategies. *Land Degrad. Amp. Dev.* **2013**, *24*, 47–56. [CrossRef]
5. Peng, L.; Xu, X.J.; Liao, X.F. Ampelocalamus luodianensis (Poaceae), a plant endemic to karst, adapts to resource heterogeneity in differing microhabitats by adjusting its biomass allocation. *Glob. Ecol. Conserv.* **2023**, *41*, e02374. [CrossRef]
6. Wang, Y.; Chen, J.; Wang, D. A systematic review on the composition, storage, processing of bamboo shoots, focusing the nutritional and functional benefits. *J. Funct. Foods* **2020**, *71*, 104015. [CrossRef]
7. Persin, Z.K.; Stana-Kleinschek, T.J.; Foster, J. Challenges and opportunities in polysaccharides research and technology: The EPNOEviews for the next decade in the areas of materials, food and health care. *Carbohydr. Polym.* **2011**, *84*, 22–32. [CrossRef]
8. Zhang, J.; Li, S.; Maratab, A.L.I. Effects of UV-B treatment on controlling lignification and quality of bamboo (*Phyllostachys prominens*) shoots without sheaths during cold storage. *J. Integr. Agric.* **2020**, *19*, 1387–1395. [CrossRef]
9. Li, D.; Limwachiranon, J.; Li, L. Hydrogen peroxide accelerated the lignification process of bamboo shoots by activating the phenylpropanoid pathway and programmed cell death in postharvest storage. *Postharvest Biol. Technol.* **2019**, *153*, 79–86. [CrossRef]
10. Lata, D.; Homa, F.; Nayyer, M.A. Effect of postharvest hydrogen sulphide on lignification and biochemical markers of pointed gourd. *Plant Biol.* **2022**, *24*, 704–710. [CrossRef]
11. Zhang, W.; Jiang, H.; Cao, J. UV-C treatment controls brown rot in postharvest nectarine by regulating ROS metabolism and anthocyanin synthesis. *Postharvest Biol. Technol.* **2021**, *180*, 111613. [CrossRef]
12. Liu, J.; Huang, Q.; Kang, P. Lignin Accumulation in Three Pumelo Cultivars in Association with Sucrose and Energy Depletion. *Biomolecules* **2019**, *9*, 701. [CrossRef]
13. Tobimatsu, Y.; Schuetz, M. Lignin polymerization: How do plants manage the chemistry so well? *Curr. Opin. Biotechnol.* **2019**, *56*, 75–81. [CrossRef]
14. Rabiei, V.; Kakavand, F.; Zaare-Nahandi, F. Nitric oxide and γ-aminobutyric acid treatments delay senescence of cornelian cherry fruits during postharvest cold storage by enhancing antioxidant system activity. *Sci. Hortic.* **2019**, *243*, 268–273. [CrossRef]
15. Cheng, S.; Yu, Y.; Guo, J. Effect of 1-methylcyclopropene and chitosan treatment on the storage quality of jujube fruit and its related enzyme activities. *Sci. Hortic.* **2020**, *265*, 109281. [CrossRef]
16. Dias, C.; Ribeiro, T.; Rodrigues, A.C. Improving the ripening process after 1-MCP application: Implications and strategies. *Trends Food Sci. Technol.* **2021**, *113*, 382–396. [CrossRef]
17. Luo, Z.; Xu, X.; Cai, Z. Effects of ethylene and 1-methylcyclopropene (1-MCP) on lignification of postharvest bamboo shoot. *Food Chem.* **2007**, *105*, 521–527. [CrossRef]
18. Blankenship, S.M.; Dole, J.M. 1-Methylcyclopropene: A review. *Postharvest Biol. Technol.* **2003**, *28*, 1–25. [CrossRef]
19. Wu, X.; An, X.; Yu, M. 1-Methylcyclopropene treatment on phenolics and the antioxidant system in postharvest peach combined with the liquid chromatography/mass spectrometry technique. *J. Agric. Food Chem.* **2018**, *66*, 6364–6372. [CrossRef]
20. Liu, H.; Cao, J.; Jiang, W. Changes in phenolics and antioxidant property of peach fruit during ripening and responses to 1-methylcyclopropene. *Postharvest Biol. Technol.* **2015**, *108*, 111–118. [CrossRef]

21. Langer, S.E.; Marina, M.; Francese, P. New insights into the cell wall preservation by 1-methylcyclopropene treatment in harvest-ripe strawberry fruit. *Sci. Hortic.* **2022**, *299*, 111032. [CrossRef]
22. Dhiman, A.; Suhag, R.; Thakur, D. Current status of Loquat (*Eriobotrya japonica* Lindl.): Bioactive functions, preservation approaches, and processed products. *Food Rev. Int.* **2022**, *38* (Suppl. 1), 286–316. [CrossRef]
23. Dukare, A.S.; Paul, S.; Nambi, V.E. Exploitation of microbial antagonists for the control of postharvest diseases of fruits: A review. *Crit. Rev. Food Sci. Nutr.* **2019**, *59*, 1498–1513. [CrossRef]
24. Joradol, A.; Uthaibutra, J.; Lithanatudom, P. Induced expression of NOX and SOD by gaseous sulfur dioxide and chlorine dioxide enhances antioxidant capacity and maintains fruit quality of 'Daw' longan fruit during storage through H_2O_2 signaling. *Postharvest Biol. Technol.* **2019**, *156*, 110938. [CrossRef]
25. Yuan, Y.; Wei, J.; Xing, S. Sulfur dioxide (SO_2) accumulation in postharvest grape: The role of pedicels of four different varieties. *Postharvest Biol. Technol.* **2022**, *190*, 111953. [CrossRef]
26. Domingues, A.; Roberto, S.; Ahmed, S. Postharvest Techniques to Prevent the Incidence of Botrytis Mold of 'BRS Vitoria' Seedless Grape under Cold Storage. *Horticulturae* **2018**, *4*, 17. [CrossRef]
27. Sabir, F.K.; Unal, S.; Sabir, A. Postharvest Aloe vera Gel Coatings Delay the Physiological Senescence of 'Alphonse Lavallée' and 'Red Globe' Grapes during Cold Storage as an Alternative to SO_2. *Erwerbs-Obstbau* **2022**, *1*, 1–10.
28. Pires, J.; Sousa, S.; Pereira, A. Management of air quality monitoring using principal component and cluster analysis—Part I: SO_2 and PM10. *Atmos. Environ.* **2008**, *42*, 1249–1260. [CrossRef]
29. Ahmed, S.; Roberto, S.; Domingues, A. Effects of Different Sulfur Dioxide Pads on Botrytis Mold in 'Italia' Table Grapes under Cold Storage. *Horticulturae* **2018**, *4*, 29. [CrossRef]
30. Zhang, D.; Xu, X.; Zhang, Z. 6-Benzylaminopurine improves the quality of harvested litchi fruit. *Postharvest Biol. Technol.* **2018**, *143*, 137–142. [CrossRef]
31. Xu, D.; Zuo, J.; Fang, Y. Effect of folic acid on the postharvest physiology of broccoli during storage. *Food Chem.* **2021**, *339*, 127981. [CrossRef]
32. Gheysarbigi, S.; Mirdehghan, S.H.; Ghasemnezhad, M. The inhibitory effect of nitric oxide on enzymatic browning reactions of in-package fresh pistachios (*Pistacia vera* L.). *Postharvest Biol. Technol.* **2020**, *159*, 110998. [CrossRef]
33. Li, C.; Suo, J.; Xuan, L. Bamboo shoot-lignification delay by melatonin during low temperature storage. *Postharvest Biol. Technol.* **2019**, *156*, 110933. [CrossRef]
34. Fukushima, R.S.; Kerley, M.S.; Ramos, M.H. The acetyl bromide lignin method accurately quantitates lignin in forage. *Anim. Feed Sci. Technol.* **2021**, *276*, 114883. [CrossRef]
35. Shi, J.; Zuo, J.; Zhou, F.; Gao, L. Low-temperature conditioning enhances chilling tolerance and reduces damage in cold-stored eggplant (*Solanum melongena* L.) fruit. *Postharvest Biol. Technol.* **2018**, *141*, 33–38. [CrossRef]
36. Karimi, F.; Hamidian, Y.; Behrouzifar, F. An applicable method for extraction of whole seeds protein and its determination through Bradford's method. *Food Chem. Toxicol.* **2022**, *164*, 113053. [CrossRef]
37. Xu, L.; Fang, X.; Wu, W. Effects of high-temperature pre-drying on the quality of air-dried shiitake mushrooms (*Lentinula edodes*). *Food Chem.* **2019**, *285*, 406–413. [CrossRef]
38. Shu, P.; Min, D.; Ai, W. L-Arginine treatment attenuates postharvest decay and maintains quality of strawberry fruit by promoting nitric oxide synthase pathway. *Postharvest Biol. Technol.* **2020**, *168*, 111253. [CrossRef]
39. Du, S.; Zhang, J.; Chen, S. The combined effect of 1-methylcyclopropene and citral suppressed postharvest grey mould of tomato fruit by inhibiting the growth of Botrytis cinerea. *J. Phytopathol.* **2019**, *167*, 123–134.
40. Bal, E. Effect of melatonin treatments on biochemical quality and postharvest life of nectarines. *J. Food Meas. Charact.* **2021**, *15*, 288–295. [CrossRef]
41. Yang, X.; Zhao, Y.; Gu, Q. Effects of Naringin on Postharvest Storage Quality of Bean Sprouts. *Foods* **2022**, *11*, 2294. [CrossRef]
42. Petriccione, M.; Pagano, L.; Forniti, R. Postharvest treatment with chitosan affects the antioxidant metabolism and quality of wine grape during partial dehydration. *Postharvest Biol. Technol.* **2018**, *137*, 38–45. [CrossRef]
43. Xia, L.; Lu, Y.; Qian, Y. Cod peptides inhibit browning in fresh-cut potato slices, A potential anti-browning agent of random peptides for regulating food properties. *Postharvest Biol. Technol.* **2018**, *146*, 36–42.
44. Kahramanoğlu, İ.; Chen, C.; Chen, Y. Improving storability of "nanfeng" mandarins by treating with postharvest hot water dipping. *J. Food Qual.* **2020**, *2020*, 8524952. [CrossRef]
45. Qi, X.; Zhao, J.; Jia, Z. Potential metabolic pathways and related processes involved in pericarp browning for postharvest pomegranate fruits. *Horticulturae* **2022**, *8*, 924. [CrossRef]
46. Wu, L.; Ma, N.; Jia, Y.C. An ethylene-induced regulatory module delays flower senescence by regulating cytokinin content. *Plant Physiol.* **2017**, *173*, 853–862. [CrossRef]
47. Li, C.; Yu, W.; Liao, W. Role of Nitric Oxide in Postharvest Senescence of Fruits. *Int. J. Mol. Sci.* **2022**, *23*, 10046. [CrossRef]
48. Wu, Z.; Tu, M.; Yang, X. Effect of cutting on the reactive oxygen species accumulation and energy change in postharvest melon fruit during storage. *Sci. Hortic.* **2019**, *257*, 108752. [CrossRef]
49. Xuan, J.J.; Lin, H.; Shi, J. Effects of a novel chitosan formulation treatment on quality attributes and storage behavior of harvested litchi fruit. *Food Chem.* **2018**, *252*, 134–141.
50. Zhang, W.; Zhao, H.; Zhang, J. Different molecular weights chitosan coatings delay the senescence of postharvest nectarine fruit in relation to changes of redox state and respiratory pathway metabolism. *Food Chem.* **2019**, *289*, 160–168. [CrossRef]

51. Zheng, J.; Li, S.; Xu, Y. Effect of oxalic acid on edible quality of bamboo shoots (*Phyllostachys prominens*) without sheaths during cold storage. *Lebensm. Wiss. Und Technol.* **2019**, *109*, 194–200. [CrossRef]
52. Ali, M.; Raza, M.A.; Li, S. 1-MCP regulates ethanol fermentation and GABA shunt pathway involved in kiwifruit quality during postharvest storage. *Hortic. Plant J.* **2021**, *7*, 23–30. [CrossRef]
53. Sortino, G.; Farina, V.; Gallotta, A. Effect of low SO_2 postharvest treatment on quality parameters of 'Italia' table grape during prolonged cold storage. *Acta Hortic.* **2018**, *1194*, 695–700. [CrossRef]
54. Li, Y.; Ishikawa, Y.; Satake, T. Effect of active modified atmosphere packaging with different initial gas compositions on nutritional compounds of shiitake mushrooms (*Lentinus edodes*). *Postharvest Biol. Technol.* **2014**, *92*, 107–113. [CrossRef]
55. Li, T.; Shi, D.; Wu, Q. Sodium para-aminosalicylate delays pericarp browning of litchi fruit by inhibiting ROS-mediated senescence during postharvest storage. *Food Chem.* **2019**, *278*, 552–559. [CrossRef]
56. Zhang, J.; Murtaza, A.; Zhu, L. High pressure CO_2 treatment alleviates lignification and browning of fresh-cut water-bamboo shoots (*Zizania latifolia*). *Postharvest Biol. Technol.* **2021**, *182*, 111690. [CrossRef]
57. Ali, S.; Khan, A.S.; Malik, A.U. Modified atmosphere packaging delays enzymatic browning and maintains quality of harvested litchi fruit during low temperature storage. *Sci. Hortic.* **2019**, *254*, 14–20. [CrossRef]
58. Luo, S.; Hu, H.; Wang, Y. The role of melatonin in alleviating the postharvest browning of lotus seeds through energy metabolism and membrane lipid metabolism. *Postharvest Biol. Technol.* **2020**, *167*, 111243. [CrossRef]
59. Chen, J.; Xu, Y.; Yi, Y. Regulations and mechanisms of 1-methylcyclopropene treatment on browning and quality of fresh-cut lotus (*Nelumbo nucifera* Gaertn.) root slices. *Postharvest Biol. Technol.* **2022**, *185*, 111782. [CrossRef]
60. Aguiar, A.C.; Higuchi, M.T.; Ribeiro, L.T.M. Bio-based and SO_2-generating plastic liners to extend the shelf life of 'Benitaka' table grapes. *Postharvest Biol. Technol.* **2023**, *197*, 112217. [CrossRef]
61. Toivonen, P.M.A.; Brummell, D.A. Biochemical bases of appearance and texture changes in fresh-cut fruit and vegetables. *Postharvest Biol. Technol.* **2008**, *48*, 1–14. [CrossRef]
62. Dong, T.; Cao, Y.; Li, G. A novel aspartic protease inhibitor inhibits the enzymatic browning of potatoes. *Postharvest Biol. Technol.* **2021**, *172*, 111353. [CrossRef]
63. Li, X.; Li, C.; Cheng, Y. Postharvest application of acibenzolar-S-methyl delays the senescence of pear fruit by regulating reactive oxygen species and fatty acid metabolism. *J. Agric. Food Chem.* **2020**, *68*, 4991–4999. [CrossRef]
64. Chen, C.; Chen, J.; Nie, Z. Chitosan coating alleviates postharvest juice sac granulation by mitigating ROS accumulation in harvested pummelo (*Citrus grandis* L. Osbeck) during room temperature storage. *Postharvest Biol. Technol.* **2020**, *169*, 111309.
65. Dunnill, C.; Patton, T.; Brennan, J. Reactive oxygen species (ROS) and wound healing: The functional role of ROS and emerging ROS-modulating technologies for augmentation of the healing process. *Int. Wound J.* **2017**, *14*, 89–96. [CrossRef]
66. Chen, C.; Nie, Z.; Wan, C. Preservation of Xinyu Tangerines with an Edible Coating Using Ficus hirta Vahl. Fruits Extract-Incorporated Chitosan. *Biomolecules* **2019**, *9*, 46. [CrossRef]
67. Wang, H.; Chen, G.; Shi, L. Influences of 1-methylcyclopropene-containing papers on the metabolisms of membrane lipids in Anxi persimmons during storage. *Food Qual. Saf.* **2020**, *4*, 143–150. [CrossRef]
68. Juan, P.Z. Hairline, a postharvest cracking disorder in table grapes induced by sulfur dioxide. *Postharvest Biol. Technol.* **2008**, *47*, 90–97.
69. Silva, W.B.; Silva, G.M.C.; Santana, D.B. Chitosan delays ripening and ROS production in guava (*Psidium guajava* L.) fruit. *Food Chem.* **2018**, *242*, 232–238. [CrossRef]
70. Lwin, W.W.; Srilaong, V.; Boonyaritthongchai, P. Electrostatic atomised water particles reduces postharvest lignification and maintain asparagus quality. *Sci. Hortic.* **2020**, *271*, 109487. [CrossRef]
71. Liu, W.; Zhang, J.; Jiao, C. Transcriptome analysis provides insights into the regulation of metabolic processes during postharvest cold storage of loquat (*Eriobotrya japonica*) fruit. *Hortic. Res.* **2019**, *6*, 11. [CrossRef]
72. García-Rojas, M.; Meneses, M.; Oviedo, K.; Carrasco, C.; Defilippi, B.; González-Agüero, M.; León, G.; Hinrichsen, P. Exogenous gibberellic acid application induces the overexpression of key genes for pedicel lignification and an increase in berry drop in table grape. *Plant Physiol. Biochem.* **2018**, *126*, 32–38. [CrossRef]
73. Boudet, A.M. Lignins and lignification: Selected issues. *Plant Physiol. Biochem.* **2000**, *38*, 81–96. [CrossRef]
74. Yang, B.; Fang, X.; Han, Y. Analysis of lignin metabolism in water bamboo shoots during storage. *Postharvest Biol. Technol.* **2022**, *192*, 111989. [CrossRef]
75. Xie, G.; Feng, Y.; Chen, Y. Effects of 1-Methylcyclopropene (1-MCP) and Ethylene on Postharvest Lignification of Common Beans (*Phaseolus vulgaris* L.). *ACS Omega* **2020**, *5*, 8659–8666. [CrossRef]
76. Rahimi, J.; Jhalegar, J.; Meti, S. Effect of ozone and sulphur dioxide as postharvest treatment to control Rhizopus rot and quality maintenance of table grape (*Vitis vinifera* L.). *Int. J. Chem. Res.* **2020**, *8*, 1816–1824. [CrossRef]

MDPI
St. Alban-Anlage 66
4052 Basel
Switzerland
www.mdpi.com

Agronomy Editorial Office
E-mail: agronomy@mdpi.com
www.mdpi.com/journal/agronomy

www.ingramcontent.com/pod-product-compliance
Lightning Source LLC
Chambersburg PA
CBHW040818250926
51850CB00007BA/1201